Plasma Science and Technology

Edited by Aamir Shahzad

Published in London, United Kingdom

IntechOpen

Supporting open minds since 2005

Plasma Science and Technology
http://dx.doi.org/10.5772/intechopen.95256
Edited by Aamir Shahzad

Contributors
Eduard V. Rostomyan, Muhammad Asif Shakoori, Maogang He, Aamir Shahzad, Misbah Khan, Santanu Karmakar, Jagadish C. Mudiganti, Hui Hu, Cem Kolbakir, Haiyang Hu, Yang Liu, Vladimir I. Sotnikov, Irina Litovko, Alexey Goncharov, Andrew Dobrovolskii, Iryna Naiko, Heremba Bailung, Rakesh Ruchel Khanikar, Sukhmander Singh, Bhavna Vidhani, Ashish Tyagi, Weihao Liu, Zijia Yu, Zhi Tao, Sachin Kaothekar, Palash Jyoti Jyoti Boruah, Parismita Kalita, Aadil Mahboob, Muhammad Kashif, Alina Manzoor, H.E. Maogang, Madiha Naheed

First published in London, United Kingdom, 2022 by IntechOpen
IntechOpen is the global imprint of INTECHOPEN LIMITED, registered in England and Wales, registration number: 11086078, 5 Princes Gate Court, London, SW7 2QJ, United Kingdom
Printed in Croatia

British Library Cataloguing-in-Publication Data
A catalogue record for this book is available from the British Library

Additional hard and PDF copies can be obtained from orders@intechopen.com

Plasma Science and Technology
Edited by Aamir Shahzad
p. cm.
Print ISBN 978-1-83969-623-7
Online ISBN 978-1-83969-624-4
eBook (PDF) ISBN 978-1-83969-625-1

For EU product safety concerns:
IN TECH d.o.o., Prolaz Marije Krucifikse Kozulić 3, 51000 Rijeka, Croatia, info@intechopen.com or visit our website at intechopen.com.

Meet the editor

Aamir Shahzad has more than seventeen years of experience in university research and teaching both at home and abroad. He received his doctoral and postdoctoral degrees from Xi'an Jiaotong University (XJTU), China, in 2012 and 2015, respectively. His research interests include computational physics, complex fluids/plasmas, plasma oncology, and bio- and energy materials. Currently, Dr. Shahzad is an associate professor in the Department of Physics, the Government College University Faisalabad (GCUF), Pakistan. He is also a member of the ThermoPhysical Society of XJTU, GCUF Physics Society, and the *University of Agriculture Faisalabad* (UAF), Pakistan.

Contents

Preface **XI**

Section 1
Plasma Developments **1**

Chapter 1 **3**
Cold Atmospheric Pressure Plasma Technology for Biomedical Application
by Rakesh Ruchel Khanikar and Heremba Bailung

Chapter 2 **25**
An Experimental Investigation on the Thermodynamic Characteristics of DBD Plasma Actuations for Aircraft Icing Mitigation
by Cem Kolbakir, Haiyang Hu, Yang Liu and Hui Hu

Chapter 3 **49**
In-Liquid Plasma: A Novel Tool for Nanofabrication
by Palash Jyoti Boruah, Parismita Kalita and Heremba Bailung

Section 2
Transport Developments **65**

Chapter 4 **67**
Polarized Thermal Conductivity of Two-Dimensional Dusty Plasmas
by Aamir Shahzad, Madiha Naheed, Aadil Mahboob, Muhammad Kashif, Alina Manzoor and H.E. Maogang

Chapter 5 **81**
Studies of Self Diffusion Coefficient in Electrorheological Complex Plasmas through Molecular Dynamics Simulations
by Muhammad Asif Shakoori, Maogang He, Aamir Shahzad and Misbah Khan

Chapter 6 **99**
Transverse Thermal Instability of Radiative Plasma with FLR Corrections for Star Formation in ISM
by Sachin Kaothekar

Section 3
Wave and Instabilities 113

Chapter 7 115
Parametric Interaction of VLF and ELF Waves in the Ionosphere
by Vladimir I. Sotnikov

Chapter 8 129
Free-electron Driven Terahertz Wave Sources Based on
Simth-Purcell Effect
by Weihao Liu, Zijia Yu and Zhi Tao

Chapter 9 149
New Types of Dissipative Streaming Instabilities
by Eduard V. Rostomyan

Section 4
Hall Thrusters and Tokamak 171

Chapter 10 173
Numerical Investigations of Electromagnetic Oscillations
and Turbulences in Hall Thrusters Using Two Fluid Approach
by Sukhmander Singh, Bhavna Vidhani and Ashish Tyagi

Chapter 11 195
The Emerging Field Trends Erosion-Free Electric Hall
Thrusters Systems
*by Iryna Litovko, Alexey Goncharov, Andrew Dobrovolskyi
and Iryna Naiko*

Chapter 12 215
Gyrotron: The Most Suitable Millimeter-Wave Source for Heating
of Plasma in Tokamak
by Santanu Karmakar and Jagadish C. Mudiganti

Preface

Plasma science and technology (PST) is a discipline investigating fundamental transport behaviors, interaction physics, and reaction chemistry of plasma, which is the fourth fundamental state of matter, and its applications in different technologies and fields. Plasma has uses in refrigeration, biotechnology, health care, microelectronics and semiconductors, nanotechnology, and space and environmental sciences, and more. Modern semiconductor devices and thin film solar cell technologies can be fabricated through plasma processes and PST to be economically viable. Plasma technology is a key advancement in modern medicine and health care. It can be used to manufacture bioengineering devices that depend on plasma processes to harden artificial joints and plan biocompatible surfaces on tissue scaffolding. In addition, PST can be used to create systems that expel plasma as a propellant for spacecraft.

This book provides a comprehensive overview of PST, including information on different types of plasma, basic interactions of plasma with organic materials, plasma-based energy devices, low-temperature plasma for complex systems, and more.

Chapter 1 introduces the exciting field of cold atmospheric pressure (CAP) plasma and its applications in the biomedical field. Chapter 2 discusses the thermal effects induced by dielectric barrier discharge (DBD) plasma actuation to suppress dynamic ice accretion over the surface of an airfoil/wing model for aircraft icing mitigation. Chapter 3 focuses on synthesizing nanomaterials using an emerging technology called in-liquid plasma. In Chapter 4, the authors investigate the effects of polarization on thermal conductivity for a wide range of plasma parameters using a homogeneous non-equilibrium molecular dynamics method. In Chapter 5, the authors employ a molecular dynamics simulation method to investigate three-dimensional electrorheological complex (dusty) plasmas. Chapter 6 examines the impact of porosity, rotation, and finite ion Larmor radius (FLR) corrections on thermal instability of immeasurable homogeneous plasma and the corresponding effects of radiative heat-loss function and thermal conductivity. Chapter 7 analyzes a non-linear parametric interaction between very low frequency (VLF) and extremely low frequency (ELF) waves in the ionosphere. Chapter 8 discusses the basic theory and latest developments of terahertz radiation schemes using a free-electron beam to interact with periodic electromagnetic structures. Chapter 9 introduces new types of dissipative streaming instabilities. Chapter 10 reports the shortcomings of Hall thrusters and their erosion problems of the channel walls, figures of merit, and a comparison with chemical propulsion. This chapter also discusses the various waves and electromagnetic instabilities propagating in a Hall thruster magnetized plasma. Chapter 11 discusses the Hall-type accelerator with closed electron drift and open walls both theoretically and experimentally. Finally, Chapter 12 discusses gyro devices as the most suitable millimeter-wave sources for heating plasma in tokamaks.

Dr. Aamir Shahzad
Professor (Associate),
Modeling and Simulation Laboratory,
Department of Physics,
Government College University Faisalabad,
Faisalabad, Pakistan

Section 1

Plasma Developments

Chapter 1

Cold Atmospheric Pressure Plasma Technology for Biomedical Application

Rakesh Ruchel Khanikar and Heremba Bailung

Abstract

Cold plasma generated in an open environment with a temperature nearly around room temperature has recently been a topic of great importance. It has unlocked the door of plasma application in a new direction: biomedical applications. Cold atmospheric pressure (CAP) plasma comprises various neutral and charged reactive species, UV radiations, electric current/fields etc., which have several impactful effects on biological matter. Some of the significant biological effects of CAP plasma are inactivation of microorganism, stimulation of cell proliferation and tissue regeneration, destruction of cells by initializing apoptosis etc. Although the detailed mechanism of action of plasma on biomaterials is still not completely understood, some basic principles are known. Studies have indicated that the reactive oxygen species and nitrogen species (ROS, RNS) play a crucial role in the observed biological effects. In this perspective, this chapter first provides a brief discussion on the fundamentals of CAP plasma and its generation methods. Then a discussion on the optical diagnostics methods to characterize the plasma is provided. Optical emission spectroscopy (OES) is used to identify the reactive species and to measure their relative concentration. Other important plasma parameters such as gas temperature, electron/excitation temperature and electron density measurement methods using OES have also been discussed. Then a discussion on the application of CAP plasma in biomedical field is provided. A thorough understanding of biochemical reaction mechanisms involving highly reactive plasma species will further improve and extend CAP plasma technology in biomedical applications.

Keywords: Cold atmospheric plasma, plasma jet, dielectric barrier discharge, reactive oxygen and nitrogen species, plasma medicine, wound healing, cancer therapy

1. Introduction

In physical sciences, the term "plasma" often refers to the fourth state of matter consisting of electrons, ions and neutral particles; while in biology, it refers to the yellowish non-cellular liquid portion of the blood. The Nobel prize-winning American chemist Irving Langmuir first used the term plasma in 1927 to an ionized gas - an electrified fluid carrying electrons and ions, in analogy to the biological plasma carrying blood corpuscles, germs etc. Despite this historical connection, there had been no real correlation between the two plasmas until the emergence of the plasma medicine field recently [1].

Plasma can exist in a variety of forms and can be created in several ways. More than 99% of the visible universe is considered to be in the plasma state. The twinkling of stars, nebulas, auroras in the night sky are some examples of plasma that we observe away from us, and so is our Sun. On the other hand, flashes of lightning, fluorescent tubes, neon signs along our city streets are some other plasma examples that we encounter in our everyday life. Plasma can be classified in different ways based on thermodynamic equilibrium, ionization degree, density etc. Based on the thermal equilibrium between the electrons and the heavy particles, plasma can be categorized into thermal or high temperature plasma and nonthermal or cold plasma [2]. The distinction between thermal and nonthermal plasmas is very important in the context of this chapter. In all plasmas produced by applying an external electric field, the energy transfer to the electrons is much faster than that to the heavier ions. Due to their very high mobility, electrons have the opportunity to heat up to several thousands of degrees of Kelvin before the surrounding environment heats up or even without heating them at all. In thermal plasma, the electrons and heavy particles (neutrals and ions) reach a local thermodynamically equilibrium state, i.e., energy transfer from electrons to the heavy particles equilibrates the energy transfer from heavy particles to the environment, and all the species in the environment remain at almost the same temperature. Because of this, this type of plasma is also called equilibrium plasma. Thermal plasma can reach a temperature up to 10^8 K, as found in the solar core. On the other hand, in nonthermal plasma, cooling of heavier particles is more efficient than the energy transfer from electrons to them, and the gas temperature remains low. Therefore nonthermal plasma is also called non-equilibrium plasma or cold plasma.

With the advent of atmospheric pressure plasma discharges in the early 1990s, various industrial and environmental applications that do not require low pressure operating conditions became possible [3]. Among these, the use of low-temperature atmospheric pressure plasma for biomedical application took center stage. The ability of these cold atmospheric pressure (CAP) plasma discharges to produce enhanced gas phase chemistry at low gas temperature has led to their widespread application in fields that require low temperatures, such as biomedical applications and material processing [4]. In recent years, different devices have been designed to generate cold plasma in atmospheric pressure and have been investigated for their ability to use in biomedical applications [5]. It is demonstrated that CAP plasma could interact with organic substances without causing any electrical/thermal damage. The early results have indicated the great potential of these CAP plasma devices for biomedical applications. These devices can produce plasma at nearly room temperature (less than 40°C) at the contact zone, which is essential for direct application on the human or animal body as well as for sterilization of some medical devices [6]. **Figure 1** shows a typical photograph of a cold plasma jet extending out of a 7 mm quartz tube.

In the mid-1990s, a few researchers, for the very first time, demonstrated the efficient bactericidal property of the CAP plasma. This has opened up a new field of research in science and technology, combining plasma physics and biology called plasma medicine. Since then, CAP plasma devices have been successfully utilized in various applications ranging from sterilization to wound healing to killing cancer cells [3, 7]. From the very beginning, it was expected that the reactive species generated by the CAP plasma play a crucial role in the observed biological effects. Even if many details regarding the mechanism of interaction of plasma with biological matter are still not clear, some basic principles are known, and our depth of knowledge is growing very fast in this field. By the beginning of the second decade of the 2000s, clinical trials on patients started with some success [8]. Several applications have reached the clinical trial stage, and some of the CAP plasma devices have already been certified as medical devices. Woedkte et al. list three clinical trials

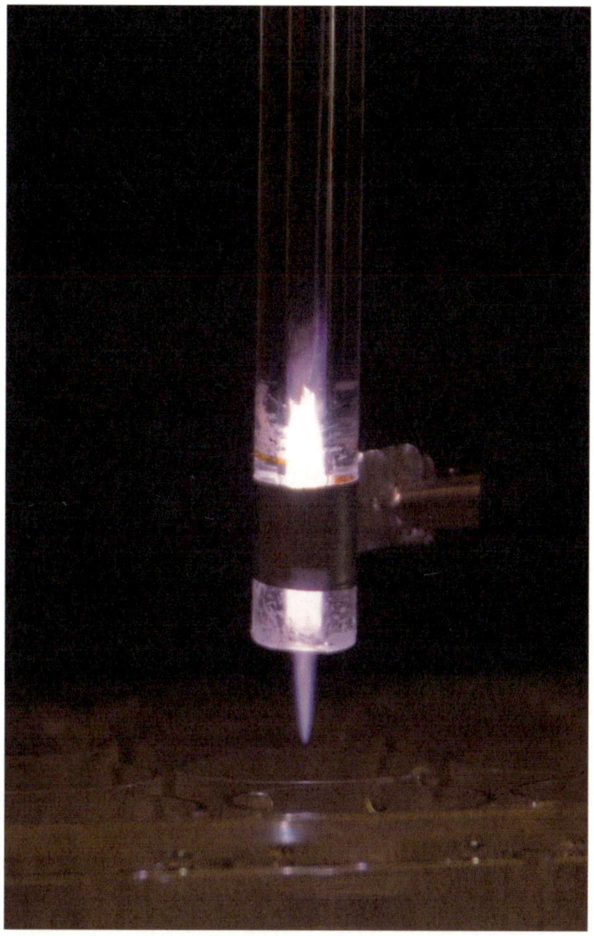

Figure 1.
Photograph of a CAP plasma jet.

using CAP plasma sources conducted in Germany, and two devices got CE marking as medical devices in 2013 [9]. Another device named SteriPlas (Adtec Ltd., London, UK) is the latest one to be certified for use as a medical device [10]. Several other devices have been tested under experimental and laboratory environments and are expecting possible clinical use certification. In 2019, the Food & Drug Administration (FDA) in the USA also approved for their first clinical trials of CAP plasma for cancer treatment [11].

It should be pointed out here that the early studies on plasma application in the biomedical field were concentrated on the thermal effects of plasma [12]. One such successful application was argon plasma coagulation (APC). It has been used to cut tissue in endoscopic applications. These devices operate by heating the tissue using electric current. Therefore their effects are mainly thermal. On the other hand, cold plasma transfers little heat, and its effects are primarily nonthermal.

These remarkable achievements of CAP plasma applications took only about 25 years from the initial discovery to the fundamental scientific investigation stage and finally to applications on actual patients.

In this chapter, we shall concentrate mainly on the nonthermal or cold plasmas produced at atmospheric pressure, their production methods, diagnostics, and their various applications in the biomedical field. After the introductory portion in Section 1, the fundamentals of nonthermal plasma is discussed in Section 2.

In Section 3, various CAP plasma generation methods will be discussed. These include dielectric barrier discharge (DBD), atmospheric pressure plasma jet (APPJ) and corona discharge. Then in Section 4, we shall discuss the diagnostics methods of CAP plasma. Due to the small size of the CAP plasma, generally passive, non-contact diagnostic methods are utilized for characterization. Optical emission spectroscopy (OES) is one such very popular non-invasive diagnostic tool for CAP plasma characterization. Then in Section 5, we shall discuss the interaction of CAP plasma with biomaterials and their biological effects. Section 6 discusses various significant biomedical applications of CAP plasma ranging from sterilization to wound healing to killing cancers. The final section then summarizes the application of CAP plasma technology in biomedical applications and their future outlook.

2. Fundamentals of nonthermal plasma

The term nonthermal plasma refers to a plasma that is not in thermodynamic equilibrium, meaning that the temperature of electrons, ions and neutrals are not equal. In this type of plasma, the electrons remain at a very high temperature (up to a few eV, 1 eV $\approx$ 11,600 K), whereas the temperature of heavy particles is quite low. Because of this reason, they are also termed as non-equilibrium or cold plasma. The high energetic electrons provide the unique reaction chemistry of the cold plasma by facilitating excitation, ionization and chemical dissociation of atoms and molecules at a very low gas temperature. The cold plasma generated at atmospheric pressure produces a myriad of reactive and charged species, including electrons, ions, free radicals, neutral or excited atoms, UV photons etc. These exciting properties of cold plasma have led to their extensive use in various technological fields such as material processing, environmental remediation, nanomaterial synthesis, textile industry, food processing and biomedical applications etc. [13].

Plasmas can be generated by supplying electrical energy to a gas in the form of an electric field. When the applied electric field between the two electrodes is high enough to initiate a breakdown, plasma is formed. Electrons can rapidly gain energy from the applied electric field because of their tiny mass and high mobility. Then they transmit the energy to the neutral atoms and molecules through collisions, providing energy for ionization, excitation, dissociation and other chemical processes. Two types of collisions occur in plasmas [14]:

Elastic collisions: These type of collisions raise the kinetic energy of the neutral species but do not change their internal energies. They increase the temperature of the heavy particles.

Inelastic collisions: These type of collisions between electrons and heavy particles are excitative or ionizing. They modify the electronic structure of the neutral species. When the electronic energy is high enough, it can create excited species or ions. Most of the excited species of plasma have a very short lifetime. They come down to the ground state by emitting a photon. The metastable species are also excited states, but they can decay only by energy transfer through collisions as there are no allowed transitions. Hence, they have a longer lifetime. These collisions do not raise the temperature of heavy particles.

2.1 Paschen's law

The voltage necessary to start a discharge in a gas between two electrodes is given by Paschen's law. It is named after Freidrich Paschen, who discovered it empirically in 1889. The breakdown voltage depends on the electrode spacing d and the pressure p and is given by the formula [15]:

$$V_b = \frac{B(p.d)}{\ln\left[A(p.d)\right] - \ln\left[\ln\left(1 + \frac{1}{\gamma_{se}}\right)\right]}$$

Here, A and B are constants determined experimentally, and γ_{se} is the secondary electron emission coefficient of the cathode. **Figure 2** shows the dependence of breakdown voltage of various gases on the product of electrode spacing and pressure. It is seen that for a constant electrode spacing, the voltage required to ionize a gas is high towards higher pressure, which implies that a narrow gap is necessary to have a reasonable breakdown voltage at atmospheric pressure.

2.2 Current-Voltage characteristics

The plasma behavior inside a discharge is determined by the values of current and voltage between the electrodes. A typical figure that almost every plasma physics textbook discusses is the current–voltage characteristic of a low-pressure (~ 1 mTorr) DC discharge, which describes different gas discharge regimes as shown in **Figure 3**. Arc discharged is characterized by a very high current and a low voltage between the anode and the cathode. Glow discharge occurs at a low current (typically in mA range) and a high voltage. The corona discharge is characterized by a very low current (few μA) and a very high voltage. For low-temperature atmospheric pressure applications, arc discharge is not acceptable as it produces a very high gas temperature. Therefore, a special setup is necessary to create a cold plasma and keep the plasma current low so that discharge remains in glow and corona regime.

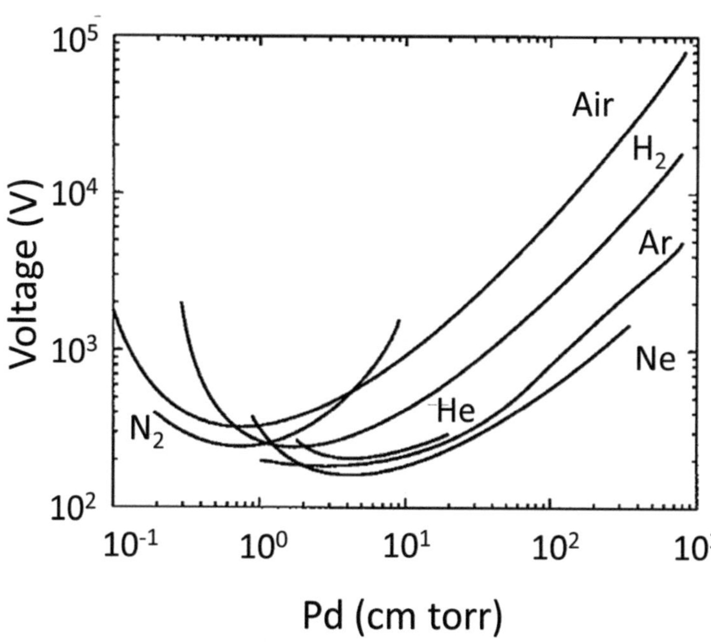

Figure 2.
Breakdown voltage in various gases as a function of the product of pressure, P and gap distance, d for plane parallel electrode.

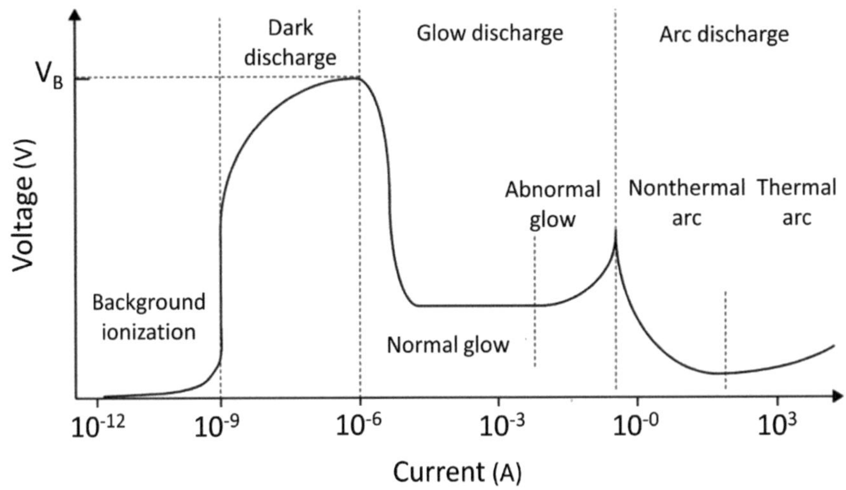

Figure 3.
Current–voltage characteristics of a low pressure DC discharge showing transitions from Townsend to glow and arc discharge.

The necessary condition for a plasma to be suitable for biomedical application is that the plasma has to be produced at atmospheric pressure and the gas temperature has to be near room temperature to avoid thermal damage of biomaterials (tissue etc.) at the contact zone. For this purpose, the plasma needs to be near glow mode. However, glow discharge is generally produced at low pressure. At higher pressure, glow discharge is unstable, and a glow to arc transition can always occur. Therefore, a special electrode arrangement is required to maintain the discharge near glow and corona regime at atmospheric pressure. One general method of producing CAP plasma is to place a dielectric barrier between the two electrodes, and the resulting plasma is known as dielectric barrier discharge (DBD). The role of the dielectric is to limit the discharge current and thus keeps the plasma temperature low. The different types of CAP plasma generation methods are discussed in the next section.

3. Methods of CAP plasma production

Production of atmospheric pressure nonthermal plasma is quite challenging due to high electron-neutral collision frequency, and low applied electric field. Fortunately, several methods have been developed over the years to overcome these challenges. Different production methods have been reported to produce cold plasma in the open environment. These include Dielectric Barrier Discharge (DBD), Atmospheric Pressure Plasma Jet (APPJ), corona discharge etc. Several different working gases such as Helium, Argon, Nitrogen, Heliox (a mix of helium and oxygen), air etc., are used to produce CAP plasma. This section gives a brief overview of the commonly used CAP plasma generation techniques.

3.1 Dielectric barrier discharge (DBD)

One of the most widely used techniques for generating CAP plasma is the dielectric barrier discharge (DBD) using alternating or pulsed electric field. As the name itself suggests, a dielectric cover is used at least at one of the two electrodes for producing the discharge. The function of the dielectric layer is to suppress the spark or arc transition by limiting the discharge current. DBDs are also called "silent" discharges as it produces no sound during discharge. Typical electrode gap distance in a DBD varies from 0.1 mm to several centimeters. Different dielectric materials such as glass, quartz, ceramics and polymers etc., are used in DBDs. To avoid a spark or arc transition, sufficient breakdown strength of the dielectric layer is necessary for insulation of discharge current. But a thicker layer requires a higher voltage, so a compromise must be made here. The electrode arrangement is generally enclosed in a chamber to introduce various gas mixtures between the two electrodes [16]. High voltages sources with frequencies in the kHz range generally drive DBDs. There are many different configurations of DBD are available, but the concept behind them all is the same. These include planar, parallel plates separated by a dielectric or a cylinder, or coaxial plates with a dielectric tube between them. Some basic DBD electrode configurations are shown in **Figure 4**.

More recently, Fridman et al. developed a floating electrode DBD (FE-DBD) [17]. It is similar to the original DBD and consists of two electrodes: an insulated high voltage electrode and an active electrode. The difference between FE-DBD and DBD is that the second electrode is active, meaning it is not grounded. The second electrode can be human skin, a sample, or any other target. Here, the powered electrode needs to be close to the surface of the second electrode to create the discharge.

The discharge in a DBD at atmospheric pressure is generally non-uniform filamentary type which can result in non-uniform treatment of the sample. The dynamic distribution of these filaments determines the appearance of the discharge. Although DBDs usually produce filamentary plasmas, under certain conditions, homogeneous diffuse plasma can also be created. Several groups have reported successful production of diffuse homogeneous atmospheric pressure glow DBD plasmas [18–21]. The mechanism of generating a glow DBD is to initiate a Townsend breakdown instead of a streamer breakdown [22]. To form an avalanche under a lower electric field and avoid growing a large number of positive space charges, sufficient initial seed electrons should exist in the gap before breakdown.

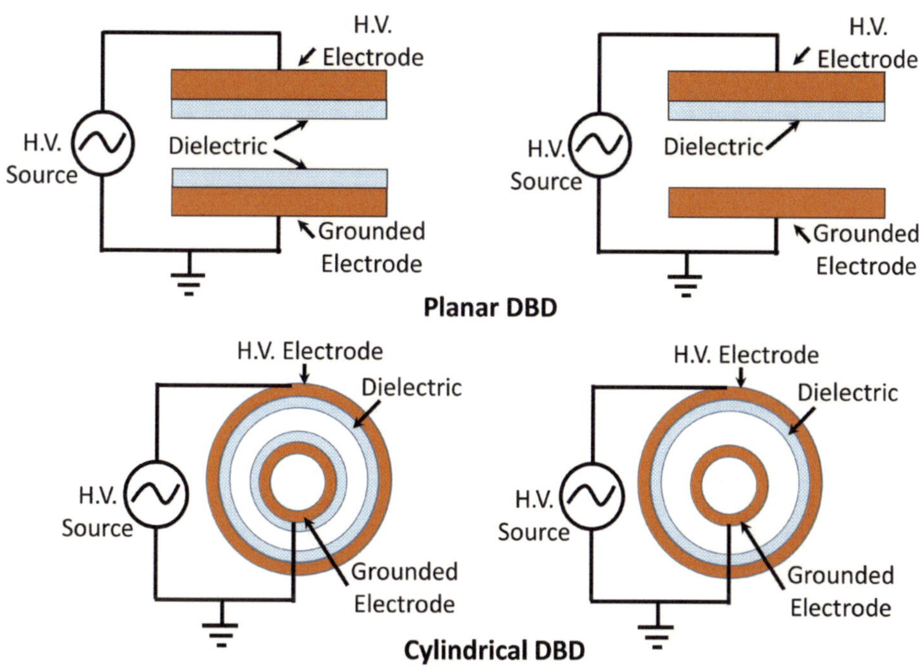

Figure 4.
Schematics of DBD with different electrode configurations.

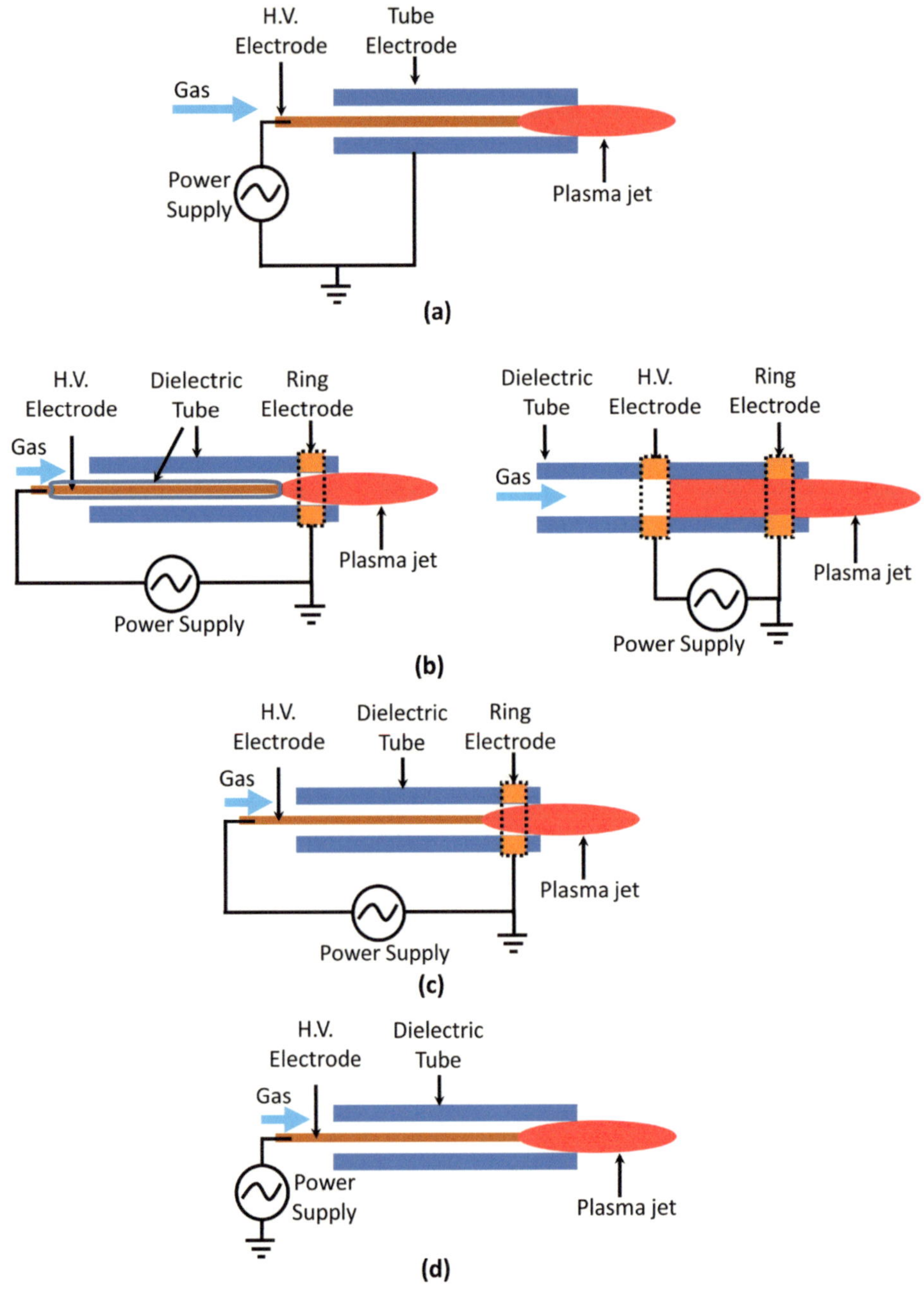

Figure 5.
Typical electrode configurations of (a) DFE jet, (b) DBD jet, (c) DBD-like jet and (d) SE jet. H.V. = high voltage.

In DBDs, residual species from the previous half period of the applied voltage provide the seed electrons or enhance the initial field for the next discharge cycle. This is the so-called memory effect [19].

3.2 Atmospheric pressure plasma jet

One of the most versatile techniques for the generation of CAP plasma for biomedical application is the nonthermal atmospheric pressure plasma jets. Because of their practical capability to produce plasmas that are not spatially bound or confined by electrodes, they can be used for direct treatment on any object irrespective of their shape and size. As a result, they can deliver the plasma generated essential short lifetime active radicals and charged particles to the sample to be treated. Many plasma jets with different configuration have been reported in the literature to generate CAP plasma. There exist various classifications schemes for CAP plasma jets. Some authors classify CAP plasma jets according to the power sources' excitation frequency to generate the plasma. This frequency range can vary from DC to GHz. Accordingly, they are named DC plasma jet, pulsed-dc plasma jet, KHz operated plasma jet, RF operated plasma jet and Microwave driven plasma jet [16, 23–25]. Some authors also use the names 'plasma flame', 'plasma plume', 'plasma gun', 'plasma stream', 'plasma pencil' etc., for plasma jets [24]. The plasma jets are operated with inert gases such as helium, argon etc. or a mixture of inert gas and a few percent of reactive gases of interest. The earliest known CAP plasma jet was developed by Koinuma et al. in 1992. It was powered by an RF source [26].

A comprehensive collection of various types of nonthermal plasma jet arrangements has been discussed in detail by Lu et al. [27]. They are classified into four different groups, namely dielectric-free electrode (DFE) jets, dielectric barrier discharge (DBD) jets, DBD-like jets and single electrode (SE) jets. The DFE jets consists of an inner powered electrode and an outer grounded electrode, as shown in **Figure 5(a)**. There is no dielectric material between the two electrodes. The gas temperature of this jet is relatively high, and cooling water is needed for continuous operation and to keep the temperature low. There is always a risk of arcing in this jet when standard operating conditions are not met. The DFE jet is not suitable for direct biomedical applications. However, it is effective for surface sterilization. It is operated by an RF power source.

In DBD jets, a dielectric layer is present between the two electrodes, and the plasma is not in contact with any electrode. The power consumed by this plasma jet is very less (of the order of few watts). Due to the presence of dielectric, these plasma jets are relatively safe as there is no risk of arcing and is ideal for biomedical applications. The DBD jets can be powered by a KHz ac source or by a pulsed-dc source. **Figure 5(b)** shows the typical electrode configurations of DBD jets.

In DBD-like plasma jets, the discharge is more or less DBD-like when plasma is not in contact with any object. There is no dielectric material between the live electrode and the object to be treated. In this type of plasma jets, easily more power can be delivered to the plasma, and the plasma can be very reactive. As long as arcing can be avoided, this type of jets has their own advantages. For biomedical applications, this kind of devices should be handled carefully because of the risk of arcing. These plasma jets can be powered by KHz ac, RF or pulsed dc sources. The typical configuration of a DBD-like plasma jet is shown in **Figure 5(c)**.

The SE jets are similar to DBD-like jets, except there is no electrode outside the dielectric tube. These jets can be operated by dc, KHz ac, RF or pulsed dc power sources. These kind of jets are not suitable for biomedical application due to the risk of arcing. **Figure 5(d)** shows the basic electrode configuration of a SE plasma jet.

Although the plasma produced by CAP plasma jets looks homogeneous to the naked eye, it is actually discrete in nature when observed by using fast imaging. The plasma volume consists of some "bullet "-like structure, with a propagation speed of more than ten kms^{-1}. This discrete nature of plasma jet was first reported by Teschke et al. using an RF-driven plasma jet [28] and by Lu et al. using a pulsed dc plasma jet [29].

3.3 Corona discharge

A corona discharge is a well-known non-equilibrium discharge that occurs around a pin or thin wire electrode where the electric field is higher near the

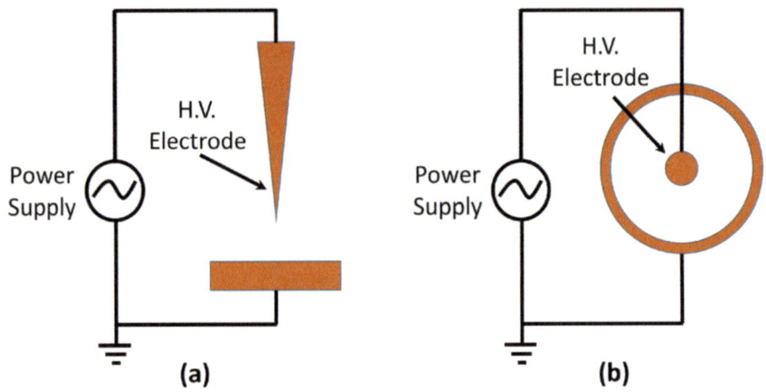

Figure 6.
Typical electrode configurations of corona discharge. (a) Point to plane and (b) wire to cylinder configuration.

electrode edge but decreases quickly with increase of distance [18, 30]. Due to this highly non-uniform electric field, the gas breakdown occurs near the pointed electrode. The electric field strength is high enough to form a conductive region but insufficient to cause electrical breakdown to nearby objects. This type of non-uniform electric field can be created using an asymmetric electrode pair arrangement such as point to plane or wire to cylindrical electrodes, as shown in **Figure 6**.

The corona discharge can be classified into two types depending on the polarity of the HV electrode. The physics of these positive and negative corona discharge is considerably different. This happens due to the vast difference in the mass of electrons and ions. In the positive corona, the electrons are attracted to the HV electrode, and the positive ions are repelled. The secondary electrons are created by photoionization in the gas near the electrode. The electrons are then attracted towards the electrode, which begins the process of further electron avalanche through inelastic collision with neutral gas molecules.

On the other hand, in the negative corona, the electrons move away from the HV electrode. In this case, the secondary electrons are primarily generated by the photoelectric effect from the electrode surface itself. The process is similar to the Townsend breakdown. The electron avalanche then multiplies through impact ionization. As we go away from the electrode, positive ion accumulation occurs, and the electric field becomes weak. As a result, ionization diminishes there.

A corona discharge can be driven by direct-current (DC), alternating-current (AC), or pulsed voltage. It has widespread applications in various fields, such as ozone synthesis, material processing, water purification, electrophotography, copier machine, bacterial inactivation, wound healing and medical surface preparation etc. This type of plasma provides substantial flexibility in treating various products and materials used in the medical industry, for example, syringe barrels, pill bottles, catheter tubing, IV tubes and surgical gowns etc.

4. Diagnostics of CAP plasma

Due to the small size and transient discharge behavior of CAP plasma, plasma diagnostics is very challenging. The use of invasive diagnostic techniques such as Langmuir probe is not suitable as they significantly disturb the plasma and, as a result, yield incorrect values. Therefore, various non-invasive optical diagnostic techniques are the choices of interest for determining the plasma characteristic of CAP plasma [31]. One such most widely used technique is Optical emission spectroscopy (OES).

It is a relatively simple and easy to implement method for determining various plasma properties. The light emitted by the plasma due to deexcitation contains various valuable information regarding the plasma. An optical emission spectrometer can capture this radiation, from which one can extract information on the different species present in the plasma. Also, using the emission spectrum, one can estimate various plasma parameters such as electron/excitation temperature, neutral or heavy particles gas temperature, electron density, concentration of different reactive excited species etc. **Figure 7** shows a typical emission spectrum obtained from an argon CAP plasma jet.

The emission spectrum of CAP plasma often contains emission from molecular species like N_2, N_2^+, OH etc. The highly energetic electrons of plasma can easily transfer their energy to the low lying molecular rotational and vibrational states. In atmospheric pressure condition, the rotational and translational degrees of freedom of gas molecules remain in equilibrium through collisions. Consequently, the rotational temperature gives the value of gas temperature. Generally, the OH rotational band around 306–309 nm, the second positive system of N_2 and the first negative system of N_2^+ are used to obtain the gas temperature. From the best fit between the experimental spectrum and a simulated theoretical spectrum of a particular molecular band, the rotational temperature or the gas temperature of the plasma can be determined. The simulated spectrum can be calculated using software like *Specair* [32] and *Lifbase* [33]. The Boltzmann distribution of the rotational levels is assumed to obtain the temperature.

Another important plasma parameter, the electron/excitation temperature, can be obtained using the Boltzmann plot technique [34]. In this technique, the spectral line intensity (I) and the excitation temperature (T_{exc}) is related by the formula: $\ln(I\lambda / g_k A) = -E_k / kT_{exc} + C$, where λ is the wavelength of the line, g_k is the upper state degeneracy, E_k is the upper level energy and A is the transition probability. If the Boltzmann law is satisfied, the plot of $\ln(I\lambda / g_k A)$ vs. E_k becomes a straight line, and the inverse of the slope gives the excitation temperature. Typically, the electron temperature is found to be near 1 eV in CAP plasmas. In low-temperature plasmas, the low energy electron number is much higher than that of high energy electrons. So the bound electrons on the higher excited levels can be in collisional equilibrium with the free electrons because the energy difference between higher excited levels and the ionization energy is small. So they can satisfy Boltzmann law.

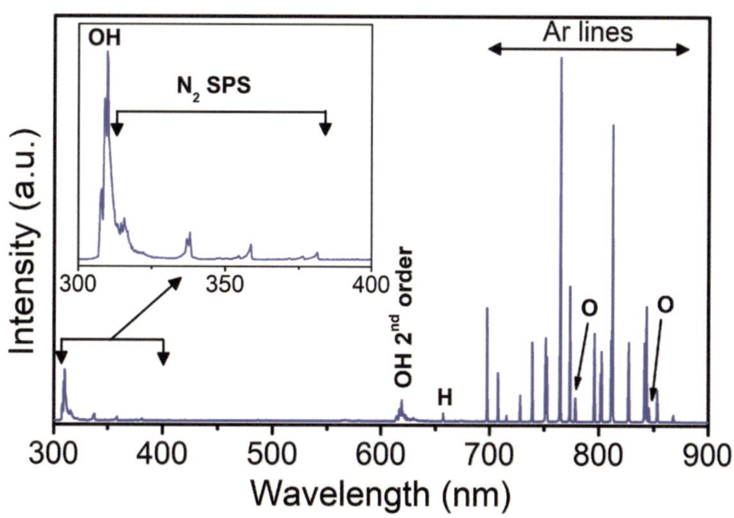

Figure 7.
Emission spectrum of an argon CAP plasma jet.

The population of electrons on the lower excited levels usually do not satisfy the Boltzmann law because they are not in collisional equilibrium with the free electrons. The excitation temperature can also be measured using another well-established method called the line intensity ratio method [35].

The electron density can be measured from the study of spectral line broadening. Spectral lines are always affected by various broadening mechanisms, such as Stark, Van-der-Waals, instrumental, Doppler broadening etc. By extracting the Stark part from the total broadening, the electron density can be determined. The popular lines used for this measurement is the hydrogen Balmer lines. These lines can appear as an impurity from the moisture, or hydrogen can be added in a small amount to the discharge for diagnostic purposes. The H_β 486.13 nm line is most widely used because of very strong Stark broadening and less self-absorption. It is also not much affected by broadening due to ion dynamics and temperature variations. Electron density as low as 5×10^{13} cm^{-3} can be measured using this method [31]. The H_α line at 656.3 nm can also be used for this purpose. However, the accuracy of the electron density value obtained from this line is relatively less. Other non-hydrogenic atomic lines can also be used to determine the electron density using this technique. For a detailed discussion on electron density measurement from the Stark broadening, an interested reader can go through the references [36, 37]. Typically, the electron density can vary between 10^{10} and 10^{14} cm^{-3} in CAP plasmas [31].

Apart from OES, there are many other techniques to study CAP plasmas. The active laser spectroscopy techniques such as laser induced fluorescence (LIF), two-photon absorption laser induced fluorescence (TALIF) can be used to obtain information on the ground state and long-lived, nonradiative excited atoms, molecules or radicals [2]. This technique has been used for many years for plasma diagnostics. Popular laser sources used in LIF are the Nd:YAG laser, dye lasers, excimer lasers, and ion lasers. Other well-known techniques such as Thompson scattering can give direct information on the electron density and temperature. Rayleigh and Raman scattering can provide information on the gas density and temperatures. The optical absorptions spectroscopy and cavity ring down spectroscopy (CRDS) can determine the absolute densities of certain plasma species. Other techniques known from chemical analysis such as UV and FTIR absorption spectroscopy, mass spectrometry, gas chromatography or electron paramagnetic resonance spectroscopy etc., are also used to identify and quantify ions and reactive species in the plasma and to track its transition from plasma phase to liquid phase [6, 18, 38].

5. Effects of CAP plasma on biomaterials

The bactericidal property of plasma was first demonstrated in the mid-1990s, and that started a new research field combining plasma physics and medicine, i.e. plasma medicine [39]. From the very beginning, it was realized that the plasma generated reactive species play a pivotal role in the observed biological effects of CAP plasma. With time, our knowledge of the mechanism of plasma action on cells and tissues has started growing significantly. The basic understanding of the mechanisms of CAP plasma effects on biomaterials is crucial to establish plasma technology application in the biomedical field.

Notwithstanding the use of different plasma sources, working gases, microorganism stains, cell types etc., some general biological plasma effects have been mentioned repeatedly in the basic research of plasma biomaterial interactions such as [9, 18]:

1. Lethal plasma effects

    a. Inactivation/killing of microorganisms

    b. Inactivation or destruction of cells by initialization of apoptosis in mammalian cells

2. Non-lethal plasma effects

    a. Stimulation of microorganism metabolism

    b. Detachment of cells from the cell cluster

    c. Influence on angiogenesis and cell proliferation and consequently promote wound healing and tissue regeneration

    d. Influence on cell migration, expression of cell surface proteins

The possibility to inactivate microorganisms on sensitive surfaces of living structures like intact or wounded skin has attracted the very early interest of physicians. The *in vitro* and *in vivo* results of plasma assisted wound healing showed that plasma acts in a two-stage process. The first one is the antiseptic effect to restore the physiological potential of the wound area by decreasing bacterial load, and the next one is additional stimulation of the healing processes by tissue regeneration independent from antiseptic effects.

Based on the current state of knowledge on the mechanism of plasma biomaterial interaction, it can be deduced that the biological effects of CAP plasma are based on two principles:

1. Biological plasma effects are primarily initiated by plasma induced changes of the liquid environment of tissue and cells.

2. Reactive oxygen species (ROS) and reactive nitrogen species (RNS) generated in or transferred into liquid phases play a vital role in plasma-induced biological responses.

The biologically important ROSs include superoxide ($O_2^-\bullet$), hydrogen peroxide (H_2O_2), hydroxyl radical ($\bullet OH$), singlet oxygen ($^1O_2$), ozone (O_3) etc. The RNSs include nitric oxide ($\bullet NO$), nitrogen dioxide ($\bullet NO_2$), nitrogen trioxide (NO_3), peroxynitrite ($ONOO-$) etc. [7]. These reactive species are formed either by plasma–liquid interaction or by plasma–air interaction. These reactive species act on cells and tissues in the same way as that occur in the body's regular biochemical and physiological processes [40]. Based on this fundamental insight, the field of redox biology can now be used to explain the biological plasma effects. For example, hydroxyl radical causes peroxidation of unsaturated fatty acids present in lipids constituting the cell membrane. The strong oxidative properties of hydrogen peroxide affect proteins, lipids and DNA. Nitric oxide is known to affect the regulation of collagen synthesis, cell proliferation, regulation of immune deficiencies, induction of phagocytosis and angiogenesis etc. In cancerous cells, it is suspected that the action of CAP plasma increases intracellular ROSs, which can lead to cell cycle arrest at the S-phase, DNA breaks, and induction of apoptosis (programmed cell death). Researchers have also shown that the plasma generated ROS and RNS can penetrate biological tissues up to more than 1 mm depth.

Therefore they can interact with the cells on the surface as well as with those beneath [41]. As these ROSs and RNSs regularly occur in cell biological processes, mammalian cells have mechanisms to protect themselves from over-concentration of these species, which might otherwise lead to oxidative stress with severe biological consequences such as genotoxic DNA changes. Detailed investigations till now indicate that the application of CAP plasma does not increase the risk of genotoxicity [6].

Other plasma components such as UV radiation, electric field/electric current also play an additional role. However, the role of UV in direct biological effects is estimated to be very low due to the low dose of UV in these plasma devices. But its supporting role in reactive species generation is essential. Electric fields or currents have varying direct biological effects on living tissue, and it strongly depends on the type of discharge. The magnitude of these electric fields can be as high as several kV/cm, and they are suspected of playing a part, such as in cellular electroporation, which may help larger molecules to enter the cells. Besides these, the plasma generated charged species such as electrons and ions are also assumed to play some roles in the observed biological effects. More research is needed to identify the role of these plasma components for their possible part in biological and medically relevant plasma action, above all reactive species.

6. Applications of cold atmospheric plasma in medicine and healthcare

CAP plasma can be employed in two different ways in biomedical applications. The first one is termed direct exposure or direct treatment, where plasma comes in direct contact with the biological sample to be treated. In this mode, all the plasma generated species, both long lived and short lived species, come in direct contact with the sample and work synergistically. The second mode is termed indirect exposure or indirect treatment. In this method, plasma is used to activate a liquid medium, and then the plasma activated liquid (PAL) is used for treatment. In this case, only the long lived chemical species such as nitrates, nitrites, and H_2O_2, which diffuse and solvate into the aqueous state, play a role. One advantage of PAL is that it can be stored and used at a later time and the composition may be tuned. Both types of CAP plasma treatment have shown significant success in biomedical applications. However, the amount of research reported using direct treatment is more than that using indirect treatment method.

Nowadays, CAP plasma has been successfully applied in various biomedical applications, including inactivation of microorganisms, sterilization of infected tissues, blood coagulation, skin regeneration, tooth bleaching, wound healing, cancer therapy etc. CAP plasma treatment for regenerative processes, such as wound healing, is one of the most advanced applications of plasma technology in the biomedical field. Three plasma devices have already been CE certified for medical use. The very first one is the medical device kINPen® MED (INP Greifswald/neoplas tools GmbH, Greifswald, Germany). The second one is PlasmaDerm® VU-2010 (CINOGY Technologies GmbH, Duderstadt, Germany), and the latest one is the SteriPlas medical device (Adtec Ltd., London, United Kingdom). Their specific purpose is wounds and skin diseases treatment. However, several more plasma devices are awaiting possible certification for clinical application [10, 42].

6.1 Use of CAP in disinfection, wound healing and dermatology

In the beginning, CAP plasma in medicine was applied to the treatment of chronic wounds [43]. Isbary et al. in 2012 first reported about the clinical trials of CAP plasma treatment on chronic ulcer wounds [44]. The authors reported significant infection reduction without any side effect. After that, several clinical trials have proven that CAP plasma action sufficiently reduces the bacterial load on wounds and improve chronic ulcer healing [42]. CAP treatment is also found to accelerate the rate of wound closure at early stages after wounding. Various studies have shown that CAP plasma is an effective tool for disinfection of a variety of bacteria and fungi on the skin and wound pathogens such as Methicillin-resistant *Staphylococcus aureus* (MRSA), *Escherichia coli*, *Pseudomonas aeruginosa* etc. [45].

Besides wound healing, the CAP plasma has also been investigated in dermatological applications. These include treatment of infective and inflammatory skin diseases like atopic eczema herpes, zoster, athlete's foot, acne and others [42, 45]. Some studies have reported positive effects of CAP plasma, but still, more research is required to understand the biochemical processes involved in dermatological applications.

6.2 Use of CAP plasma in cancer therapy

Another most exciting and promising area of CAP plasma application in medicine is cancer treatment. Cancer cells appear to be more vulnerable to CAP plasma than healthy cells. Researchers have shown that CAP plasma can induce programmed cell death (apoptosis) in cancerous cells while leaving their nearby healthy cells essentially unaffected. Up until various *in vitro* and *in vivo* studies have been performed to study the CAP plasma effect on different cancer cells. These studies have repeatedly shown the anti-cancer capacity of CAP plasmas.

It has been reported that CAP plasma treatment increases intracellular ROS concentration. This, in turn, creates a severe redox imbalance in cancer cells as they are already under oxidative stress. Then, the redox imbalance leads to mitochondrial dysfunction, DNA damage, advanced state of oxidation of proteins, caspase activation, etc., and ultimately leads to death of cancer cells [11].

The early animal studies performed had shown the promising potential of CAP plasma in cancer therapy [46, 47]. Since then, many *in vivo* studies have been conducted, and similar positive results have been reported for various types of cancer cells injected under the skin of mice. To date, CAP plasma treatment has demonstrated a significant anti-cancer effect on approximately twenty cancer types *in vivo*. These cancer cell lines include skin, brain, head and neck, breast, leukemia, hepatoma, colorectal, bladder, cervical, lung etc. [48]. In 2019, Metelmann et al. reported the first clinical trials on patients with advanced head and neck cancer [49]. The trial demonstrated the clinical relevance of CAP plasma in cancer treatment and reported an overall positive effect. Some other researchers have shown that some radiation-resistant and chemo-resistant cancer cells are also sensitive to plasma treatment. Some studies have demonstrated CAP plasma as an intra-operative adjuvant treatment. It can be used to inactivate the remaining cancer cells after a surgery [42, 50]. In the USA, the Food & Drug Administration (FDA) approved the first clinical trials of CAP plasma to treat the cancer tumors remaining after surgery in 2019 [11].

6.3 Use of CAP plasma in dentistry

Another long time studied field of CAP plasma application is in dental medicine. The most predominant oral diseases are caries and periodontitis, which are initiated by dysbiotic biofilms. The application of plasma primarily aims to reduce these biofilms on tooth substances and surrounding tissues. Also, instead of using an antimicrobial solution to oral cavity sites for disinfection, CAP plasma treatment can eliminate the unpleasant side effects from anti-microbial solution use. There is a broad spectrum of research going on the possible oral and dental application of CAP plasma ranging from disinfection of root canals, inactivation and removal of biofilm on teeth, treatment of infections and wounds of oral mucosa, dentures and on dental implants, tooth whitening, decontamination and coating of dental prosthesis, cleaning and optimization of tooth and implant surfaces to improve bonding of dental fillings [10, 42].

6.4 Use of CAP plasma in other biomedical areas

Another field explored from the beginning of plasma medicine research is CAP plasma use for hemostasis and blood coagulation [42]. Application of CAP plasma leads to blood coagulation in a much localized manner without damaging the tissue. Therefore it can be a valuable supporting technique in surgery. Aside from these large fields, the potential of CAP plasma in ophthalmology [51, 52] and neurology [53, 54] is also under investigation.

7. Summary and future outlook

The application of CAP plasma technology in the biomedical field has opened up new frontiers in science and technology. It has reached new heights of scientific progress in recent years and has been successfully applied in numerous applications, ranging from sterilization to wound healing to killing cancers. Three CAP devices have already been certified for clinical use for the treatment of chronic and infected wounds. Even if many details regarding the mechanism of interaction of plasma with biological matter are still not clear, some basic principles are known. The biological and medically beneficial plasma effects are primarily triggered by the plasma generated reactive oxygen species (ROS) and reactive nitrogen species (RNS). While CAP plasma has already reached standard medical care status in some areas, such as sterilization, disinfection, wound treatment etc., only primary and pre-clinical data for its effects are available in some other areas. A better understanding of the mechanism of its action will allow further improvement and extensions of the CAP plasma technology to achieve its full therapeutic potential. Also, the development of new plasma devices and modifications of the existing ones will open up new opportunities. In this case, international standardization of the methods to characterize the plasma devices is required to allow better comparability of results obtained from different findings.

It cannot be hypothetically omitted that CAP plasma application does not have some minimal adverse consequences at the molecular level. All these findings and the odds are subject to ongoing investigations, but the current results indicate that the various benefits of CAP plasma outweigh the unproven negative effects. What we all foresee is that the CAP plasma is on its way to clinical routine. Essentially a multidisciplinary research platform is growing with experts from many different fields like plasma physics, biochemistry, molecular biology, medicine etc. to address the issues with synergetic approach.

In the 20th century, laser technology caused a medical revolution with swift development. It successfully got integrated into medicine and created its own medical field called laser medicine. Now, CAP plasma has the chance to repeat this history and to be at the forefront of scientific and technological progress in medicine of the 21st century.

Author details

Rakesh Ruchel Khanikar and Heremba Bailung*
Plasma Application Laboratory, Physical Sciences Division, Institute of Advanced Study in Science and Technology, Guwahati, Assam, India

*Address all correspondence to: hbailung@yahoo.com

References

[1] Fridman G, Friedman G, Gutsol A, Shekhter A B, Vasilets V N and Fridman A 2008 Applied plasma medicine Plasma Process. Polym. **5** 503-533

[2] Chu P K and Lu X 2014 *Low Temperature Plasma Technology: Methods and Applications* (London: CRC Press)

[3] Laroussi M 2014 From killing bacteria to destroying cancer cells: 20 years of plasma medicine Plasma Process. Polym. **11** 1138-1141

[4] Laroussi M and Akan T 2007 Arc-free atmospheric pressure cold plasma jets: A review Plasma Process. Polym. **4** 777-788

[5] Kolb J F, Mohamed A A H, Price R O, Swanson R J, Bowman A, Chiavarini R L, Stacey M and Schoenbach K H 2008 Cold atmospheric pressure air plasma jet for medical applications Appl. Phys. Lett. **92** 1-3

[6] Weltmann K D and Von Woedtke T 2017 Plasma medicine - Current state of research and medical application Plasma Phys. Control. Fusion **59** 014031

[7] Graves D B 2014 Low temperature plasma biomedicine: A tutorial review Phys. Plasmas **21** 080901

[8] Isbary G, Morfill G, Schmidt H U, Georgi M, Ramrath K, Heinlin J, Karrer S, Landthaler M, Shimizu T, Steffes B, Bunk W, Monetti R, Zimmermann J L, Pompl R and Stolz W 2010 A first prospective randomized controlled trial to decrease bacterial load using cold atmospheric argon plasma on chronic wounds in patients Br. J. Dermatol. **163** 78-82

[9] Von Woedtke T, Metelmann H R and Weltmann K D 2014 Clinical Plasma Medicine: State and Perspectives of in Vivo Application of Cold Atmospheric Plasma Contrib. to Plasma Phys. **54** 104-117

[10] Braný D, Dvorská D, Halašová E and Škovierová H 2020 Cold atmospheric plasma: A powerful tool for modern medicine Int. J. Mol. Sci. **21** 2932

[11] Laroussi M 2020 Cold Plasma in Medicine and Healthcare: The New Frontier in Low Temperature Plasma Applications Front. Phys. **8** 74

[12] Keidar M, Shashurin A, Volotskova O, Ann Stepp M, Srinivasan P, Sandler A and Trink B 2013 Cold atmospheric plasma in cancer therapy Phys. Plasmas **20** 057101

[13] Bruggeman P J, Iza F and Brandenburg R 2017 Foundations of atmospheric pressure non-equilibrium plasmas Plasma Sources Sci. Technol. **26** 123002

[14] Braithwaite N S J 2000 Introduction to gas discharges *Plasma Sources Sci.* Technol. **9** 517-527

[15] Raizer Y P 1991 *Gas Discharge Physics* (New York: Springer-Verlag)

[16] Hoffmann C, Berganza C and Zhang J 2013 Cold Atmospheric Plasma: Methods of production and application in dentistry and oncology *Med.* Gas Res. **3** 1-15

[17] Fridman G, Peddinghaus M, Ayan H, Fridman A, Balasubramanian M, Gutsol A, Brooks A and Friedman G 2006 Blood coagulation and living tissue sterilization by floating-electrode dielectric barrier discharge in air Plasma Chem. Plasma Process. **26** 425-442

[18] Lu X, Naidis G V., Laroussi M, Reuter S, Graves D B and Ostrikov K 2016 Reactive species in non-equilibrium atmospheric-pressure plasmas: Generation, transport, and biological effects Phys. Rep. **630** 1-84

[19] Kogelschatz U 2002 Filamentary, patterned, and diffuse barrier

discharges IEEE Trans. Plasma Sci. **30** 1400-1408

[20] Okazaki S, Kogoma M, Ueharaÿ M and Kimura Y 1993 Appearance of stable glow discharge in air, argon, oxygen and nitrogen at atmospheric pressure using a 50 hz source J. Phys. D. Appl. Phys. **26** 889-892

[21] Massines F, Segur P, Gherardi N, Khamphan C and Ricard A 2003 Physics and chemistry in a glow dielectric barrier discharge at atmospheric pressure: diagnostics and modelling *Surf.* Coatings Technol. **175** 8-14

[22] Becker K H, Kogelschatz U, Schoenbach K H and Barker R J 2004 *Non-Equilibrium Air Plasmas at Atmospheric Pressure* (London: Institute of Physics Publishing)

[23] Laroussi M and Akan T 2007 Arc-free atmospheric pressure cold plasma jets: A review Plasma Process. Polym. **4** 777-788

[24] Winter J, Brandenburg R and Weltmann K-D 2015 Atmospheric pressure plasma jets: an overview of devices and new directions Plasma Sources Sci. Technol. **24** 064001

[25] Nokhandani A M, Mahsa S, Otaghsara T and Abolfazli M K 2015 A Review of New Method of Cold Plasma in Cancer Treatment *Sch.* Acad. J. Biosci. **3** 222-230

[26] Koinuma H, Ohkubo H, Hashimoto T, Inomata K, Shiraishi T, Miyanaga A and Ihayashi S 1992 Development and application of a microbeam plasma generator Appl. Phys. Lett. **60** 816-817

[27] Lu X, Laroussi M and Puech V 2012 On atmospheric-pressure non-equilibrium plasma jets and plasma bullets Plasma Sources Sci. Technol. **21** 034005

[28] Teschke M, Kedzierski J, Finantu-Dinu E G, Korzec D and Engemann J 2005 High-speed photographs of a dielectric barrier atmospheric pressure plasma jet IEEE Trans. Plasma Sci. **33** 310-311

[29] Lu X and Laroussi M 2006 Dynamics of an atmospheric pressure plasma plume generated by submicrosecond voltage pulses J. Appl. Phys. **100** 063302

[30] Schutze A, Jeong J Y, Babayan S E, Park J, Selwyn G S and Hicks R F 1998 The atmospheric-pressure plasma jet: a review and comparison to other plasma sources Plasma Sci. IEEE Trans. **26** 1685-1694

[31] Laroussi M, Lu X and Keidar M 2017 Perspective: The physics, diagnostics, and applications of atmospheric pressure low temperature plasma sources used in plasma medicine J. Appl. Phys. **122** 020901

[32] Laux C O 2002 Radiation and nonequilibrium collisional–radiative models *Physico-Chemical Model. High Enthalpy Plasma Flows (Rhode-Saint-Genèse, Belgium, 4-7 June 2002) ed D Fletcher al (von Karman Inst. Spec. Course) www.specair-radiation.net*

[33] Luque J M and Crosley D R 1999 LIFBASE: Database and Spectral Simulation Program (Version 2.1) *SRI* Int. Rep. MP 99-009

[34] Staack D, Farouk B, Gutsol A and Fridman A 2008 DC normal glow discharges in atmospheric pressure atomic and molecular gases Plasma Sources Sci. Technol. **17** 025013

[35] Khanikar R R, Boruah P J and Bailung H 2020 Development and optical characterization of an atmospheric pressure non-thermal plasma jet for superhydrophobic surface fabrication Plasma Res. Express **2** 045002

[36] Konjević R and Konjević N 1997 On the use of non-hydrogenic spectral line profiles for electron density diagnostics

of inductively coupled plasmas Spectrochim. Acta - Part B At. Spectrosc. **52** 2077-2084

[37] Konjević N, Ivković M and Sakan N 2012 Hydrogen Balmer lines for low electron number density plasma diagnostics Spectrochim. Acta - Part B At. Spectrosc. **76** 16-26

[38] Magureanu M and Lukes P 2012 *Plasma Chemistry and Catalysis in Gases and Liquids* vol 66 (Singapore: Willey-VCH)

[39] Laroussi M 1996 Sterilization of contaminated matter with an atmospheric pressure plasma IEEE Trans. Plasma Sci. **24** 1188-1191

[40] Graves D B 2012 The emerging role of reactive oxygen and nitrogen species in redox biology and some implications for plasma applications to medicine and biology J. Phys. D. Appl. Phys. **45** 263001

[41] Laroussi M 2018 Plasma Medicine: A Brief Introduction Plasma **1** 47-60

[42] Von Woedtke T, Emmert S, Metelmann H R, Rupf S and Weltmann K D 2020 Perspectives on cold atmospheric plasma (CAP) applications in medicine Phys. Plasmas **27** 070601

[43] Kramer A, Hübner N-O, Weltmann K-D, Lademann J, Ekkernkamp A, Hinz P and Assadian O 2008 Polypragmasia in the therapy of infected wounds - conclusions drawn from the perspectives of low temperature plasma technology for plasma wound therapy. *GMS Krankenhhyg. Interdiszip.* **3** 13

[44] Foster J, Sommers B S, Gucker S N, Blankson I M and Adamovsky G 2012 Perspectives on the interaction of plasmas with liquid water for water purification IEEE Trans. Plasma Sci. **40** 1311-1323

[45] Bernhardt T, Semmler M L, Schäfer M, Bekeschus S, Emmert S and Boeckmann L 2019 Plasma Medicine: Applications of Cold Atmospheric Pressure Plasma in Dermatology ed N K Kaushik *Oxid. Med. Cell. Longev.* **2019** 3873928

[46] Keidar M, Walk R, Shashurin A, Srinivasan P, Sandler A, Dasgupta S, Ravi R, Guerrero-Preston R and Trink B 2011 Cold plasma selectivity and the possibility of a paradigm shift in cancer therapy Br. J. Cancer **105** 1295-1301

[47] Vandamme M, Robert E, Pesnel S, Barbosa E, Dozias S, Sobilo J, Lerondel S, Pape A Le and Pouvesle J M 2010 Antitumor effect of plasma treatment on u87 glioma xenografts: Preliminary results Plasma Process. Polym. 7 264-273

[48] Yan D, Sherman J H and Keidar M 2017 Cold atmospheric plasma, a novel promising anti-cancer treatment modality Oncotarget **8** 15977-15995

[49] Metelmann H R, Seebauer C, Miller V, Fridman A, Bauer G, Graves D B, Pouvesle J M, Rutkowski R, Schuster M, Bekeschus S, Wende K, Masur K, Hasse S, Gerling T, Hori M, Tanaka H, Ha Choi E, Weltmann K D, Metelmann P H, Von Hoff D D and Woedtke T von 2018 Clinical experience with cold plasma in the treatment of locally advanced head and neck cancer *Clin.* Plasma Med. **9** 6-13

[50] Yoon Y J, Suh M J, Lee H Y, Lee H J, Choi E H, Moon I S and Song K 2018 Anti-tumor effects of cold atmospheric pressure plasma on vestibular schwannoma demonstrate its feasibility as an intra-operative adjuvant treatment Free Radic. Biol. Med. **115** 43-56

[51] Martines E, Brun P, Brun P, Cavazzana R, Deligianni V, Leonardi A, Tarricone E and Zuin M 2013 Towards a plasma treatment of corneal infections Clin. Plasma Med. **1** 17-24

[52] Nikmaram H, Rezaei Kanavi M, Ghoranneviss M, Balagholi S, Ahmadieh H, Roshandel D and Amini

M 2018 Cold atmospheric pressure plasma jet for the treatment of Aspergillus keratitis *Clin.* Plasma Med. **9** 14-18

[53] Yan X, Ouyang J, Zhang C, Shi Z, Wang B and Ostrikov K 2019 Plasma medicine for neuroscience - An introduction *Chinese Neurosurg.* J. **5** 1-8

[54] Xiong Z 2018 Cold Atmospheric Plasmas: A Novel and Promising Way to Treat Neurological Diseases Trends Biotechnol. **36** 582-583

Chapter 2

An Experimental Investigation on the Thermodynamic Characteristics of DBD Plasma Actuations for Aircraft Icing Mitigation

Cem Kolbakir, Haiyang Hu, Yang Liu and Hui Hu

Abstract

We report the research progress made in our research efforts to utilize the thermal effects induced by DBD plasma actuation to suppress dynamic ice accretion over the surface of an airfoil/wing model for aircraft icing mitigation. While the fundamental mechanism of thermal energy generation in DBD plasma discharges were introduced briefly, the significant differences in the working mechanisms of the plasma-based surface heating approach from those of conventional resistive electric heating methods were highlighted for aircraft anti−/de-icing applications. By leveraging the unique Icing Research Tunnel available at Iowa State University (*i.e.*, ISU-IRT), a comprehensive experimental campaign was conducted to quantify the thermodynamic characteristics of a DBD plasma actuator exposed to frozen cold incoming airflow coupled with significant convective heat transfer. By embedding a DBD plasma actuator and a conventional electrical film heater on the surface of the same airfoil/wing model, a comprehensive experimental campaign was conducted to provide a side-by-side comparison between the DBD plasma-based approach and conventional resistive electrical heating method in preventing ice accretion over the airfoil surface. The experimental results clearly reveal that, with the same power consumption level, the DBD plasma actuator was found to have a noticeably better performance to suppress ice accretion over the airfoil surface, in comparison to the conventional electrical film heater. A duty-cycle modulation concept was adopted to further enhance the plasma-induced thermal effects for improved anti−/de-icing performance. The findings derived from the present study could be used to explore/optimize design paradigm for the development of novel plasma-based anti−/de-icing strategies tailored specifically for aircraft icing mitigation.

Keywords: DBD plasma actuation, Thermodynamics of DBD plasma discharges, Aircraft icing mitigation and protection

1. Introduction

Aircraft icing is widely recognized as one of the most serious weather hazards to flight safety [1–3]. Ice accumulation has been found to induce large-scale flow separation over airframe surfaces, thereby, degrading the aerodynamic performance of an airplane significantly [4]. Ice accretion over airframe surfaces can make the aircraft to roll or pitch uncontrollably, and even causes crashes [5]. While considerable research progresses have been made in recent years to provide a better understanding about aircraft icing phenomena, preventing the loss of control due to ice accretion over airframe surfaces still remains an important unsolved problem at the top of National Transportation Safety Board (NTSB)'s most wanted list of aviation safety improvements as highlighted at https://www.ntsb.gov/safety/mwl/Pages/mwl5-2017-18.aspx.

It should be noted that, while anti-icing refers to the prevention of ice buildup on an airframe surface, de-icing denotes the scenario where ice has already formed on an airframe surface, which is removed subsequently. While a number of anti–/de-icing systems have been developed and implemented for aircraft icing mitigation in recent years, all aircraft anti–/de-icing systems can generally be classified into two categories: active and passive methods. While active methods rely on energy input from an external system for the anti–/de-icing operation, passive methods take advantage of the physical properties of the airframe surfaces (e.g., surface wettability) to prevent/delay ice formation and accretion. Current active anti–/de-icing strategies for aircraft icing mitigation suffer from various drawbacks. For example, spraying aqueous solutions of propylene and ethylene glycol (minimum of 50% concentration) along with other chemical additives are widely used for ground anti–/de-icing at airports before aircraft takeoff. Propylene and ethylene glycol, although readily biodegradable, exert an extremely high biochemical oxygen demand on aquatic systems that result in killing fish and other aquatic creatures due to the depletion of dissolved oxygen [6]. There has been an increasing concern of the environmental impacts from the aircraft de-icing fluid swept away with storm and melt water runoff at airports to ground water and nearby waterways [7]. Pneumatic de-icing systems with rubber boots have been used to break off ice chunks accreted at airfoil/wing leading edge for aircraft in-flight icing protection, but they are usually quite heavy and sometime unreliable [8]. Ultrasonic and mechanical de-icing solutions are not easily integrated into existing aircraft and pose foreign object damage (FOD) hazards to aero-engines [8]. While electric resistant heating l or hot air bleeding systems have been used to melt out ice by heating airframe surfaces, they are usually very inefficient and have demanding power requirements and can also cause damage to composite materials from overheating. Furthermore, the melt water may simply run back and re-freeze at a downstream location to cause uncontrolled ice accretion [8]. Passive anti-icing approaches with hydro–/ice-phobic surface coatings have also been suggested as viable strategies for aircraft icing mitigation [9–11]. However, none of the passive approaches are found to be able to eliminate/prevent ice accretion over airframe surfaces completely, especially in the critical regions (e.g., near the airfoil leading edges) [12, 13]. Thus, it is highly desirable and important to develop novel and effective anti–/de-icing strategies to ensure safer and more efficient operation of aircraft under atmospheric icing conditions.

Dielectric barrier discharge (DBD) plasma actuators, which are fully electronic devices without any moving parts, have been studied extensively in the aerospace engineering community [14–16]. A DBD plasma actuator usually features two electrodes attached asymmetrically on the opposite side of a dielectric barrier layer. When a high voltage (i.e., either in alternating current (AC) or nanosecond pulses), is applied to the electrodes, the air over the encapsulated electrode will be ionized to generate a streak of plasma discharges. For AC-DBD plasma discharge, powered by an AC electric field, ionized air molecules is formed in the discharge region above the covered electrode inducing a fluid velocity adding momentum to the boundary layer [17–20]. For the cases with the applied high voltages in nano-second pulses (i.e., ns-DBD plasma actuation), it would induce an ultrafast gas heating mechanism, leading to the generation of a shockwave [17–20]. The use of DBD plasma actuators has gained significant interest in the aerospace engineering community as a promising flow control tool to suppress airfoil stall [21–23] and eliminate separation of laminar boundary layer flows [24, 25] for improved aerodynamic performances. It should be noted that, even though DBD plasma actuators have been widely used for various flow control applications [26, 27], the electro-mechanical efficiency of DBD-plasma- based approach (e.g., the ratio of the energy used to induce ionic wall jet flows for flow control to the total energy consumed by the actuator) was found to be usually very low (i.e., no more than 0.20%) [28], and majority of the energy consumed by the plasma actuators would be dissipated via gas heating and dielectric heating [29].

As revealed clearly by Stanfield et al. [30] and Dong et al. [31], the rotational temperature of the gas above the grounded electrode of a DBD actuator during the plasma actuation can be increased up to 200°C, while the vibrational temperatures were observed to be an order of magnitude higher than the rotational temperature. It was also found that the primary mechanism for the heating of the dielectric layer is through heat transfer from the plasma, i.e., through direct injection, convection, and radiation. To further characterize the thermal effects of DBD plasma discharges, Tirumala et al. [32] used an infrared thermography technique to measure the surface temperature over a DBD plasma actuator, and found that the predominant mechanism of dielectric heating is due to the heat transfer from the plasma to the gas, which then heats up the dielectric surface through forced convection. The increase of the surface temperature was found to have linear relationship with both the applied voltage and the input frequency. By adopting the significant thermal effects of DBD plasma actuators, Cai et al. [33] conducted an explore study to demonstrate the feasibility of using plasma-induced thermal effects for anti–/de-icing operations by embedding an AC-DBD plasma actuator on an ice accreting cylinder model. The thermal effects of AC-DBD plasma actuation were found to be effective for both anti-icing and de-icing operations.

For the flow control applications on aircraft, DBD plasma actuators are usually designed to be mounted in the aerodynamically delicate regions where the aerodynamic characteristics would alter greatly as incoming flow changes (e.g., leading edges of wings and inlet lips of aeroengines) [14, 34]. It should be noted that, such aerodynamically delicate regions are usually also the preferential sites for ice formation and accretion [33]. Since DBD plasma actuation has been found to induce significant surface heating effects along with the ionic wind generation [30, 31], DBD plasma actuators can also be used promising candidates for aircraft icing mitigation. By leveraging icing research tunnels to generate icing conditions to simulate the dynamic ice accretion process over airfoil/wing surfaces, a series of experimental studies were conducted recently to demonstrate the feasibility of utilizing the plasma-induced thermal effects to suppress dynamic ice accretion process over the surfaces of airfoil/wing models for aircraft icing mitigation [29, 35–39].

In the present study, we report the research progress made in our research efforts to utilize the plasma-induced thermal effects to suppress dynamic ice accretion process over the surfaces of airfoil/wing models for aircraft icing mitigation. In the context that follows, while the fundamental mechanism of thermal energy generation in DBD plasma actuation is introduced briefly, the significant differences in the working mechanism of the plasma-based surface heating approach from those of conventional resistive electric heating methods for aircraft anti–/de-icing applications are highlighted. By leveraging the unique Icing Research Tunnel available at Iowa State University (*i.e.*, ISU-IRT), a comprehensive experimental campaign is conducted to quantify the thermodynamic characteristics of an DBD plasma actuator embed over the surface an airfoil/wing model exposed to frozen cold incoming airflow with significant convective heat transfer in the context of aircraft anti–/de-icing. By embedding both a DBD plasma actuator and a conventional electrical film heater onto the surface of the same airfoil/wing model, an experimental investigation is also conducted to provide a side-by-side comparison between the DBD plasma actuator and the electrical film heater in preventing ice formation and accretion over the airfoil surface under a typical icing condition. While a high-speed camera is used to capture the transient details of the dynamic ice accretion and water transport processes over the airfoil surface, an infrared thermal imaging system is utilized to map the surface temperature evolutions during the dynamic ice accretion process or anti–/de-icing process with the AC-DBD plasma and the electrical film heater turned on. The temporally-synchronized-and-resolved IR thermal imaging results are correlated with the acquired ice accretion images to elucidate the underlying physics for a better understanding of the fundamentals of the DBD plasma-based approach for aircraft icing mitigation.

2. Mechanism of surface heating due to DBD plasma actuation

As that shown schematically in **Figure 1**, when a high voltage (i.e., either in alternating current (AC) or nanosecond pulses) is applied to the electrodes of an DBD plasma actuator, the air over the encapsulated electrode would be ionized to generate a streak of plasma charges. It has been reported that DBD plasma actuation would have significant thermal effects [24, 30–32, 40]. While substantial thermal energy is generated along with the formation of ionic airflow for AC-DBD plasma actuation [40], ns-DBD plasma discharge was found to induce an ultra-fast gas heating (FGH), which can dramatically affect the kinetics of chemical reactions, leading to the development of shockwaves in the near-surface gas layer [41–43].

It is well known that, when the high-voltages are applied to the electrodes, a high-intensity electric field would be generated between the exposed electrode and the grounded electrode separated by the dielectric layer. Driven by the electric field, the free electrons and ions in the air are responsible for energy transmission from the external power source to gas heating [32, 42]. As suggested by Popov [44] and Aleksandrov et al. [20], the dynamic gas heating during the plasma discharge is mainly caused by the complex collisions, reactions, and interactions between electrons, ions, and the excited molecules in the electrical field, as summarized in **Figure 2**.

The free electrons get energy from the electric field through acceleration, and then collide with neutrals and ions in the air. If an elastic collision occurs, there is an immediate, but only a rather small portion of total energy release, while in inelastic collisions, ionized particles and excited molecules can be produced, which are the main sources of energy heating the gas. Collision between ions and neutrals and

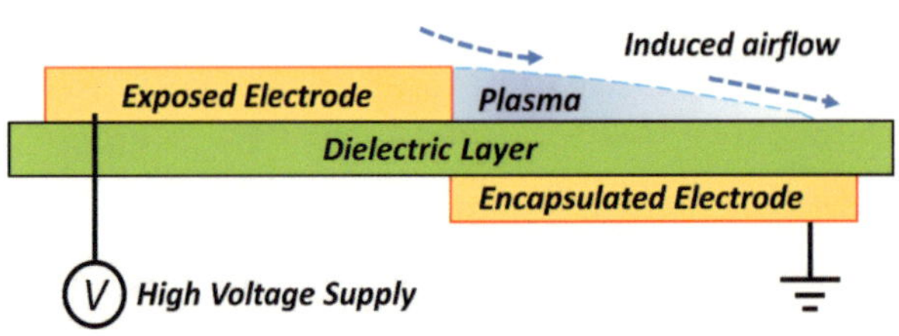

Figure 1.
Schematic of a typical DBD plasma actuator.

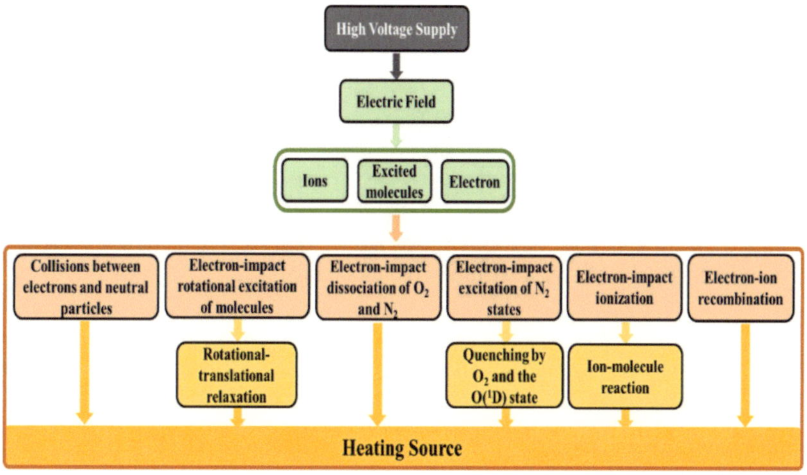

Figure 2.
Heating mechanisms of DBD plasma actuation.

electrons is another source that should be taken into consideration [45]. **Figure 3** summarizes the primary reactions contributing to the gas heating in DBD plasma actuation. When the free electrons impact the molecules in the air (e.g., N_2 and O_2), these molecules would be excited from the ground states to the electronic states. Then, dissociation of the excited molecules would occur, which can generate a significant amount of thermal energy [44]. When electrons impact with the molecular ions in the electrical field, recombination would also occur [20], in which process, the energy would be released between the electronic and translational degrees of freedom of the produced atoms [44]. During the dissociation processes of the electronically excited molecules, while the energy released in the collisions is expended on the rotational excitation of molecules and gas heating, the rotational energy is relaxed into the translational degrees of freedom during the multiple collisions,

Molecule excitation from the ground state to the electronic states	$e+\mathrm{N_2}\rightarrow e+\mathrm{N_2^*}$
	$e+\mathrm{O_2}\rightarrow e+\mathrm{O_2}\left(B^3\Sigma_u^-,v\right)$
	$e+\mathrm{O_2}\rightarrow e+\mathrm{O_2}\left(A^3\Sigma_u^+,C^3\Delta_u\right)$
Dissociation of the excited molecules	$\mathrm{N_2^*}\rightarrow \mathrm{N}\left({}^4S\right)+\mathrm{N}\left({}^2D\right)+\varepsilon_R$
	$\mathrm{O_2}\left(B^3\Sigma_u^-,v\right)\rightarrow \mathrm{O}\left({}^3P\right)+\mathrm{O}\left({}^1D\right)+\varepsilon_R$
	$\mathrm{O_2}\left(A^3\Sigma_u^+,C^3\Delta_u\right)\rightarrow \mathrm{O}\left({}^3P\right)+\mathrm{O}\left({}^3P\right)+\varepsilon_R$
Recombination when electrons impact with the molecular ions	$e+\mathrm{O_2^+}\rightarrow \mathrm{O}\left({}^3P\right)+\mathrm{O}\left({}^1D\right)+\varepsilon_R$
	$e+\mathrm{NO^+}\rightarrow \mathrm{O}\left({}^3P\right)+\mathrm{N}\left({}^2D\right)+\varepsilon_R$
	$e+\mathrm{N_2^+}\rightarrow \mathrm{N}\left({}^4S\right)+\mathrm{N}\left({}^2D\right)+\varepsilon_R$
Quenching of the excited atoms	$\mathrm{O}\left({}^1D\right)+\mathrm{N_2}\rightarrow \mathrm{O}\left({}^3P\right)+\mathrm{N_2}(v)+\varepsilon_R$
	$\mathrm{O}\left({}^1D\right)+\mathrm{O_2}\rightarrow \mathrm{O}\left({}^3P\right)+\mathrm{O_2}\left(b^1\Sigma_g^+,v=2\right)$
Reaction of VT relaxation	$\mathrm{N_2}(v)+\mathrm{O}\left({}^3P\right)\rightarrow \mathrm{N_2}(v-1)+\mathrm{O}\left({}^3P\right)+\varepsilon_R$

Figure 3.
Primary reactions for gas heating in DBD plasma actuation.

which is termed as quenching of the excited molecules. The kinetic energy produced in the quenching processes is rapidly converted into gas heating [44]. It should be noted that, a large number of excited oxygen atoms are produced in the dissociation-recombination-quenching reactions. These excited atoms can also be quenched by the molecules in the air, i.e., N_2 and O_2. It was suggested about 70% of the excitation energy of the excited atoms $O(^1D)$ is expended on gas heating [44]. Along with the above dissociation and quenching reactions, the excited oxygen atoms $O(^3P)$ would also lead to the reaction of VT relaxation, which is considered to be a significant reaction contributing to the gas heating [44].

3. Differences in the working mechanisms of the plasma-based approach from conventional resistive electric heating methods for aircraft inflight icing mitigation

Since the impingement of supercooled water droplets onto an airframe surface is the precursor for the ice accretion over the airframe surface, a better understanding about the heat transfer mechanisms during the impinging process of the water droplets onto the surface of a plasma actuator against that of a conventional electrical film heater is very helpful to elucidate the underlying physics to reveal the significant differences in the working mechanisms of DBD plasma-based approaches from those of conventional resistive electric heating methods for aircraft icing mitigation.

Figure 4 shows the schematics to reveal the great differences in the heating mechanisms as a water droplet impinging onto airfoil surfaces protected by using two different anti–/de-icing systems (*i.e.*, conventional electrical heating method *vs.* DBD plasma-based approach). For the case with a conventional electrical film heater, the thermal energy is generated on the heater surface through resistive electric heating as supplied from the electrical power source. While a portion of the thermal energy may be dissipated to the airflow above the heater surface due to the development of thermal boundary layer via convective heat transfer, which could warm up the impinging water droplet before it is in contact with the heater surface, the dominating mechanism for heating the water droplet would be through heat conduction after the dynamic impinging process (*i.e.*, droplet impacting, splashing, and receding), as

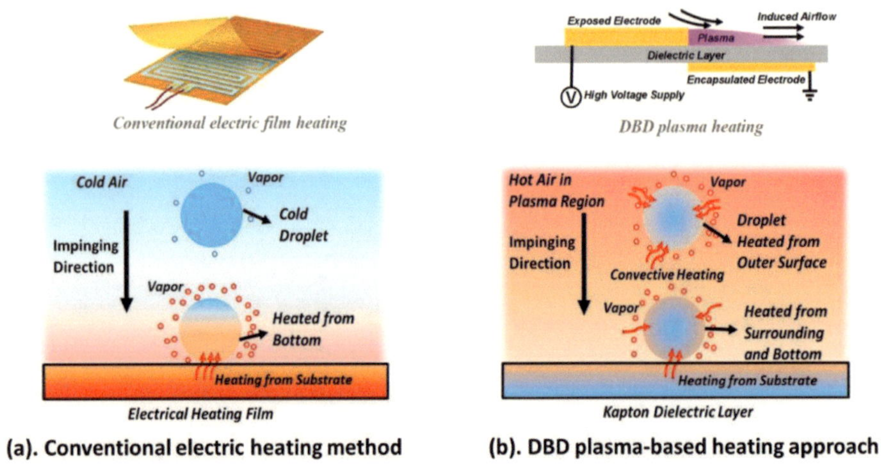

Figure 4.
Comparison of different heating mechanisms as water droplet impinging onto the surface of an electrical film heater against that of a DBD plasma-based actuator.

shown schematically in **Figure 4(a)**. Due to the significant temperature differences between the impinging supercooled water droplet and the heater surface, the thermal energy would be transferred from the heater surface to the water droplet, which can keep the droplet warm up (*i.e.*, above the freezing point) or even being evaporated after the water droplet impacted onto the surface of the electric film heater.

However, as shown schematically in **Figure 4(b)**, the situation would become much different for the case as the water droplet impinging onto the airfoil surface protected by a DBD plasma actuator. As described in Tirumala et al. [32], the primary heating mechanism for plasma discharges is through heat transfer from the plasma to the ambient gas, which then heats up the dielectric surface through direct injection, convection and radiation. This is a reverse thermal path in comparison with that of the scenario of the conventional electrical heating method. Therefore, as the water droplet impinging onto the airfoil surface protected by the DBD plasma actuator, the water droplets would not only be heated up through heat conduction after impacted onto the hot dielectric surface, more importantly, but also be effectively heated up through forced convective heat transfer as the droplet traveling through the hot air in the plasma region even before becoming in contacting with the surface of the plasma actuator, as shown clearly in **Figure 4(b)**.

As described in Li *et al*. [46], the transient temperature of an in-flight droplet can be calculated by using equation of

$$T = \left[T_i - T_e + T_e \exp\left[\left(6t_f h\right)/\left(\rho c_p D\right)\right]\right] / \exp\left[\left(6t_f h\right)/\left(\rho c_p D\right)\right] \tag{1}$$

where T is the transient temperature of the in-flight droplet, h is the convection coefficient of air around the surface of the in-flight droplet, T_i is the initial temperature of the droplet, T_e is the air temperature surrounding the droplet, ρ is the density of the droplet, c_p is the specific heat of the droplet; t_f is the time of flight of the droplet in the convective air flow, and D is the diameter of the flying droplet. It is obvious that, with the same flight time, a higher temperature of the surrounding air would imply a higher transient temperature of the inflight water droplet. Since DBD plasma actuation would induce a significant gas heating above the surface of the plasma actuator, the temperature of the water droplet before impacting on the surface of the plasma actuator would become much higher than that of the case above the electrical film heater.

4. Evaluation of anti–/de-icing performance of the DBD plasma-based approach against conventional resistive electric heating methods for aircraft icing mitigation

As described above, even though both the DBD plasma-based approach and conventional electrical heating method utilize thermal energy to prevent the impinging supercooled water droplets from being frozen to cause ice accretion on the airframe surfaces, the fundamental working mechanisms of the two strategies are quite different for aircraft icing mitigation. It should also be noted that, the conventional electrical heating heaters usually have almost 100% energy efficiency in the sense that all the input electric energy would be converted to thermal energy, while the heating efficiency of DBD plasma actuation was found to vary from 50–90% under different operating conditions [47]. It is highly desirable to evaluate the overall effectiveness of the two different methods for aircraft anti–/de-icing applications. By embedding both a DBD plasma actuator and a conventional electrical film heater onto the surface of the same airfoil/wing model, a comprehensive experimental campaign was conducted to provide a side-by-side comparison between the DBD plasma actuator and the conventional electrical film heater in preventing the ice formation and accretion over the airfoil surface.

4.1 Test model and experimental setup

The experimental study was performed in the Icing Research Tunnel available at Aerospace Engineering Department of Iowa State University (i.e., ISU-IRT). As shown schematically in **Figure 5**, ISU-IRT is a research-grade, multi-functional icing research tunnel with a test section of 2.0 m in length × 0.4 m in width × 0.4 m in height and four transparent side walls. It has the capacity of generating a maximum wind speed of 60 m/s in the test section and an airflow temperature down to −25°C. An array of eight pneumatic atomizer/spray nozzles are installed at the entrance of the contraction section of ISU-IRT to inject micro-sized water droplets (10 ~ 100 μm in size) into the airflow. By manipulating the pressure and flow rate supplied to the atomizer/spray nozzles, the liquid water content (*LWC*) in ISU-IRT is adjustable (i.e., *LWC* ranging from 0.1 g/m^3 to 5.0 g/m^3). In summary, ISU-IRT can be used to simulate various atmospheric icing phenomena over a range of icing conditions (i.e., from dry *rime* to wet *glaze* ice conditions). In the present study, a typical glaze icing condition was generated in ISU-IRT with the freestream airflow velocity of U_∞ = 40 m/s, temperature of T_∞ = −5°C and liquid water content level (LWC) of LWC = 1.0 g/m^3.

Figure 5 also gives the schematic of the airfoil/wing model used in the present study, which has a NACA0012 airfoil profile in the cross section and a chord length of 150 mm (i.e., C = 150 mm). A resistive electrical film heater (*i.e.*, Kapton® Polyimide Film insulated heater), which was selected due to its outstanding operational performance among the electrical film heaters available on the market, was

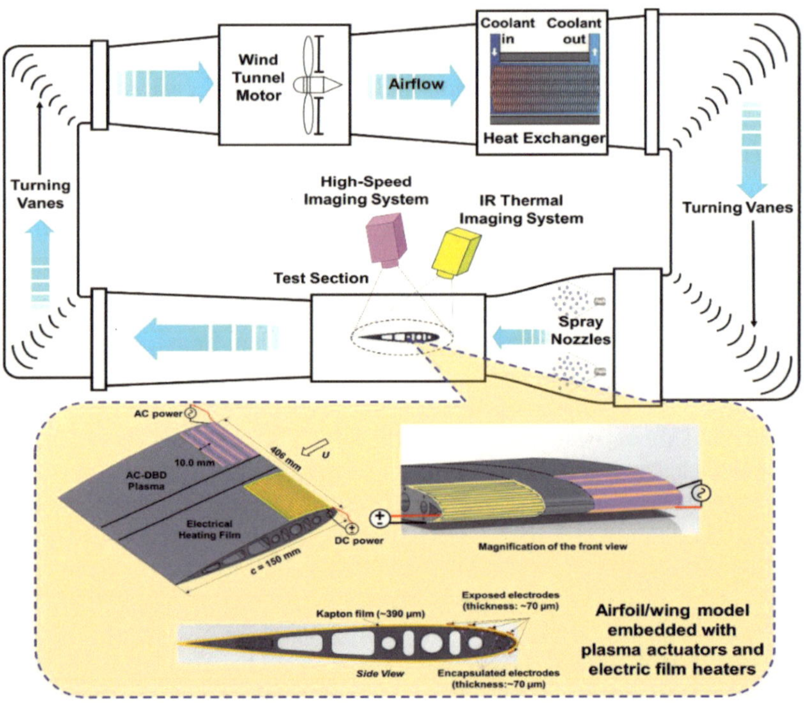

Figure 5.
Schematic of ISU-IRT and the airfoil/wing model used for the experimental study.

embedded over one side of the airfoil surface. The electric film heater consists of an etched foil element of 0.013 mm thickness that is encapsulated between two layers of 0.05 mm Polyimide Film and 0.025 mm FEP adhesive tape. The coverage area of the film heater is 50.8 mm × 101.6 mm. A DC power source was used to power the electrical film heater for the anti–/de-icing operation during the experiments.

A DBD plasma actuator was embedded over the other half surface of the airfoil/wing model for a side-by-side comparison of the anti–/de-icing methods. The DBD plasma actuator consist of four encapsulated electrodes and five exposed electrodes, with the same electrode thickness of about 70 μm. Three layers of Kapton film (*i.e.*, 130 μm for each layer) were integrated to serve as the dielectric barrier to separate the encapsulated electrodes from the exposed electrodes. Ranging from the airfoil leading-edge to about 27% chord length downstream, four encapsulated electrodes were distributed evenly over the airfoil model with a separation distance of 3.0 mm. The length of the encapsulated electrodes was about 350 mm, and the width was 10.0 mm (except the one at the leading edge which was 5.0 mm). As reported by Waldman and Hu [48], since most of the ice would accrete around the leading edge of the airfoil/wing model, the width of the first encapsulated electrode was reduced to 5.0 mm in order to generate more plasma discharges near the airfoil leading-edge for a successful anti–/de-icing operation in the region, while the encapsulated electrodes were attached symmetrically around the leading edge of airfoil model. The exposed electrodes (*i.e.*, 96 mm in length and 3.0 mm in width) were placed right above the encapsuled electrodes with zero overlap between the exposed and encapsulated electrodes. The DBD plasma actuator were wired to a high-voltage AC power supply (Nanjing Suman Co., CTP-2000 K), which is capable of providing a maximum 30 kV peak-to-peak sinusoidal voltage with a center frequency of 10 kHz. During the experiments, while the AC current applied to the plasma actuator was measured by using a high response current probe (Pearson Electronics, Inc., Pearson 2877), the high-amplitude voltage was measured by using a high voltage probe (*i.e.*, P6015A from Tektronix). The electric voltage supplied to the electrodes was manipulated with a variable voltage transformer at a constant frequency of 10 kHz. In order to quantitatively compare the anti–/de-icing performance of the DBD plasma actuator against the electrical film heater under the pre-selected icing conditions, the applied power (i.e., in the term of the applied power density, P_d) to the DBD plasma actuator was adjusted to be same as that applied to the electrical film heater.

During the experiments, in addition to use a high-speed, high-resolution camera (PCO Tech, Dimax) with a 60 mm lens (Nikon, 60 mm Nikkor f/2.8) to record the dynamic ice accreting or anti–/de-icing process over the airfoil surface, an infrared (IR) thermal imaging system (FLIR A615) was also used to map the surface temperature of the ice accreting airfoil surface via an infrared window (i.e., FLIR IR Window-IRW-4C with optic material of Calcium Fluoride) flush mounted on the top wall of the ISU-IRT test section. An in-situ calibration was performed to validate the IR thermal imaging results against the measured surface temperature data with a high-accuracy RTD probe. The measurement uncertainty for the IR thermal imaging system was found to be within ±0.5°C. The high-speed video camera and the IR thermal imaging system were connected to a digital delay generator (Berkeley Nucleonics, model 575) that synchronized the timing between the two systems.

4.2 Thermodynamic characteristic of DBD plasma actuations under frozen cold conditions with significant heat convection pertinent to aircraft anti–/de-icing

As described above, while a number of investigations have been conducted to characterize the thermal effects of DBD plasma discharges [30–32, 40], almost all the previous studies were conducted in quiescent air at room temperature without considering frozen cold conditions and significant convective heat transfer pertinent to aircraft icing phenomena. With the experimental set up described above, a comprehensive investigation was conducted to characterize the thermodynamic characteristics of DBD plasma actuations as a function of relevant controlling parameters (e.g., applied voltage, frequency, and power input, etc…) under frozen-cold test conditions coupled with significant convective heat transfer pertinent to aircraft anti–/de-icing.

Figure 6 shows one example of the experimental results to reveal the time evolution of the measured temperature distribution over the airfoil surface protected by the DBD plasma actuator a dry test condition (i.e., without turning on the water spray system of the ISU-IRT). For the experiment results, while the incoming airflow was set at of U_∞ = 40 m/s and T_∞ = −5.0°C, the AC-DBD plasma actuators were supplied by AC voltage of $V_{\text{p-p}}$ = 12.5 kV and f = 10 kHz, with the corresponding applied power density of P_d = 7.8. kW/m^2. It is clearly seen that, after the plasma actuator was switched on, the temperatures over the airfoil surface were found to increase rapidly, with the local surface temperatures at the edges of the exposed electrodes being raised from −5°C to more than 25°C in less than 5 seconds. As shown clearly in **Figure 6(a)**, the surface heating was first initiated at the edges of the exposed electrodes with evident local temperature peaks (i.e., as indicated by the white strips in the temperature map over the airfoil surface). As the time goes by, more and more thermal energy was generated during the plasma discharges, as seen from the measurement results shown in **Figure 6(b)–(d)**. It should be noted that the maximum temperatures were always found to be located at the edges of the exposed electrodes, which agrees with the findings reported in the previous studies [40]. Meanwhile, the temperature over the exposed electrodes (i.e., copper tap) appeared to be much higher than that over the dielectric layer (i.e., Kapton film). The temperature differences between the electrode surfaces and the surface of the dielectric layer were believed to be caused by the significant difference in the thermal conductivity between the copper tape and the Kapton film (i.e., 385.0 W/m·K for copper tape *vs.* 1.57 W/m·K for Kapton film). The measured surface temperature was found to be relatively low near the airfoil leading edge in general, and increased gradually at further downstream locations, which was correlated well with the chordwise development of the convective heat transfer over the airfoil surface (i.e., the heat convection would be maximum at the airfoil leading edge, and decrease gradually in the downstream [49]).

Based on the measured temperature distributions given in **Figure 6**, the spanwise-averaged temperature profiles along the airfoil chord can be extracted, and the extracted results are given in **Figure 7**. It can be seen clearly that the spanwise-averaged temperature profiles at the different time instances have a very similar distribution pattern, i.e., the surface temperatures were always found to reach the local peak values at the edges of the exposed electrodes, and then decrease gradually

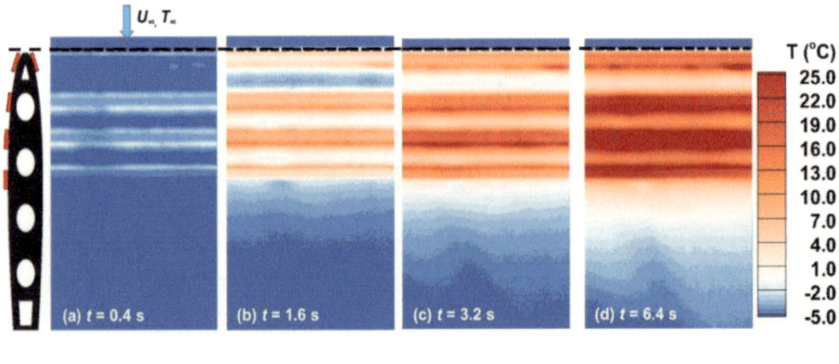

Figure 6.
Measured temperature distribution over the airfoil surface with the DBD plasma actuator operating under a dry test condition of U_∞ = 40 m/s and T_∞ = −5°C.

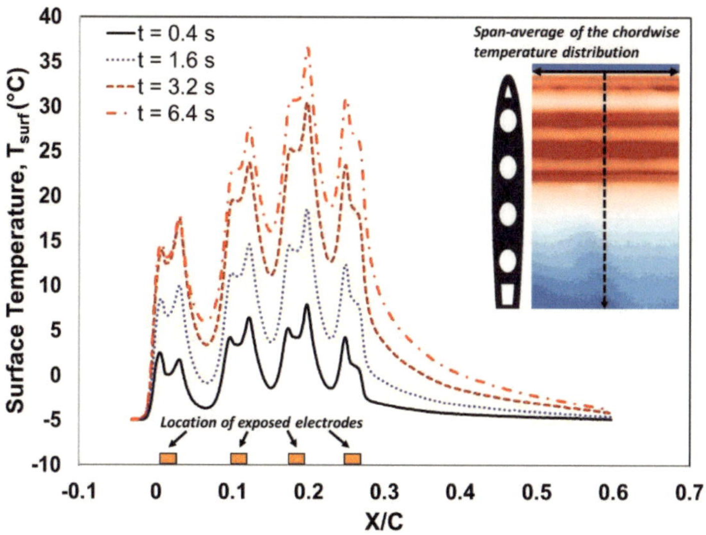

Figure 7.
Spanwise-averaged temperature profiles along the airfoil chord with the DBD plasma actuators operating under a dry test condition of U_∞ = 40 m/s and T_∞ = −5°C.

with the increasing distance away from the edges of the exposed electrodes. For example, at the time instance of t = 0.4 s, while the surface temperatures over the dielectric layer were found to have almost no change (i.e., still being frozen cold at $T_{surface}$ = −5°C), the local temperature peaks at the electrode edges were found to increase rapidly and become higher than +5.0°C. Along with the rapid temperature rise at the edges of the exposed electrodes, the temperatures over the surfaces of copper-based electrodes were also found to become much higher in comparison to those over the dielectric layer. It was suggested that the predominant heating mechanism in plasma discharges is due to the heat transfer from the plasma to the gas, which then heats up the surface of the plasma actuator through forced convection [32]. With the plasma actuator embedded over the airfoil surface was exposed in the frozen cold airflow coupled with significant convective heat transfer, the hot air originally heated by plasma discharges would not only be in contact with the dielectric layer, but also convect over the copper-based exposed electrodes. It caused the temperature rise over the surfaces of the exposed electrodes. Due to the much higher thermal conductivity of the copper-based exposed electrodes, the electrode surfaces were found to have a much faster thermal response (*i.e.*, rapid temperature increases) in comparison to that of the Kapton-based dielectric layer, as shown quantitatively from the measured temperature profiles given in **Figure 6**. As the time goes on, the temperatures over the airfoil surface were found to increase rapidly, with the maximum temperature raised to more than 35°C at 6.4 seconds after turning on the DBD plasma actuator. It can also be seen clearly that, the surface temperature around the third exposed electrode was found to be always higher than those at other locations, which is believed to be a result of the development of the thermal boundary layer over the airfoil surface, i.e., due to the effects of the significant convective hear transfer over the airfoil surface [50].

4.3 Comparison of the anti–/de-icing performance of the DBD plasma-based approach against the convention electrical surface heating methods

With the experimental setup given in **Figure 5**, a comprehensive experimental campaign was conducted to provide a side-by-side comparison of the DBD plasma-based approach against conventional electrical heating methods in preventing the ice formation and accretion over the airfoil surface. In performing the ice accretion experiments, ISU-IRT was operated at a prescribed frozen-cold temperature level (e.g., T_∞ = −5°C for the present study) for at least 30 minutes in order to ensure ISU-IRT reaching a thermal steady state. Then, the DBD plasma actuator and the electrical film heater embedded over the airfoil/wing surface were switched on simultaneously for about 60 seconds to achieve a thermal equilibrium state before turning on the water spray system of ISU-IRT. After the water spray system was switched on at $t = t_0$, the super-cooled water droplets carried by the incoming airflow would impinge onto the surface of the airfoil/wing model to start the ice accretion process. During the experiments, the high-speed imaging system and IR thermal imaging system were synchronized to record the dynamic ice accretion or anti–/de-icing process and map the corresponding surface temperature distributions over the ice accreting airfoil/wing model simultaneously.

Figure 8 shows the typical snapshots of the dynamic ice accretion process over the airfoil surface with the same electric power supplied to the DBD plasma actuator and the electrical film heater (i.e., P_d = 7.8 kW/m^2) for the anti–/de-icing operation. The box in red dashed lines in the acquired images indicates the measurement window of the IR thermal imaging system. Since very similar features were observed for all the test cases, only the measurement results obtained under the test conditions of U_∞ = 40 m/s, T_∞ = −5°C and *LWC* = 1.0 g/m^3 were shown and analyzed here for conciseness. As shown clearly in **Figure 8(a)**, right after starting the ice accretion experiments (*i.e.*, t = 10.0 s), since both the DBD plasma actuator and the electrical film heater had already been switched on for a while to make the surface temperatures of the airfoil/wing model being well above the freezing point of water, the supercooled water droplets were found to be heated up rapidly, upon impacting onto the heated airfoil surface. Therefore, the front surface of the airfoil model protected by the plasma actuator and the electrical film heater (*i.e.*, from the leading edge to ~30% chord length) was found to be totally ice free, with evident water runback flow observed over the airfoil surface. Driven by the boundary layer airflow over the airfoil surface, the unfrozen water was found to run back in the form of film/rivulet flows. The runback water over the airfoil surface on the electrical film heater side was found to be refrozen into ice eventually to form rivulet-shaped ice structures at the downstream region of X/C ≈ 60%. In comparison,

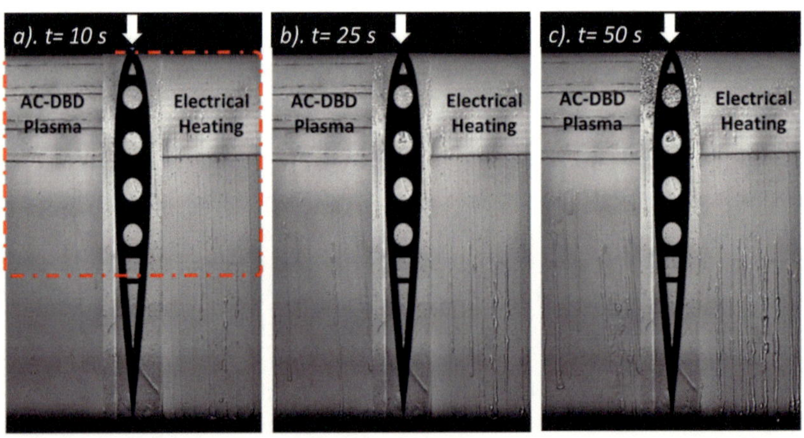

Figure 8.
Acquired images to reveal the dynamic ice accretion process over the airfoil surface with the electric power supplied to the plasma actuator and the film heater being P_d *= 7.8 kW/m*^{2}.

much less runback water was observed over the airfoil surface on the DBD plasma actutor side with much fewer ice structures formed at the downstream locations.

The experimental observation suggests that, with the same electric powere inputs, the DBD plasam actuator seems to have a better anti–/de-icing performance in comparison to the conventioneal electric film heater. This can be explained by the facts that, since the airflow over the region covered by the DBD plasma actuator would be sufficiently heated due to the gas heating effects in the plasma acuatuation, a portion of the airbrone water droplets would be warming up rapidly and even evaporated as they flying through the plasam region before impacting onto with airfloil surface, resulting in the less water mass collected over the airfoil surface proptected by the DBD plasma actutor. However, with the supercooled water droplets impinging onto the airfoil surface protected by conventional electrical film heater, the thermal energy was mainly transferred from the heating elements to the airfoil surface via heat conduction. Since the input power for this cases was not sufficient to instantly evaporate the impacted water droplets (i.e., P_d = 7.8 kW/m^2), the impacted water droplets were found to coalesce quickly on the airfoil surface to form rivulets/film flows to transport the impacted water mass to further downstream locations, as driven by the airflow over the airfoil surface. Due to the intense convective heat transfer between the surface water and the frozen cold incoming airflow over the airfoil surface, the runback water was found to be refrozen into ice eventually, result in the formation of rivulet-shaped ice strcutures at further downstream locations.

As the time goes on, more and more super-cooled water droplets would impinge onto the airfoil surface. As a result, more water mass was found to be collected over the airfoil surface to cause the formation of more rivulet-shaped runback ice accreted at the downstream locations of the airfoil surface. As shown in **Figure 8(b)** and **(c)**, while electric power supplied to the two systems were set to be the same value of P_d = 7.8 kW/m^2, the ice structures accreted over the airfoil surface on the plasma actuator side were always found to be less than those on the electric film heater side.

In order to achieve a better anti–/de-icing performance, the electric power supplied to the plasma actuator and the electrical film heater were increased by a factor of two (*i.e.*, P_d = 15.6 kW/m^2) to generate more thermal energy for the anti–/de-icing. The typical snapshots of the dynamic water runback/ice accretion process with elevated power input are shown in **Figure 9**. It can be clearly seen that, with the higher power input of P_d = 15.6 kW/m^2, the airfoil surface was found to become completely free of ice on both sides of the airfoil surface. It can also be seen that, similar to that observed for the case with relatively lower power input described above, the DBD plasma side of the airfoil surface appeared to have much less water runback in comparison to that on the electric film heater side. The rapid evaporation of the airborne water droplets as flying through the plasma region described above is believed to be the reason to cause the much less water mass collected on the plasma actuator side of the airfoil surface.

The corresponding IR thermal imaging results can reveal more details on the different working mechanisms of the two system for the anti–/de-icing operation. While **Figure 10** shows the time evolution of the measured temperature distributions over the airfoil surface before and after starting the ice accretion process, **Figure 11** gives the corresponding surface temperatures at different chordwise locations (*i.e.*, locations of *A, B, C* and *D* at X/C = 2.0%, 10% 18% and 45% chord respectively, as indicated in **Figure 10**) as a function of the time on the two sides of the airfoil surface (*i.e.*, plasma actuator side *vs.* electric film heater side). It can be seen clearly that, after the DBD plasma actuator was switched on for 10 seconds (i.e., at t = 10s), the temperatures over the exposed electrodes of the plasma

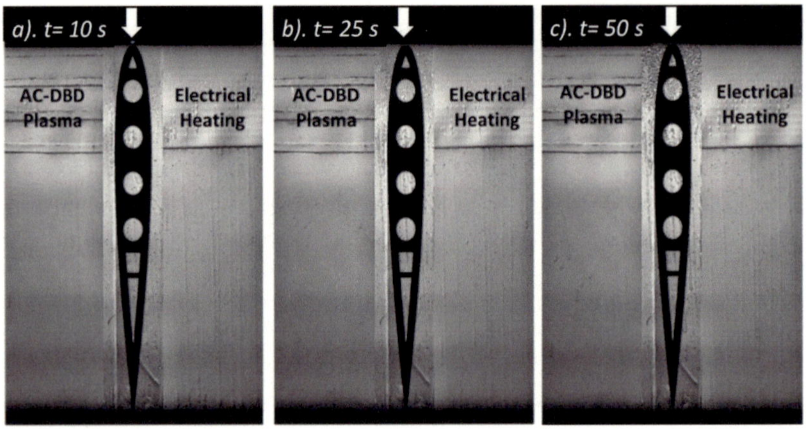

Figure 9.
Acquired images to reveal the dynamic ice accretion process over the airfoil surface with the electric power supplied to the plasma actuator and the film heater being P_d = 15.6 kW/m².

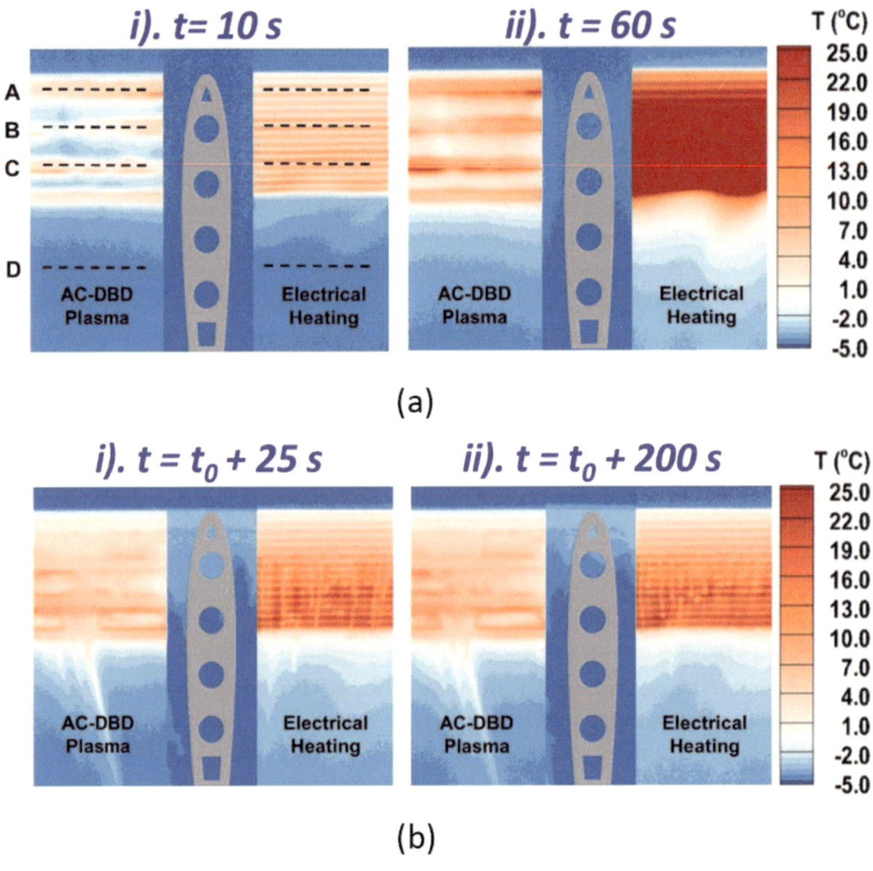

Figure 10.
Time evolution of the measured temperature distributions over the airfoil surface before and after starting the ice accretion process.

actuator were found to increase to about 10°C, while the temperatures over the dielectric surface (*i.e.*, in the spacings between the electrodes) were still quite low (i.e., below the freezing point of water), which has been discussed in the previous section. Since the electrical film heater was also switched on simultaneously, a

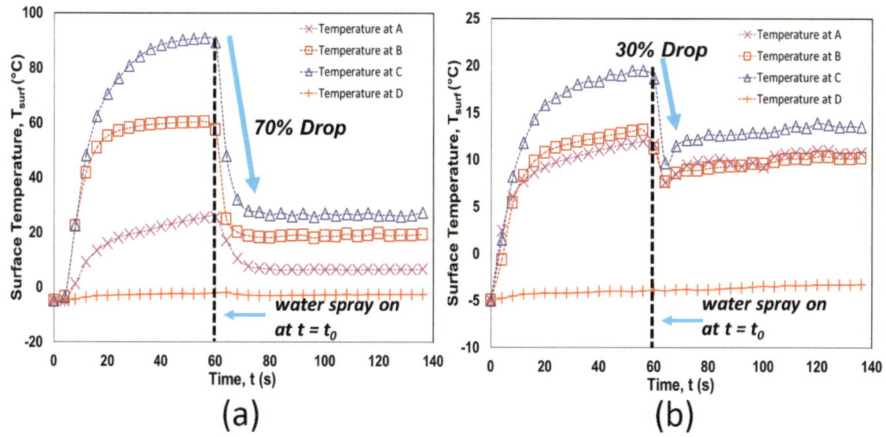

Figure 11.
Measured surface temperatures at different chordwise locations on the airfoil surface before and after starting the ice accretion process.

strip-patterned temperature distribution was found over the airfoil surface on electric film heater side at t = 10 seconds. Such a strip-like temperature distribution was due to the configuration of the etched foil resistance element encapsulated between the Polyimide films in the electric film heater.

As the time goes by, more and more thermal energy would be generated on both the plasma actuator side and the electric film heater side of the airfoil surface. At about 50 second after turning on the plasma actuator and the electric film heater, a thermal equilibrium state was found to achieve on both sides of the airfoil surface, as indicated by the flattening surface temperature profiles shown in **Figure 11(a)** and **(b)**. It can be seen clearly that, the surface temperatures over the electrical film heater were much higher than those over the DBD plasma actuator as they reached the thermal equilibrium state. The temperatures at the downstream location "C" (*i.e.*, X/D = 18%) were found to be the maximum on both sides with the measured values becoming 20°C and 90°C on the plasma actuator side and the electric film heater side, respectively. As described above, while the thermal energy generated by the electrical film heater is mainly at the heater surface through resistive heating, the primary heating mechanism in DBD plasma actuation is through gas heating and then heating up the dielectric/electrodes surfaces through direct injection, convection and radiation [32]. Therefore, with the same power input, the measured surface temperatures on the electric film heater surface were found to be much higher than those over the DBD plasma actuator. It was also found that, after the thermal equilibrium state was achieved, the temperature was higher at the locations further away from the airfoil leading edge (i.e., X/C ≈ 18%) as clearly shown in **Figures 10** and **11**. Existence of such a temperature gradient over the airfoil surface was believed to be caused by the development of the thermal boundary layer over the airfoil surface, i.e., due to the significant convective heat transfer over the airfoil surface with the maximum heat convection locating at the airfoil leading edge and decreasing gradually in the downstream region.

As indicated by the dashed line in **Figure 11**, the water spray system of ISU was switched on at t = 60 s to start the ice accretion process, i.e., at 60 seconds after turning on the plasma actuator and the electrical film heater. This time instant was also defined as t_0, as given in **Figure 10(b)**. It was found that, after the super-cooled water droplets impinged onto the airfoil surface, for the time instance at $t = t_0 + 25$ s, while the surface temperature on the electric film heater was found to decrease significantly, the temperature on the surface of the plasma actuator only dropped slightly as shown in **Figure 10**. As the time goes on, more and more impinged water would be collected on the airfoil surface. Since the power input to the plasma actuator and the electric film heater were sufficiently high to prevent ice accretion over the airfoil surfaces (i.e., P_d = 15.6 kW/m^2), the mass transport and energy transfer on both sides of the airfoil surface were found to reach an equilibrium state, as indicated by the almost unchanged temperature distributions. More quantitatively, after the water droplets impinged on the airfoil surface, while the temperature drop on the airfoil surface protected by the plasma actuator was about 33% (*i.e.*, the temperature dropped from 12.0°C to 8.0°C at the location A and B, and dropped from 18.0°C to 12.0°C at the location C), the corresponding temperature decrease son the airfoil surface protected by the electrical film heater were found to be around 70% (*i.e.*, the temperature decreased from 25.0°C to 6.0°C at the location A, from 60.0°C to 20.0°C at the location B, and from 90.0°C to 25.0°C at the location C). Such significant differences in the surface temperature changes before and after the impingement of the super-cooled water droplets can be explained by the different heating mechanisms discussed in the previous section. For the electrical film heater case, since the thermal energy was mainly generated at the heater surface, and then transferred into the impinged supercooled water droplets, the measured surface temperature, therefore, appeared to drop significantly due to the great temperature differences between the heater surface and the impinged water droplets. However, for the case with the water droplets impinging onto the surface of the DBD plasma actuator, the water droplets had already been effectively warmed up through the forced heat convection as they were flying through the hot air above the plasma actuator. Since the temperatures of the water droplets would be increased substantially before impacting onto the dielectric/electrodes surface, it results in the much smaller surface temperature drops upon the impacting of the water droplets onto the airfoil surface protected by the plasma actuator, as revealed quantitatively in **Figure 11(b)**.

4.4 Further improve the anti–/de-icing performance of the DBD plasma-based approach with a duty-cycle modulation technique

As described above, with the same power input, the DBD plasma-based approach was demonstrated to be more effective in preventing ice accretion over the airfoil surface, in comparison with the conventional electrical heating method. The anti–/de-icing performance of the DBD plasma-based approach can be further improved through optimization of the design paradigms of the plasma actuation in the terms of plasma actuation modes (i.e., AC-DBD vs. ns-DBD plasma actuation) [38], the layout design of the plasma actuator over the airfoil surface [36], the applied voltage and frequency, etc. [51]. As an example of the attempts to optimize the DBD plasma actuation for improved anti–/de-icing performance, a duty-cycled modulation concept was utilized by leveraging the unique feature of fast response time for the DBD plasma actuation (*i.e.*, on the order of 10 ~ 100 ms) in terms of momentum transfer [28] and thermal effects [47] induced by plasma discharges. **Figure 12** shows a schematic of the duty-cycle modulation from a continuous DBD plasma actuation to a duty-cycled DBD plasma actuation. Various frequencies of the duty-cycled DBD plasma actuation were examined to evaluate the effects of the duty cycle frequency on the thermal characteristics of the DBD plasma actuation [39].

Figure 13 shows the comparison of the acquired ice accretion images and the corresponding IR thermal imaging results with the DBD plasma actuators being operated at different duty-cycled frequencies. The experimental study was conducted under the same icing test condition of the test conditions of U_∞ = 40 m/s, T_∞ = −5°C and *LWC* = 1.0 g/m^3. While the maximum instantaneous power inputs

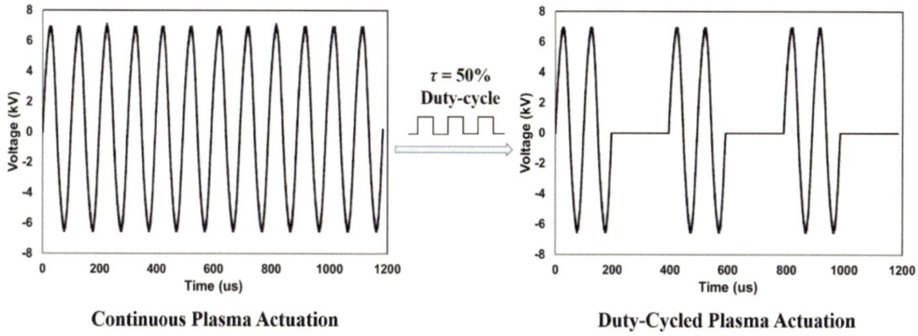

Figure 12.
A schematic of the modulation from a continuous plasma actuation to a duty-cycled plasma actuation for improved anti–/de-icing performance.

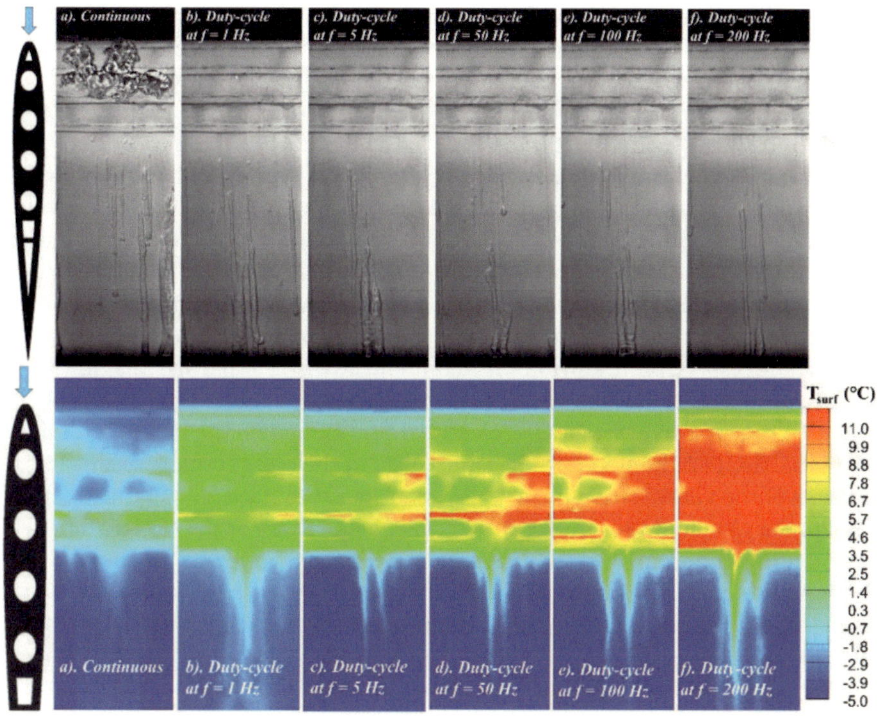

Figure 13.
Comparison of duty-cycled DBD plasma actuations modulated with different duty-cycle frequencies.

supplied to the plasma actuator were different for the cases with different duty-cycle modulation frequencies, the total power consumptions over a given period of time were kept at the same level for all the compared cases. It is clearly seen that, the cases with duty-cycled plasma actuations show much better anti–/de-icing performance (i.e., with the plasma region being completely free of ice), in comparison to that of the continuous plasma actuation (i.e., the baseline case shown at the most left side of **Figure 13**). As the duty cycle frequency increases, much less rivulets-shaped ice features were found to form over the airfoil surface. Since the surface temperatures for the cases with the duty cycled DBD plasma actuations are much higher than that of the baseline case with continuous plasma actuation, the increase in the duty cycle frequency was found to further enhance the thermal effects of the DBD plasma actuation, resulting in higher temperatures over the airfoil surface, as revealed quantitatively from the acquired IR thermal imaging results. The enhanced thermal effects of the duty-cycled DBD plasma actuation at higher duty cycle frequencies was demonstrated to be able to further improve the anti–/de-icing performance of the DBD-plasma-based approach for aircraft icing mitigation.

5. Conclusion

In the present study, we report the research progress made in our efforts to utilize the thermal effects induced by DBD plasma actuation to suppress dynamic ice accretion process over the surface of an airfoil/wing model for aircraft icing mitigation. While the fundamental mechanisms for thermal energy generation in DBD plasma actuation were introduced briefly, the significant differences in the working mechanism of the DBD-plasma-based surface heating approach from those of conventional resistive electric heating methods were highlighted for aircraft anti–/de-icing applications. By leveraging the unique Icing Research Tunnel available at Iowa State University (*i.e.*, ISU-IRT), a comprehensive experimental campaign was conducted to quantify the thermodynamic characteristics of an DBD plasma actuator embedded over the surface of an airfoil model exposed to frozen cold incoming airflow coupled with significant convective heat transfer in the context of aircraft anti–/de-icing. By integrating a DBD plasma actuator and a conventional electrical film heater onto the same airfoil/wing model, an experimental investigation was conducted to provide a side-by-side comparison between the DBD plasma-based approach and conventional resistive electrical heating method in preventing ice formation and accretion over the airfoil surface under a typical aircraft icing condition. While a high-speed camera was used to capture the transient details of the dynamic ice accretion and water transport processes over the airfoil surface, an infrared (IR) thermal imaging system was utilized to map the corresponding temperature over the airfoil surface protected by the DBD plasma actuator and the electrical film heater during the anti–/de-icing operation. Based on the side-by-side comparison of the measurement results (*i.e.*, snapshots of the visualization images and quantitative surface temperature distributions) on the plasma side of the airfoil surface against those of the electric film heater side under the same icing test condition, the effectiveness of using the thermal effects induced by DBD plasma actuation and the conventional electrical heating in preventing ice formation and accretion over the airfoil surface was evaluated and analyzed in details.

It was found that, with the same input power density, the surface temperature on the electric film heater was much higher than that over the surface of the DBD plasma actuator before the water droplets impingement, which was essentially due to the different heating mechanisms of the two methods. For the conventional electrical film heater, the thermal energy was mainly generated at the heater surface. For the case of the DBD plasma actuation, the heating path is through heat transfer from the plasma discharges to the ambient gas at first, and then heating up the surfaces of the dielectric layer and electrodes through direct injection, convection, and radiation. Upon the impingement of the super-cooled water droplets onto the airfoil surface, while the surface temperature over the electrical film heater was found to descend significantly due to the instant heat transfer from the electric film heater surface to the impinged water mass, the decrease in the measured surface temperature over the airfoil surface on the plasma actuator side appeared to be much less since the airborne water droplets were pre-heated greatly as flying through the hot air above the DBD plasma actuator before being in contacting with the heated airfoil surface. As a result, the DBD plasma-based method showed a more promising performance in preventing ice formation and accretion over the airfoil surface, in comparison with that of the conventional electrical heating method.

An explorative study was also conducted to further improve the anti–/de-icing performance of the DBD plasma-based method by adopting a duty-cycle modulation concept. It was found that the implementation of duty-cycled modulation to the DBD plasma actuation can significantly enhance the thermal effects induced by the DBD plasma actuation. It was demonstrated clearly that, under the same icing condition and the same total power input, the duty-cycled plasma actuation has a better anti–/de-icing performance in comparison to the continuous plasma actuation. The findings derived from the present study could be used to explore/optimize design paradigm for the development of novel DBD-plasma-based anti–/de-icing strategies tailored specifically for aircraft icing mitigation.

Acknowledgements

The research work is partially supported by Iowa Energy Center for Wind Turbine Icing Study under the IEC Competitive Grant # 312350 and National Science Foundation (NSF) under award numbers of CBET-1935363 and CBET-1916380.

Author details

Cem Kolbakir[1,2], Haiyang Hu[1], Yang Liu[1,3] and Hui Hu[1]*

1 Department of Aerospace Engineering, Iowa State University, Ames, IA, USA

2 Department of Aerospace Engineering, Samsun University, Samsun,Turkey

3 Department of Engineering, East Carolina University, Greenville, USA

*Address all correspondence to: huhui@iastate.edu

References

[1] Korkan, K. D., Dadone, L., and Shaw, R. J. "Helicopter Rotor Performance Degradation in Natural Icing Encounter." *Jorunall of Aircraft*, Vol. 21, No. 1, 1984, pp. 84-85. https://doi.org/10.2514/3.48226.

[2] Thomas, S. K., Cassoni, R. P., and MacArthur, C. D. "Aircraft Anti-Icing and de-Icing Techniques and Modeling." *Journal of Aircraft*, Vol. 33, No. 5, 1996, pp. 841-854. https://doi.org/10.2514/3.47027.

[3] Cebeci, T., and Kafyeke, F. "AIRCRAFT ICING." *Annual Review of Fluid Mechanics*, Vol. 35, No. 1, 2003, pp. 11-21. https://doi.org/10.1146/annurev.fluid.35.101101.161217.

[4] Bragg, M. B., Broeren, A. P., and Blumenthal, L. A. "Iced-Airfoil Aerodynamics." *Progress in Aerospace Sciences*, Vol. 41, No. 5, 2005, pp. 323-362. https://doi.org/10.1016/j.paerosci.2005.07.001.

[5] Heinrich, A., Ross, R., Zumwalt, G., Provorse, J., and Padmanabhan, V. *Aircraft Icing Handbook. Volume 2*. Gates LwarJet Corp, Wichita, KS, 1991.

[6] Gray, L. Review of Aircraft Deicing and Anti-Icing Fluid Storm Water Runoff Control Technologies. 2013.

[7] Kent, R., and Andersen, D. "Canadian Water Quality Guidelines for Glycols–An Ecotoxicological Review of Glycols and Associated Aircraft Anti-Icing and Deicing Fluids." *Environmental Toxicology*, Vol. 14(5), 1999, pp. 481-522.

[8] Thomas, S. K., Cassoni, R. P., and MacArthur, C. D. "Aircraft Anti-Icing and de-Icing Techniques and Modeling." *Journal of Aircraft*, Vol. 33(5), 2012, pp. 841-854.

[9] Lin, Y., Chen, H., Wang, G., and Liu, A. "Coatings Recent Progress in Preparation and Anti-Icing Applications of Superhydrophobic Coatings." https://doi.org/10.3390/coatings8060208.

[10] Antonini, C., Innocenti, M., Horn, T., Marengo, M., and Amirfazli, A. "Understanding the Effect of Superhydrophobic Coatings on Energy Reduction in Anti-Icing Systems." *Cold Regions Science and Technology*, Vol. 67, Nos. 1-2, 2011, pp. 58-67. https://doi.org/10.1016/j.coldregions.2011.02.006.

[11] Zhang, Z., Ma, L., Liu, Y., and Hu, H. "An Experimental Study on the Durability of a Hydro-/Ice-Phobic Surface Coating for Aircraft Icing Mitigation." *2018 Atmospheric and Space Environments Conference*, 2018, pp. 1-15. https://doi.org/10.2514/6.2018-3655.

[12] Antonini, C., Innocenti, M., Horn, T., Marengo, M., and Amirfazli, a. "Understanding the Effect of Superhydrophobic Coatings on Energy Reduction in Anti-Icing Systems." *Cold Regions Science and Technology*, Vol. 67, Nos. 1-2, 2011, pp. 58-67. https://doi.org/10.1016/j.coldregions.2011.02.006.

[13] Gao, L., Liu, Y., Ma, L., and Hu, H. "A Hybrid Strategy Combining Minimized Leading-Edge Electric-Heating and Superhydro-/Ice-Phobic Surface Coating for Wind Turbine Icing Mitigation." *Renewable Energy*, Vol. 140, 2019, pp. 943-956. https://doi.org/10.1016/J.RENENE.2019.03.112.

[14] Wang, J.-J., Choi, K.-S., Feng, L.-H., Jukes, T. N., and Whalley, R. D. "Recent Developments in DBD Plasma Flow Control." *Progress in Aerospace Sciences*, Vol. 62, 2013, pp. 52-78. https://doi.org/10.1016/j.paerosci.2013.05.003.

[15] Ebrahimi, A., and Movahhedi, M. "Power Improvement of NREL 5-MW Wind Turbine Using Multi-DBD Plasma Actuators." *Energy Conversion and Management*, Vol. 146, 2017, pp. 96-106. https://doi.org/10.1016/j.enconman.2017.05.019.

[16] Meng, X., Hu, H., Yan, X., Liu, F., and Luo, S. "Lift Improvements Using Duty-Cycled Plasma Actuation at Low Reynolds Numbers." *Aerospace Science and Technology*, Vol. 72, 2018. https://doi.org/10.1016/j.ast.2017.10.038.

[17] Roupassov, D. V., Nikipelov, A. A., Nudnova, M. M., and Starikovskii, A. Y. "Flow Separation Control by Plasma Actuator with Nanosecond Pulsed-Periodic Discharge." *AIAA Journal*, Vol. 47, No. 1, 2009, pp. 168-185. https://doi.org/10.2514/1.38113.

[18] Correale, G., Michelis, T., Ragni, D., Kotsonis, M., and Scarano, F. "Nanosecond-Pulsed Plasma Actuation in Quiescent Air and Laminar Boundary Layer." *Journal of Physics D: Applied Physics*, Vol. 47, No. 10, 2014, p. 105201. https://doi.org/10.1088/0022-3727/47/10/105201.

[19] SAMIMY, M., KIM, J.-H., KASTNER, J., ADAMOVICH, I., and UTKIN, Y. "Active Control of High-Speed and High-Reynolds-Number Jets Using Plasma Actuators." *Journal of Fluid Mechanics*, Vol. 578, 2007, p. 305. https://doi.org/10.1017/S0022112007004867.

[20] Aleksandrov, N. L., Kindysheva, S. V, Nudnova, M. M., and Starikovskiy, A. Y. "Mechanism of Ultra-Fast Heating in a Non-Equilibrium Weakly Ionized Air Discharge Plasma in High Electric Fields." *Journal of Physics D: Applied Physics*, Vol. 43, No. 25, 2010, p. 255201. https://doi.org/10.1088/0022-3727/43/25/255201.

[21] Little, J., Takashima, K., Nishihara, M., Adamovich, I., and Samimy, M. "Separation Control with Nanosecond-Pulse-Driven Dielectric Barrier Discharge Plasma Actuators." *AIAA Journal*, Vol. 50, No. 2, 2012, pp. 350-365. https://doi.org/10.2514/1.J051114.

[22] Benard, N., and Moreau, E. "On the Vortex Dynamic of Airflow Reattachment Forced by a Single Non-Thermal Plasma Discharge Actuator." *Flow, Turbulence and Combustion*, Vol. 87, No. 1, 2011, pp. 1-31. https://doi.org/10.1007/s10494-011-9325-4.

[23] Little, J., Nishihara, M., Adamovich, I., and Samimy, M. "High-Lift Airfoil Trailing Edge Separation Control Using a Single Dielectric Barrier Discharge Plasma Actuator." *Experiments in Fluids*, Vol. 48, No. 3, 2010, pp. 521-537. https://doi.org/10.1007/s00348-009-0755-x.

[24] Joussot, R., Hong, D., Weber-Rozenbaum, R., and Leroy-Chesneau, A. Modification of the Laminar-to-Turbulent Transition on a Flat Plate Using DBD Plasma Actuator. In *5th Flow Control Conference*, American Institute of Aeronautics and Astronautics, 2010.

[25] Duchmann, A., Grundmann, S., and Tropea, C. "Delay of Natural Transition with Dielectric Barrier Discharges." *Experiments in Fluids*, Vol. 54, No. 3, 2013, p. 1461. https://doi.org/10.1007/s00348-013-1461-2.

[26] Corke, T. C., Enloe, C. L., and Wilkinson, S. P. "Dielectric Barrier Discharge Plasma Actuators for Flow Control." *Annual Review of Fluid Mechanics*, Vol. 42, No. 1, 2010, pp. 505-529. https://doi.org/10.1146/annurev-fluid-121108-145550.

[27] Thomas, F. O., Corke, T. C., Iqbal, M., Kozlov, A., and Schatzman, D. "Optimization of Dielectric Barrier Discharge Plasma Actuators for Active Aerodynamic Flow Control." *AIAA Journal*, Vol. 47, No. 9, 2009, pp. 2169-2178. https://doi.org/10.2514/1.41588.

[28] Benard, N., and Moreau, E. "Electrical and Mechanical Characteristics of Surface AC Dielectric Barrier Discharge Plasma Actuators Applied to Airflow Control." *Experiments in Fluids*, Vol. 55, No. 11, 2014, p. 1846. https://doi.org/10.1007/s00348-014-1846-x.

[29] Liu, Y., Kolbakir, C., Starikovskiy, A. Y., Miles, R., and Hu, H. "An

Experimental Study on the Thermal Characteristics of NS-DBD Plasma Actuation and Application for Aircraft Icing Mitigation." *Plasma Sources Science and Technology*, Vol. 28, No. 1, 2019, p. 014001. https://doi.org/10.1088/1361-6595/aaedf8.

[30] Stanfield, S. A., Menart, J., DeJoseph, C., Kimmel, R. L., and Hayes, J. R. "Rotational and Vibrational Temperature Distributions for a Dielectric Barrier Discharge in Air." *AIAA Journal*, Vol. 47, No. 5, 2009, pp. 1107-1115. https://doi.org/10.2514/1.37648.

[31] Dong, B., Bauchire, J. M., Pouvesle, J. M., Magnier, P., and Hong, D. "Experimental Study of a DBD Surface Discharge for the Active Control of Subsonic Airflow." *Journal of Physics D: Applied Physics*, Vol. 41, No. 15, 2008, p. 155201. https://doi.org/10.1088/0022-3727/41/15/155201.

[32] Tirumala, R., Benard, N., Moreau, E., Fenot, M., Lalizel, G., and Dorignac, E. "Temperature Characterization of Dielectric Barrier Discharge Actuators: Influence of Electrical and Geometric Parameters." *Journal of Physics D: Applied Physics*, Vol. 47, No. 25, 2014, p. 255203. https://doi.org/10.1088/0022-3727/47/25/255203.

[33] Cai, J., Tian, Y., Meng, X., Han, X., Zhang, D., and Hu, H. "An Experimental Study of Icing Control Using DBD Plasma Actuator." *Experiments in Fluids*, Vol. 58, No. 8, 2017, p. 102. https://doi.org/10.1007/s00348-017-2378-y.

[34] Corke, T. C., Enloe, C. L., and Wilkinson, S. P. "Dielectric Barrier Discharge Plasma Actuators for Flow Control." *Annual Review of Fluid Mechanics*, Vol. 42, No. 1, 2010, pp. 505-529. https://doi.org/10.1146/annurev-fluid-121108-145550.

[35] Zhou, W., Liu, Y., Hu, H., Hu, H., and Meng, X. "Utilization of Thermal Effect Induced by Plasma Generation for Aircraft Icing Mitigation." *AIAA Journal*, Vol. 56, No. 3, 2018, pp. 1097-1104. https://doi.org/10.2514/1.J056358.

[36] Kolbakir, C., Hu, H., Liu, Y., and Hu, H. "An Experimental Study on Different Plasma Actuator Layouts for Aircraft Icing Mitigation." *Aerospace Science and Technology*, Vol. 107, 2020, p. 106325. https://doi.org/10.1016/j.ast.2020.106325.

[37] Meng, X., Hu, H., Li, C., Abbasi, A. A., Cai, J., and Hu, H. "Mechanism Study of Coupled Aerodynamic and Thermal Effects Using Plasma Actuation for Anti-Icing." *Physics of Fluids*, Vol. 31, No. 3, 2019. https://doi.org/10.1063/1.5086884.

[38] Liu, Y., Kolbakir, C., Hu, H., Starikovskiy, A., and Miles, R. B. "An Experimental Study on the Thermal Characteristics of NS-DBD Plasma Actuation and Application for Aircraft Icing Mitigation." *Plasma Sources Science and Technology*, 2018. https://doi.org/10.1088/1361-6595/aaedf8.

[39] Liu, Y., Kolbakir, C., Hu, H., Meng, X., and Hu, H. "An Experimental Study on the Thermal Effects of Duty-Cycled Plasma Actuation Pertinent to Aircraft Icing Mitigation." *International Journal of Heat and Mass Transfer*, Vol. 136, 2019. https://doi.org/10.1016/j.ijheatmasstransfer.2019.03.068.

[40] Joussot, R., Boucinha, V., Weber-Rozenbaum, R., Rabat, H., Leroy-Chesneau, A., and Hong, D. Thermal Characterization of a DBD Plasma Actuator: Dielectric Temperature Measurements Using Infrared Thermography. 2010.

[41] Starikovskii, A. Y., Nikipelov, A. A., Nudnova, M. M., and Roupassov, D. V. "SDBD Plasma Actuator with Nanosecond Pulse-Periodic Discharge." *Plasma Sources Science and Technology*, Vol. 18, No. 3, 2009, p. 034015. https://doi.org/10.1088/0963-0252/18/3/034015.

[42] Zhu, Y., Wu, Y., Cui, W., Li, Y., and Jia, M. "Numerical Investigation of Energy Transfer for Fast Gas Heating in an Atmospheric Nanosecond-Pulsed DBD under Different Negative Slopes." *Journal of Physics D: Applied Physics*, Vol. 46, No. 49, 2013, p. 495205. https://doi.org/10.1088/0022-3727/46/49/495205.

[43] Leonov, S. B., Petrishchev, V., and Adamovich, I. V. "Dynamics of Energy Coupling and Thermalization in Barrier Discharges over Dielectric and Weakly Conducting Surfaces on μ s to Ms Time Scales." *Journal of Physics D: Applied Physics*, Vol. 47, No. 46, 2014, p. 465201. https://doi.org/10.1088/0022-3727/47/46/465201.

[44] Popov, N. A. "Investigation of the Mechanism for Rapid Heating of Nitrogen and Air in Gas Discharges." *Plasma Physics Reports*, Vol. 27, No. 10, 2001, pp. 886-896. https://doi.org/10.1134/1.1409722.

[45] Conrads, H., and Schmidt, M. "Plasma Generation and Plasma Sources." *Plasma Sources Science and Technology*, Vol. 9, No. 4, 2000, pp. 441-454. https://doi.org/10.1088/0963-0252/9/4/301.

[46] Li, H., Chen, F., and Hu, H. "Simultaneous Measurements of Droplet Size, Flying Velocity and Transient Temperature of in-Flight Droplets by Using a Molecular Tagging Technique." *Experiments in Fluids*, Vol. 56, No. 10, 2015, p. 194. https://doi.org/10.1007/s00348-015-2063-y.

[47] Rodrigues, F., Pascoa, J., and Trancossi, M. "Heat Generation Mechanisms of DBD Plasma Actuators." *Experimental Thermal and Fluid Science*, Vol. 90, 2018, pp. 55-65. https://doi.org/10.1016/J.EXPTHERMFLUSCI.2017.09.005.

[48] Waldman, R. M., and Hu, H. "High-Speed Imaging to Quantify Transient Ice Accretion Process over an Airfoil." *Journal of Aircraft*, Vol. 53, No. 2, 2015, pp. 369-377. https://doi.org/10.2514/1.C033367.

[49] Incropera, F. P. *Fundamentals of Heat and Mass Transfer*. BOOK. John Wiley & Sons, 2011.

[50] Liu, Y., Kolbakir, C., Hu, H., and Hu, H. "A Comparison Study on the Thermal Effects in DBD Plasma Actuation and Electrical Heating for Aircraft Icing Mitigation." *International Journal of Heat and Mass Transfer*, Vol. 124, 2018, pp. 319-330. https://doi.org/10.1016/J.IJHEATMASSTRANSFER.2018.03.076.

[51] Liu, Y., Kolbakir, C., Hu, H., Starikovskiy, A. Y., and Miles, R. "A Parametric Study to Explore Ns-DBD Plasma Actuation for Aircraft Icing Mitigation." 2018. https://doi.org/10.2514/6.2018-3756.

Chapter 3

In-Liquid Plasma: A Novel Tool for Nanofabrication

Palash Jyoti Boruah, Parismita Kalita and Heremba Bailung

Abstract

This chapter focuses on synthesising nanomaterials using an emerging technology called In-Liquid Plasma, i.e., plasma generation inside a liquid. The generation of various reactive species and energetic electrons in the plasma zone plays a crucial role in synthesising nanomaterials. They act as the reducing agent. Non-requirement of the toxic chemical reducing agents make In-Liquid Plasma an environmentally friendly green approach to fabricate nanomaterials. This method enables the simultaneous synthesis of nanoparticles from the electrode material and liquid precursor, which gains much importance on the single-step synthesis of nanocomposites. Moreover, it gives flexibility in controlling both the physical and chemical parameters, which provide fine-tuning required for the size, shape and composition of nanomaterials.

Keywords: In-Liquid Plasma, Plasma zone, Reactive species, Nanomaterial, Precursor solution, Size and shape transformation

1. Introduction

Plasma technology has been involved in various biomedical and environmental applications for quite a long time. From the treatment of cancer to the polluted water, its involvement increases day by day to humankind. Moreover, plasma has also been immensely engaged in the fabrication of high-quality nanomaterials [1]. Various methods have been employed to generate the plasma both in the non-thermal and thermal conditions. Amongst them, Plasma – Liquid Interaction (PLI) gains a lot of attention as it involves both the physical and chemical processes simultaneously [2]. It offers single-step, rapid and large-scale synthesis of uniform nanomaterials with different shapes and sizes [3]. One of the advantages of PLI is that it does not require any external reducing and stabilising chemical agents, so a few purification steps can be avoided before applying in any application. The reactive species form during the generation of plasma act as the reducing or oxidising agents. Moreover, it does not require water cool vacuum chambers or pumping systems. PLI offers two effective ways for nanomaterial synthesis. The first one is from the electrode material by generating plasma between two electrodes placed in a liquid. This process is termed as In-Liquid Plasma or solution plasma. Here, nanomaterials can also synthesised using specific metal precursor solutions. Hence, both the electrode material and the precursor solution can be the source of nanomaterial formation. In the second one, nanomaterials are fabricated by generating plasma above a liquid surface. Here, one electrode is placed above the liquid surface and the other is immersed into the liquid in such a way that the liquid surface acts as the counter electrode. The liquid (precursor) acts as the nanomaterial source i.e., plasma interact with the liquid solution to synthesise nanomaterials.

In-Liquid Plasma mainly consists of three zones. The central zone is the plasma region, where the temperature goes beyond thousands of kelvin. The next one is the gaseous region, which is formed due to the evaporation of water/solvent. The outermost zone is the liquid medium, where the temperature is slightly more than the room temperature. Besides these zones, there are two interfacial regions called the plasma/gas and gas/liquid interface. These are very significant regions for nanomaterial synthesis, where many physical and chemical activities occur. A broad range of active radicals such as OH^*, H_2O_2, O_3, NO, $e^-_{(aq)}$, UV radiations and shock waves are formed in these interfacial zones [4]. **Figure 1** shows the presence of the three zones (plasma, gas and liquid) and the interfacial regions (plasma/gas and gas/liquid) during the generation of plasma inside a liquid. This chapter will mainly discuss the In-Liquid Plasma, i.e., plasma generation inside liquid to synthesise various nanomaterials. A review of nanofabrication by In-Liquid Plasma over the last two decades has been discussed in the following paragraphs.

Since the beginning of the 21st century, In- Liquid Plasma has been vigorously involved in the synthesis of various nanomaterials. In the early years, researchers have mainly focused on the synthesis of carbon-based nanomaterials [5–8]. In 2000, Ishigami et al. reported the continuous synthesis of multi-walled carbon nanotubes by generating arc discharge between two graphite electrodes inside liquid nitrogen [5]. Whereas Sano et al. in 2001 investigated the synthesis of carbon onions in water [6] and in 2004, single-walled carbon nanotubes with nanohorns in liquid nitrogen [9]. Bera et al. synthesised palladium nanoparticles filled carbon nanotubes using arc discharge in palladium chloride ($PdCl_2$) solution in 2004 [10]. After these pioneering investigations, researchers began to work on the synthesis of noble and

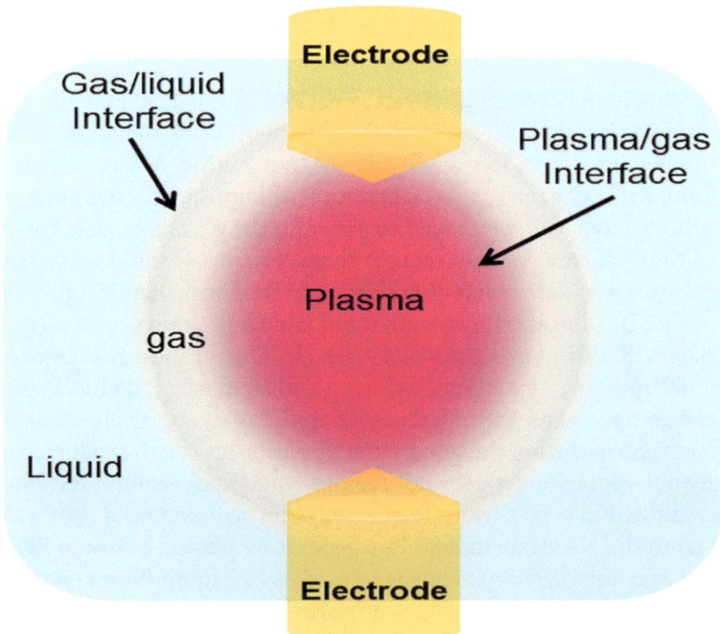

Figure 1.
Schematic representation of the generation of plasma inside a liquid showing the presence of plasma/gas and gas/liquid interfaces.

transition metal nanomaterials and their composites. Lo et al. synthesised Copper (Cu) – based nanofluids and silver nanofluids using submerged arc nanoparticle synthesis system (SANSS) in 2005 and 2007 respectively [11, 12]. Lung et al., in 2007, reported the synthesis of gold (Au) nanoparticles in water by arc discharge [13]. Ashkarran et al. synthesised tungsten trioxide (WO_3), Zinc oxide (ZnO) and Zirconium oxide (ZrO_2) nanoparticles inside water in 2008 [14], 2009 [15] and 2010 [16] respectively. In 2009, Omurzak et al. reported the synthesis of blue amorphous titanium oxide (TiO_2) and Ti_nO_{2n-1} nanoparticles by generating plasma between two titanium electrodes [17].

In the meantime, besides the erosion of electrode materials, the researchers investigated the reduction of metallic salt solutions by the In-Liquid Plasma for the synthesis of various nanomaterials. Saito et al., in 2009, synthesised Au nanoparticles by generating plasma between two tungsten electrodes inside Chloroauric acid ($HAuCl_4$) [18]. $HAuCl_4$ acts as the metal precursor for Au nanoparticles. Pootawang et al. investigated the synthesis of silver/platinum (Ag/Pt) bimetallic nanocomposites by producing plasma between a silver and a platinum electrode inside a mixture of solution containing sodium dodecylsulfonate (SDS) and sodium chloride (NaCl) using a unipolar pulse power supply in 2012 [19]. Synthesis of WO_3, Ag and Au nanoparticles by generating plasma between a metal electrode (tungsten or silver or gold) and a copper plate was investigated by Hattori et al. in 2013 [20]. Lee et al., in 2014, synthesised tin (Sn) and tin oxide (SnO_2) nanoparticles by the reduction of tin chloride dehydrate ($SnCl_2$. $2H_2O$) [21]. Fabrication of manganese (Mn) oxide/activated carbon composites was investigated by Lee et al. in 2015 [22]. They used a mixture of manganese chloride tetrahydrate ($MnCl_2.4H_2O$) and activated carbon powder as the solution. The former and latter act as the precursor for manganese oxide and carbonaceous material respectively. Synthesis of bimetallic Nickel (Ni)/Copper (Cu) nanoparticles by generating plasma inside a mixture of nickel nitrate hexahydrate ($Ni(NO_3)_2.6H_2O$) and copper nitrate tetrahydrate ($Cu(NO_3)_2.4H_2O$) solution was reported by Sun et al. in 2016 [23]. To enhance catalytic activity towards oxygen reduction reaction (ORR), Panomsuwan et al. reported the synthesis of metal-free composite of nitrogen-doped carbon nanoparticles (NCNP)/carbon nanofiber (CNF) using solution plasma in 2016 [24]. The composites were obtained by generating plasma inside a mixture of CNF and 2 – cyanopyridine ($C_6H_4N_2$), where the latter act as the source of nitrogen. Fabrication of bead-chain-like nanostructures of ZnO from the oriented attachment of spherical Zn/ZnO nanoparticles by generating DC plasma between two Zinc electrodes inside deionised water was reported by Ziashahabi et al. in 2017 [25]. Fabrication of nitrogen-doped activated carbon-supported iron oxide (Fe_2O_3) nanocomposites for supercapacitor applications was investigated by Lee et al. in 2018 [26]. They first prepared the nitrogen-doped carbon (NC) by Liquid Phase Plasma (LPP) inside a solution containing ammonium chloride (NH_4Cl) and activated carbon (AC) powder, where the former act as the precursor for nitrogen. After that, the resultant particles were mixed in iron chloride ($FeCl_2$) and Cetyltrimethyl ammonium bromide (CTAB) solution. The LPP reaction then gives the iron oxide/NC composites (IONCC). The specific capacitance and cyclic stability of NC and IONCC were found superior to the bare AC. Synthesis of Cu – Ni/CuO – NiO (CNO) nanocomposites by generating a plasma between a copper and nickel electrodes inside water using a bipolar pulse high voltage power supply was reported by Yang et al. in 2020 [27]. They found superior catalytic activity of CNO towards methanol electrocatalytic oxidation in alkaline media than the other transition metal or metal oxide based catalyst. Boruah et al. in 2021, reported a novel single-step synthesis method of Au/CuO micro/nanocomposites by generating plasma between two copper electrodes inside a solution of $HAuCl_4$ [28]. The copper electrodes acted as the source of CuO particles and the $HAuCl_4$ acted as the precursor of Au nanoparticles.

Moreover, researchers have also focused on the fabrication of nanomaterials having various defect states to enhance the catalytic activity of the materials during the last few years. Panomsuwan et al., in 2015, reported the synthesis of defect-induced black titanium oxide (H-TiO_{2-x}) nanoparticles by generating plasma between two titanium electrodes inside water [29]. They observed a higher photocatalytic performance of H-TiO_{2-x} (90%) than the commercial TiO_2 particles (18%) for the degradation of Methylene blue (MB) dye under visible light irradiation. Moreover, about 51% of MB molecules adsorbed on the surface of H-TiO_{2-x} under dark, whereas for commercial TiO_2, the adsorption was about only 9%. The same group in 2018 fabricated defect–induced heterophase anatase/brookite TiO_{2-x} nanocrystals by generating plasma inside a solution containing commercially available TiO_2 powder [30]. Active radicals present in the plasma interact with the TiO_2 particles to form defective sites. A higher gaseous photocatalytic activity towards acetaldehyde degradation to CO_2 of the plasma-treated particles (TiO_{2-x}) (91.1%) than the untreated commercial TiO_2 particles (51%) was observed. Boruah et al. in 2020 synthesised narrow bandgap tungsten oxide (WO_{3-x}) nanoparticles by generating plasma inside deionised water [31]. The reason behind the formation of narrow bandgap nanoparticles was investigated to be the presence of higher amount of oxygen vacancies. They observed higher photocatalytic performance of WO_{3-x} nanoparticles (77%) than the commercial nano WO_3 (62%) and bulk WO_3 (50%) particles under a solar simulator.

2. Experimental details and methodology

2.1 Experimental designs

The design of experimental setups mainly depends on two electrode configurations: pin-to-pin and pin-to-plane. In the first one, two pointed electrodes are placed vertically towards each other and in the second one, a pointed electrode is positioned vertically to a planar (plate) electrode. The advantage of pin-to-pin and pin-to-plane electrode configuration is to attain maximum electric field at a desired electrode gap. Various types of voltage waveforms such as DC, pulsed DC, AC, RF (radio-frequency) and microwave are used to generate plasma. **Figure 2** (a) and (b) shows the generation of plasma using the pin-to-plane and pin-to-pin electrode configurations respectively, in our laboratory. For pin-to-plane, it is observed that

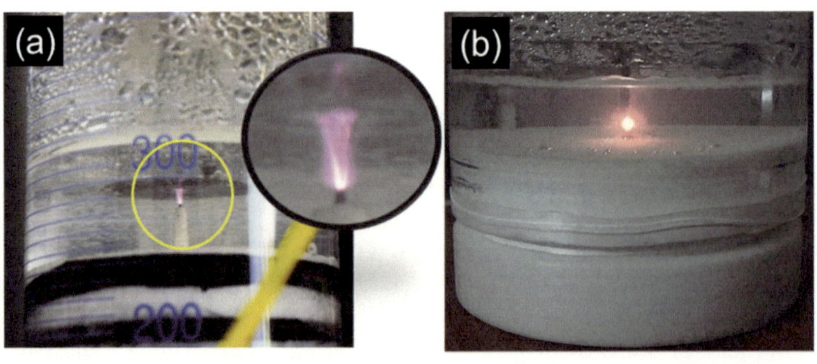

Figure 2.
Generation of plasma using (a) pin-to-plane and (b) pin-to-pin electrode configurations.

the plasma channel diverse from the pin electrode towards the plane electrode and for pin-to-pin, the plasma expands at the central region from the tip of the two electrodes. Hattori et al. employed a pin-to-plane electrode configuration to generate a plasma between a metallic electrode and a copper plate using radio-frequency plasma [20]. Whereas Lange et al. reported the use of pin-to-pin electrode configuration to generate a plasma between two vertically pointed graphite electrodes using a DC power supply [8].

Many different types of experimental setups have been designed so far for the fabrication of nanomaterials in various liquids. Ishigami et al. designed the experimental setup for the continuous production and transportation of carbon nanotubes [5]. They dipped a graphite anode and a short copper or graphite cathode having pin-to-pin electrode geometry into a vessel containing liquid nitrogen. After the generation of plasma, carbon nanotubes were formed due to the erosion of the anode material. They made funnel-shaped bottom of the vessel and sealed it with a valve to operate continuously. The valve opens periodically to transfer the nanotubes from the vessel. In a pin-to-pin or pin-to-plane electrode configuration, erosion of anode is much higher than the cathode erosion [32]. Hence, the resultant nanomaterial is made of anode material.

2.2 Mechanism of plasma generation

Usually, a pulsed high voltage power supply with a voltage rise time shorter than the Maxwellian relaxation time of the liquid is required to generate plasma inside a liquid [33]. When a high voltage is applied between the two electrodes inside a liquid, it induces a current and redistribution of the electric field. Subsequently, Joules heating at the tip of the electrodes takes place, which initiates the bubble formation due to the evaporation of the liquid. Eq. (1) provides an expression for the theoretical maximum of the electric field in pin-to-plane electrode configuration [34]:

$$E_m = \frac{2V_i}{Rln\left(2D/R\right)} \tag{1}$$

Where, V_i, R and D represent the applied voltage, the radius of curvature of the pin electrode tip and the distance between the two electrodes respectively. Therefore, to attain the maximum electric field (E_m) at a constant voltage, both the radius of curvature of the pin electrode tip and the distance between the two electrodes should be as minimum as possible. When the applied voltage is increased to a specific value, the high electric field initiates the discharge inside the bubbles. When the bubbles are bridged together, a continuous plasma channel is formed between the two electrodes. The formation of plasma or the conductive channel depends on the Joules heating. When it is larger than a threshold value, instability occurs, which stimulates the immediate evaporation of the liquid followed by thermal breakdown. Hence, plasma is generated between the two electrodes. However, when Joules heating is smaller than the threshold value, only electrolysis takes place. The mechanism of plasma discharge inside liquid also depends on its polarity. As water is a polar medium, it can conduct current and the plasma discharge occurs using the mechanism as explained above. However, for a non-polar medium, the discharge mechanism is slightly different. As non-polar medium cannot conduct electricity hence, bubble formation does not take place. Therefore, plasma generation can only be possible when the electric field between the two electrodes is high enough to trigger the dielectric breakdown of the medium. The dielectric breakdown can be defined as the sharp reduction in the electric resistance of a medium, when the electric field is higher than the dielectric constant of the medium. Li et al. compared the mechanism of plasma discharge in polar (tap water) and non-polar (benzene) solution [35]. The dielectric constant of benzene is ~10^6 V/m. They observed the discharge inside benzene after applying the voltage between the two electrodes (electrode gap 0.5 mm) reached 1.5 kV. However, for tap water, the plasma discharge channel was observed by applying less than 1 kV.

2.3 Plasma diagnostics and influence of reactive species on material fabrication

As the plasma inside the liquid is confined to a tiny region, spectroscopic diagnostics is mainly employed to determine the plasma parameters such as plasma density and temperature. Optical emission spectroscopy (OES) helps to identify the presence of various reactive species in the plasma zone. The emission spectrum obtained from the plasma zone is the superposition of the continuous spectra of electron radiation and band or line spectra of various molecules, atoms and radicals. Stark broadening of spectral lines and line intensity ratios are employed to determine the plasma density and temperatures respectively [36]. The temperature of the plasma zone is very crucial to fabricate various nanoparticles from the electrode materials. When two or more spectral lines of the same element (atom or ion) are present in the emission spectrum, then the electron/excitation temperature of the plasma from the line intensity ratio can be expressed as [37].

$$T_e = \frac{E_2 - E_1}{k \ln\left(\frac{I_1 A_2 g_2 \lambda_1}{I_2 A_1 g_1 \lambda_2}\right)} \tag{2}$$

Where, subscripts 1 and 2 denote two different spectral lines of the same element. I, E, A, λ, g, and k are the relative intensity, the energy of upper level, transition probability, the wavelength of the emission line, statistical weight and Boltzmann constant respectively. When the temperature exceeds the boiling point of the electrode material, there is a high probability of the vaporisation of the electrode material to form the nanoparticles. Dunleavy et al. observed the presence of two well-defined regions of plasma [38]. A central core having high temperature ~ (16000 ± 3500) K with high electron density $N_e \sim 5 \times 10^{17}$ cm^{-3}. The region is at local thermodynamic equilibrium (LTE). As for LTE, the minimum electron density is 2×10^{17} cm^{-3}. The other is the low density ($N_e \sim 10^{15}$ cm^{-3}) peripheral region, which is much cooler having temperature ~ 3500 K. In a unique work of deposition of anti-corrosion layer using plasma electrolytic carbonitriding on pure aluminium, the electron density and temperature were calculated to be 6×10^{15} cm^{-3} and 4000 K respectively [39].

The generation of various reactive species such as hydrogen (H^*), oxygen (O^*), hydroxyl (OH^*) and superoxide (O_2^{*-}) radicals in the plasma region can be explained using the following reactions [2]:

$$H_2O + e^- \rightarrow 2H^* + O^* + e^- \tag{3}$$

$$H_2O + e^- \rightarrow H^* + OH^* + e^- \tag{4}$$

$$O^* + O^* \rightarrow O_2 \tag{5}$$

$$H^* + O_2 \rightarrow HO_2^* \leftrightarrow O_2^{*-} + H^+ \tag{6}$$

During the formation of metal nanoparticles from the electrolytic solution, the expression for the reduction of metal ions (M^{n+}) dissolved in the solution by energetic electrons in the plasma zone is as follows:

$$M^{n+} + ne^- = M \tag{7}$$

The mechanism of the formation of Au nanoparticles using the gold precursor solution ($HAuCl_4$) can be understood by considering the Eqs. (8)-(12) reported by Ashkarran et al. [40] and Bratescu et al. [41]. Here $0 < j < 4$ and the replacement of Cl^- by OH^- depends on the pH of the solution.

$$HAuCl_4 \rightarrow H^+ + AuCl_4^- \tag{8}$$

$$AuCl_4^- + 3e^- \rightarrow Au^0 + 4Cl^- \tag{9}$$

$$AuCl_4^- + jOH^- \leftrightarrow AuCl_{4-j}^-(OH)_j^- + jCl^- \tag{10}$$

$$AuCl_4^- + 3H^* \rightarrow Au^0 + 4Cl^- + 3H^+ \tag{11}$$

$$Au(OH)_4^- + 3H^* \rightarrow Au^0 + OH^- + 3H_2O \tag{12}$$

Klapkiv et al. reported a simulation study on the synthesis of Al_2O_3 by considering the plasma channel into three zones [42]. In the central zone of the plasma channel, the temperature ranges from 7000 to 10000 K and the density of electrons in the order of 10^{22} cm^{-3}. Here, the evaporated anodic materials (made of aluminium, Al) are partially ionised and all the other species are in a monoatomic state. In this zone, Al reacts with singlet oxygen to form AlO, Al_2O, and AlO_2 as given by the following reactions:

$$Al + O \leftrightarrow AlO \tag{13}$$

$$2Al + O \leftrightarrow Al_2O \tag{14}$$

$$Al + 2O \leftrightarrow AlO_2 \tag{15}$$

In the next zone, the temperature is about 5400 K and the reactive molecular species can exist. The following reactions are possible in this zone:

$$Al + OH \leftrightarrow AlO + \tfrac{1}{2}H_2 \tag{16}$$

$$2Al + OH \leftrightarrow Al_2O + \tfrac{1}{2}H_2 \tag{17}$$

The temperature of the third zone is around 2327 K and the formation of Al_2O_3 is possible using the following equations:

$$2Al + 3O \leftrightarrow Al_2O_3 \tag{18}$$

$$2AlO + O \leftrightarrow Al_2O_3 \tag{19}$$

$$Al_2O + 2O \leftrightarrow Al_2O_3 \tag{20}$$

$$2Al + 3OH \leftrightarrow Al_2O_3 + \tfrac{3}{2}H_2 \tag{21}$$

After this region, the temperature of the liquid medium falls to around 300 K (room temperature of the liquid). During this drastic temperature change, the polymorphic transition in the oxide phases is possible.

Optical Emission Spectroscopy also provides the emission spectra of electrode materials; therefore, one can get an idea about the formation of nanoparticles from the electrode material before going through the other material characterisation techniques. To detect the plasma species, Lu et al. used OES, where they observed the emission of Cu atoms along with the other plasma species such as OH, H_α, H_β, O and Na [43]. From the emission of Cu atoms, they suggested that at first, copper foil anode is oxidised to form Cu^{2+}, which then move towards the cathode due to the external electric field inside the plasma region. During their movement, they react with high-energy electrons and H atoms to form Cu atoms. As copper atoms are highly reactive in water, they are easily oxidised to form CuO nanoparticles. However, Saito et al. reported the synthesis of CuO nanoflower by considering a slightly different mechanism [44]. Firstly, copper hydroxide [$Cu(OH)_2$] is formed at the surface of the copper electrode. The temperature of the electrolyte covering the electrode goes beyond 260°, which is sufficient to melt the surface of the electrode. Secondly, $Cu(OH)_2$ interacts with plasma-generated OH^- to form tetrahydroxocuprate(II) anions [$Cu(OH)_4]^{2-}$. Lastly, as the temperature drops, precipitation of CuO (s) occurs by releasing H_2O and OH^-. Preferential growth of crystal plane along with a specific direction forms spindle structures.

3. Synthesis of nanomaterials by controlling process parameter

It is well known that nanoparticles have widespread applications in various field such as solar cells, photo-thermal cancer treatment, controlled drug delivery, catalysis etc., because of its tunable optical, electrical and catalytic properties which strictly depends on its structure and morphology of the particle. Therefore, nowadays, researchers have devoted substantial effort to have proper control on the shape and size of nanoparticles by simply controlling the fundamental physical and chemical parameters. One of the main advantages of In-Liquid Plasma method for nanofabrication is that it provides flexibility in controlling both the physical (plasma) as well as the chemical (solution) parameters. However, in the chemical-based synthesis, only solution parameters and in the gas phase plasma synthesis of nanomaterial, only plasma parameters can be controlled. Process parameters have a very significant role in the morphology and composition of nanomaterials.

3.1 Variation of physical parameters

In the plasma – liquid interface, due to the interaction of energetic electrons with the liquid medium, various reactive species are formed, which play a significant role as the reducing agent in nanomaterial synthesis. Interestingly, these plasma-generated reducing agents are directly related to the discharge voltage and current applied to generate plasma. Therefore, the proper understanding of the role of discharge voltage and current on nanomaterial synthesis is very essential. Saito et al. investigated the size of Au nanoparticles by varying the discharge voltage [18]. For the applied voltage of 1600 and 3200 V, after 1 min of plasma discharge, dendrite shape nanoparticles of size ~150 nm and after 5 min of discharge ~50 nm nanoparticles were observed. After 20 mins of discharge, for 1600 V applied voltage, a slight change of the size of the nanoparticles was observed. However, for 3200 V applied voltage, the particle size decreased significantly with some anisotropic shapes such as triangular, pentagonal and hexagonal. After 45 mins of discharge, the size of the nanoparticles reduced up to 20 nm. During the experiment, they also observed that the pH of the solution ($HAuCl_4$) decreased with the increase of discharge time. The change in pH explains the formation small nanoparticles. At low pH, gold nanoparticles dissolve and the reduction rate of gold ion decreases, which leads to a reduction in nanoparticle size and formation of exotic or anisotropic shapes. Ashkarran et al. reported the effect of discharge current on the size of Zirconium oxide (ZrO_2) nanoparticles [16]. For 10 A and 20 A arc current, the average size (diameter) of the spherical particles were 21 nm and 42 nm respectively, i.e. size of the particles increased with the increase in discharge current. As the smaller particles have a larger specific surface area, therefore, they observed higher photocatalytic activity of the nanoparticles synthesised at lower discharge current. However, Ziashahabi et al. reported contradictory results, where they observed a decrease in the size of the Zn/ZnO nanocomposites with the increase of discharge current [25]. By maintaining the discharge current at 20, 50, 100 and 150 A, the diameter of the particles was 60, 40, 35 and 26 nm respectively. They also observed that the shape of the nanoparticles changed from spherical to bead-like at higher discharge current. For 50, 100 and 150 A discharge current, the length of bead-like aggregates was 114, 117 and 120 nm respectively. Therefore, with the increase in discharge current, the size of nanoparticles decreased in diameter and increased in length. Jin et al. observed the effect of discharge voltage on the shape and size of Ag nanoparticles [45]. To investigate the phenomenon, they fixed the other physical and chemical parameters such as pulse frequency, pulse width, discharge duration, electrode gap and solution concentration at 20 kHz, 2 μs, 600 s, 1 mm and 0.5 mM respectively. When the discharge voltage was in the range of 800–900 V, the particles formed were mostly aggregated and had some dendritic structures. Further increase in discharge voltage to 1000 V, only dendritic structure was observed. This change in morphology of the Ag nanoparticles at higher voltage has been explained by considering the orbit – limiting charging model [46]. The model relates the particle charge with the surface potential as:

$$Q = C\Phi_s \tag{22}$$

Where, Q, C and Φ_s represents the charge, capacitance of the particle in plasma and the surface potential of the particle respectively. As the quantity of electrons is directly proportional to the particle charge. Therefore, when the discharge voltage increases, the generation of energetic electrons also increases, generating more charged particles. A considerable number of electrons surrounds the surface of these charged particles. Hence, further reduction of the Ag ion will take place quite quickly on the already nucleated negatively charged particle, which helps in forming a dendrite structure. They also investigated the effect of nanoparticle morphology by controlling the discharge duration. They also studied the impact on the morphology of the nanoparticles by varying the discharge duration and keeping the applied voltage at 1000 V. At 120 s of discharge duration, nearly spherical Ag nanoparticles of size 12.7 ± 4.4 nm in diameter were observed. When the duration was increased to 500 s, aggregated nanoparticles of a dendritic shape having branch of size 61.8 ± 21.8 nm was observed. However, further increase in discharge duration to 600 s showed an abrupt increase in the size of the dendritic structure to 153.8 ± 54.6 nm. Moreover, during the experiment, they observed the rise of solution temperature from 333 to 368 K. From this observation of temperature rise, they explained the change of morphology of the nanoparticles by considering the Brownian motion of the particles. The below equation is used to explain the relationship between the temperature and the Brownian motion of the nanoparticles:

$$D = kT/6\pi\eta r \tag{23}$$

Where, D, k, T, η and r represents the diffusion constant of particle, Boltzmann constant, temperature, viscosity and radius of particle respectively. Since, along with the discharge duration, the temperature of the solution increased. Therefore, the average kinetic energy of the nanoparticles also increases, which helps in aggregating the particles, due to continuous encounters with each other.

Sun et al. observed the variation on Ni – Cu bimetallic nanoparticle size and shape as well as the percentage of metal components with the plasma discharge duration [23]. With the increase of discharge duration, the size of the nanoparticles was observed to increase. At the initial stage of discharge, spherical bimetallic nanoparticles and after longer plasma treatment time flower-shaped nanoparticles were observed. Moreover, they observed higher copper content in the bimetallic particles. The atomic composition of the resulting bimetallic nanoparticles was 60% Cu, 30% Ni and 10% Oxygen. The observation can be explained by considering the standard potentials of copper and nickel, as shown by the following equations.

$$Cu^{2+}(aq) + 2e^{-} \rightarrow Cu(s) E^{0} = +0.34V \tag{24}$$

$$Ni^{2+}(aq) + 2e^{-} \rightarrow Ni(s) E^{0} = -0.25V \tag{25}$$

Since, copper has a higher standard potential i.e. lower ionisation tendency than nickel. Hence, copper ions are reduced faster than nickels to have more copper content in the resulting bimetallic nanoparticles. Kang et al. investigated a different phenomenon, where they observed the size and crystallinity of carbon nanospheres by varying the pulse frequency of a bipolar pulse power supply [47]. The voltage, pulse width and electrode gap were controlled at 1.3 kV, 2 μs and 1 mm respectively. Benzene was used as the precursor for carbon nanospheres. By adjusting the pulse frequency from 25 to 65 kHz, the average diameter of the carbon nanospheres was observed to be 20 to 100 nm. Moreover, during the discharge, another interesting phenomenon was observed. When the pulse frequency was adjusted from 25 to 50 kHz, amorphous carbon spheres were synthesised. On the other hand, at 65 kHz, synthesised carbon nanospheres composed continuous short-range graphite with turbostratic structure.

3.2 Variation of chemical parameters

The chemical parameters such as concentration of the solution, pH and use of surfactant plays a significant role in the morphological and chemical compositions of the nanomaterials. In a recent work on the synthesis of Au/CuO micro/nanocomposites, we have reported the shape transformation of CuO particles by simply varying the concentration of the gold precursor ($HAuCl_4$) solution [28]. In the experiment, simultaneously, both the electrode (Cu) and the liquid solution act as the source of materials for the formation of Au/CuO micro/nanocomposites. At low concentration (0.1 mM $HAuCl_4$) of the solution, the shape of CuO was found to be spindle. When the concentration was increased to 0.5 mM, along with the spindle shape, several rod-like structures of CuO were also observed. However, at higher concentration (1 mM), the spindle shape of CuO completely transformed to sheet – like structure. The shape transformation of CuO is believed to be due to the presence of a large number of foreign metal (Au^{3+}) and halide (Cl^{-}) ions at a higher concentration of gold precursor solution. The pH of the gold solution could also be responsible for the shape transformation process. Moreover, with the increase of solution concentration from 0.1 to 1 mM, the size of the Au nanoparticles was found to increase from 7.73 ± 0.11 to 37.50 ± 1.50 nm. A different work reported by Saito et al. investigated the morphology of copper/copper oxide nanoparticles synthesised from the electrode material by varying the concentration of K_2CO_3 solution [44]. They observed the formation of CuO nanoflowers having sharp nanorods, where size increased with increasing the solution concentration. The pH value of the precursor solution also plays a pivotal role in controlling the structure of nanomaterials. In most cases, the pH value of the initial precursor solution is controlled by using different concentrations of NaOH solution. Bratescu et al. investigated the variation of pH on the size of the Au nanoparticles [41]. They used $HAuCl_4.3H_2O$ as the gold precursor. To adjust the pH of the solution, different amount of NaOH was used. At pH 3, 6 and 12 the average size of the Au nanoparticles was measured to be 10, 4 and 2 nm respectively. They explained the size variation of Au nanoparticles with pH by considering the standard redox potential of Eqs. (11) and (12), which occur at pH 3 and 12 respectively. The standard redox potential of the reduction of $AuCl_4^-$ to Au^0 at pH 3 is 0.95 eV, whereas for $Au(OH)_4^-$ to Au^0 at pH 12 is 0.60 eV [48]. As greater redox potential leads to the formation of a high number of atoms, hence, at pH 3 more number of Au atoms were formed, which aggregates and generate nanoparticles with sizes ~10 nm. At higher pH, less number of atoms were formed, which lead to the formation of smaller Au nanoparticles.

Although most researchers do not use any stabilising agents, a few reported the addition of surfactant to the initial solution. Surfactants mainly act as a capping agent to control the size and shape of the nanoparticles by preventing the aggregation of particles. Kim et al. reported the use of Polyvinylpyrrolidone (PVP) as the stabiliser to observe the size variation of Au nanoparticles at different concentrations [49]. For the experiment, they dissolved 0.1 mM gold precursor $HAuCl_4.3H_2O$ in ethylene glycol. After the plasma discharge, Au nanoparticles with various shapes such as triangular, square and nearly spherical, having sizes 20.85 ± 2.78 nm in diameter and a few nanorods with 10 nm in diameter and 40–45 nm in length, were observed. After that, they mixed potassium chloride (KCl) to generate high plasma density, as potassium ions have higher oxidation potential than hydrogen. For 0.05 and 0.1 M KCl, most of the nanoparticles were observed to be nearly spherical of diameters 17.1 ± 0.48 and 16.38 ± 0.48 nm respectively. Therefore, the use of KCl had a significant role in the shape of the nanoparticles. When 0.01 mM PVP was added to a mixture of 0.05 M KCl and 0.1 mM $HAuCl_4.3H_2O$ solution, the size of the Au nanoparticles was reduced to 12.32 ± 0.87 nm. Moreover, with the increase of the concentration of PVP, the size of the Au nanoparticles decreased and they have a high tendency to become spherical. It indicates that the use of PVP effectively protects the surface of the Au nanoparticles by limiting the crystal growth and results in spherical nanoparticles. Lee et al. reported the effect of a cationic and anionic surfactant on the size of nickel (Ni) nanoparticles [50]. For the cationic and anionic surfactant, Cetyltrimethyl ammonium bromide (CTAB) and Sodium dodecyl sulfate (SDS) respectively were used. Nickel chloride hexahydrate ($NiCl_2.6H_2O$) solution was used as the precursor for Ni nanoparticles. When SDS was added to the solution, formation of spherical nanoparticles at all molar ratios of SDS/$NiCl_2$ were observed. On the other hand, when CTAB was added up to 20% of the molar ratio of CTAB/$NiCl_2$, smaller spherical nanoparticles than the no surfactant case was observed. When the molar ratio was increased to 30% or greater, the formation of large polygonal or whisker-shaped particles was observed. From this experiment, it has been established that cationic surfactant play a significant role in tuning the size and shape of the nanoparticles. Use of surfactant also influenced the composition of transition metal nanoparticles, as they have very high probability to form oxides in liquid environment. For the synthesis of copper (Cu) nanoparticles, CTAB or other surfactants have to be used otherwise formation of spindle-like Cu_2O/CuO structures are frequently observed [51, 52]. Change of composition of the nanoparticles from metal (Sn) to metal oxide (SnO_2) with discharge time even after the use of surfactant has also been reported by Lee et al. [21]. They explained the observation by investigating the pH of the solution. With the increase in discharge time for 50 mins, solution pH decreased i.e., H_2O_2 and HNO_3 were formed in the solution. HNO_3 may react with Sn to form SnO_2 nanoparticles as shown by the equation:

$$Sn + 4HNO_3 \rightarrow SnO_2 + 4NO_2(g) + 2H_2O \tag{26}$$

As HNO_3 was consumed during SnO_2 synthesis, the pH of the solution again increased for the discharge duration of 50 to 60 mins.

4. Conclusion

The main concern of this chapter is to deliver a general perception on the generation of plasma inside liquid and its importance in the field of material fabrication. In-Liquid Plasma method offers single-step, rapid and large-scale synthesis of uniform nanomaterials with different shapes, sizes and compositions. Moreover, from the economic point of view, it is a cost-effective approach as it does not require any gas, water cool vacuum chambers and pumping systems. Interaction of various reactive species on the synthesis of pure metal and metal oxide nanoparticles as well as nanocomposites has been discussed in this chapter. Morphological and compositional modification of nanomaterials by controlling the physical (e.g. applied voltage, current, pulse width and discharge duration) and chemical (e.g. solution concentration, pH and surfactant) parameters have also been discussed.

Author details

Palash Jyoti Boruah, Parismita Kalita and Heremba Bailung*
Plasma Application Laboratory, Physical Sciences Division, Institute of Advanced Study in Science and Technology (IASST), Guwahati, India

*Address all correspondence to: hbailung@yahoo.com

References

[1] Dou S, Tao L, Wang R, et al (2018) Plasma-Assisted Synthesis and Surface Modification of Electrode Materials for Renewable Energy. Adv Mater 30: 1705850. https://doi.org/10.1002/adma.201705850

[2] Bruggeman PJ, Kushner MJ, Locke BR, et al (2016) Plasma-liquid interactions: A review and roadmap. Plasma Sources Sci Technol 25:053002. https://doi.org/10.1088/0963-0252/25/5/053002

[3] Chen Q, Li J, Li Y (2015) A review of plasma-liquid interactions for nanomaterial synthesis. J Phys D Appl Phys 48:424005. https://doi.org/10.1088/0022-3727/48/42/424005

[4] Lukes P, Locke BR (2005) Plasmachemical oxidation processes in a hybrid gas-liquid electrical discharge reactor. J Phys D Appl Phys 38:4074–4081. https://doi.org/10.1088/0022-3727/38/22/010

[5] Ishigami M, Cumings J, Zettl A, Chen S (2000) A simple method for the continuous production of carbon nanotubes. Chem Phys Lett 319:457–459. https://doi.org/10.1016/S0009-2614(00)00151-2

[6] Sano N, Wang H, Chhowalla M, et al (2001) Synthesis of carbon 'onions' in water. Nature 414:506–507. https://doi.org/10.1038/35107141

[7] Zhu HW, Li XS, Jiang B, et al (2002) Formation of carbon nanotubes in water by the electric-arc technique. Chem Phys Lett 366:664–669. https://doi.org/10.1016/S0009-2614(02)01648-2

[8] Lange H, Sioda M, Huczko A, et al (2003) Nanocarbon production by arc discharge in water. Carbon N Y 41:1617–1623. https://doi.org/10.1016/S0008-6223(03)00111-8

[9] Sano N, Nakano J, Kanki T (2004) Synthesis of single-walled carbon nanotubes with nanohorns by arc in liquid nitrogen. Carbon N Y 42:686–691. https://doi.org/10.1016/j.carbon.2003.12.078

[10] Bera D, Kuiry SC, McCutchen M, et al (2004) In-situ synthesis of palladium nanoparticles-filled carbon nanotubes using arc-discharge in solution. Chem Phys Lett 386:364–368. https://doi.org/10.1016/j.cplett.2004.01.082

[11] Lo CH, Tsung TT, Chen LC (2005) Shape-controlled synthesis of Cu-based nanofluid using submerged arc nanoparticle synthesis system (SANSS). J Cryst Growth 277:636–642. https://doi.org/10.1016/j.jcrysgro.2005.01.067

[12] Lo CH, Tsung TT, Lin HM (2007) Preparation of silver nanofluid by the submerged arc nanoparticle synthesis system (SANSS). J Alloys Compd 434–435:659–662. https://doi.org/10.1016/j.jallcom.2006.08.217

[13] Lung JK, Huang JC, Tien DC, et al (2007) Preparation of gold nanoparticles by arc discharge in water. J Alloys Compd 434–435:655–658. https://doi.org/10.1016/j.jallcom.2006.08.213

[14] Ashkarran AA, Zad AI, Ahadian MM, Ardakani SAM (2008) Synthesis and photocatalytic activity of WO3 nanoparticles prepared by the arc discharge method in deionized water. Nanotechnology 19:195709. https://doi.org/10.1088/0957-4484/19/19/195709

[15] Ashkarran AA, Iraji zad A, Mahdavi SM, Ahadian MM (2009) ZnO nanoparticles prepared by electrical arc discharge method in water. Mater Chem Phys 118:6–8. https://doi.org/10.1016/j.matchemphys.2009.07.002

[16] Ashkarran AA, Afshar SAA, Aghigh SM, kavianipour M (2010) Photocatalytic activity of ZrO2 nanoparticles prepared by electrical arc discharge method in water. Polyhedron 29:1370–1374. https://doi.org/10.1016/j.poly.2010.01.003

[17] Omurzak E, Mashimo T, Iwamoto C, et al (2009) Synthesis of blue amorphous TiO 2 and Ti nO 2n-1 by the impulse plasma in liquid. J Nanosci Nanotechnol 9:6372–6375. https://doi.org/10.1166/jnn.2009.1331

[18] Saito N, Hieda J, Takai O (2009) Synthesis process of gold nanoparticles in solution plasma. Thin Solid Films 518: 912–917. https://doi.org/10.1016/j.tsf.2009.07.156

[19] Pootawang P, Saito N, Takai O, Lee SY (2012) Synthesis and characteristics of Ag/Pt bimetallic nanocomposites by arc-discharge solution plasma processing. Nanotechnology 23:395602. https://doi.org/10.1088/0957-4484/23/39/395602

[20] Hattori Y, Nomura S, Mukasa S, et al (2013) Synthesis of tungsten oxide, silver, and gold nanoparticles by radio frequency plasma in water. J Alloys Compd 578:148–152. https://doi.org/10.1016/j.jallcom.2013.05.032

[21] Lee H, Park SH, Kim SJ, et al (2014) Synthesis of tin and tin oxide nanoparticles using liquid phase plasma in an aqueous solution. Microelectron Eng 126:153–157. https://doi.org/10.1016/j.mee.2014.07.014

[22] Lee H, Park SH, Kim SJ, et al (2015) Synthesis of manganese oxide/activated carbon composites for supercapacitor application using a liquid phase plasma reduction system. Int J Hydrogen Energy 40:754–759. https://doi.org/10.1016/j.ijhydene.2014.08.085

[23] Sun SH, Jung SC (2016) Facile synthesis of bimetallic Ni-Cu nanoparticles using liquid phase plasma method. Korean J Chem Eng 33:1075–1079. https://doi.org/10.1007/s11814-015-0262-0

[24] Panomsuwan G, Saito N, Ishizaki T (2016) Nitrogen-Doped Carbon Nanoparticle-Carbon Nanofiber Composite as an Efficient Metal-Free Cathode Catalyst for Oxygen Reduction Reaction. ACS Appl. Mater. Interfaces 8: 6962–6971

[25] Ziashahabi A, Poursalehi R, Naseri N (2017) Formation mechanism of bead-chain-like ZnO nanostructures from oriented attachment of Zn/ZnO nanocomposites prepared via DC arc discharge in liquid. Mater Sci Semicond Process 72:128–133. https://doi.org/10.1016/j.mssp.2017.09.032

[26] Lee H, Lee WJ, Park YK, et al (2018) Liquid phase plasma synthesis of iron oxide nanoparticles on nitrogen-doped activated carbon resulting in nanocomposite for supercapacitor applications. Nanomaterials 8:190. https://doi.org/10.3390/nano8040190

[27] Yang B, Yu Y, Qiao J, et al (2020) Solution plasma method for the preparation of Cu-Ni/CuO-NiO with excellent methanol electrocatalytic oxidation performance. Appl Surf Sci 513:145808. https://doi.org/10.1016/j.apsusc.2020.145808

[28] Boruah PJ, Khanikar RR, Bailung H (2021) Novel single-step synthesis and shape transformation of Au/CuO micro/nanocomposites using plasma-liquid interaction. Nanotechnology 32:245601. https://doi.org/10.1088/1361-6528/abecb9

[29] Panomsuwan G, Watthanaphanit A, Ishizaki T, Saito N (2015) Water-plasma-assisted synthesis of black titania spheres with efficient visible-light photocatalytic activity. Phys Chem Chem Phys 17:13794–13799. https://doi.org/10.1039/c5cp00171d

[30] Pitchaimuthu S, Honda K, Suzuki S, et al (2018) Solution Plasma Process-Derived Defect-Induced Heterophase Anatase/Brookite TiO2 Nanocrystals for Enhanced Gaseous Photocatalytic Performance. ACS Omega 3:898–905. https://doi.org/10.1021/acsomega.7b01698

[31] Boruah PJ, Khanikar RR, Bailung H (2020) Synthesis and Characterization of Oxygen Vacancy Induced Narrow Bandgap Tungsten Oxide (WO3−x) Nanoparticles by Plasma Discharge in Liquid and Its Photocatalytic Activity. Plasma Chem Plasma Process 40:1019–1036. https://doi.org/10.1007/s11090-020-10073-3

[32] Parkansky N, Glikman L, Beilis II, et al (2007) W-C electrode erosion in a pulsed arc submerged in liquid. Plasma Chem Plasma Process 27:789–797. https://doi.org/10.1007/s11090-007-9099-6

[33] Yong Yang, Young I. Cho AF (2012) Plasma Discharge in Liquid: Water Treatment and Applications. CRC Press, Taylor & Francis Group, New York

[34] Fridman A, Gutsol A, Cho YI (2007) Non-thermal atmospheric pressure plasma. Adv Heat Transf 40:1–142. https://doi.org/10.1016/S0065-2717(07)40001-6

[35] Lun Li HO, Kang J, Urashima K, Saito N (2013) Comparison between the Mechanism of Liquid Plasma Discharge Process in Water and Organic Solution. J Inst Electrostat Jpn 37:22–27

[36] Hussein RO, Nie X, Northwood DO, et al (2010) Spectroscopic study of electrolytic plasma and discharging behaviour during the plasma electrolytic oxidation (PEO) process. J Phys D Appl Phys 43:105203. https://doi.org/10.1088/0022-3727/43/10/105203

[37] Klapkiv MD, Nykyforchyn HM, Posuvailo VM (1995) Spectral analysis of an electrolytic plasma in the process of synthesis of aluminum oxide. Mater Sci 30:333–343. https://doi.org/10.1007/BF00569685

[38] Dunleavy CS, Golosnoy IO, Curran JA, Clyne TW (2009) Characterisation of discharge events during plasma electrolytic oxidation. Surf Coatings Technol 203:3410–3419. https://doi.org/10.1016/j.surfcoat.2009.05.004

[39] Wu J, Zhang Y, Liu R, et al (2015) Anti-corrosion layer prepared by plasma electrolytic carbonitriding on pure aluminum. Appl Surf Sci 347:673–678. https://doi.org/10.1016/j.apsusc.2015.04.171

[40] Ashkarran AA, Iraji zad A, Mahdavi SM, et al (2009) Rapid and efficient synthesis of colloidal gold nanoparticles by arc discharge method. Appl Phys A Mater Sci Process 96:423–428. https://doi.org/10.1007/s00339-009-5288-x

[41] Bratescu MA, Cho SP, Takai O, Saito N (2011) Size-controlled gold nanoparticles synthesized in solution plasma. J Phys Chem C 115:24569–24576. https://doi.org/10.1021/jp207447c

[42] Klapkiv MD (1999) Simulation of synthesis of oxide-ceramic coatings in discharge channels of a metal-electrolyte system. Mater Sci 35:279–283. https://doi.org/10.1007/BF02359992

[43] Lu Q, Wang X, Yu J, et al (2020) Synthesis of spindle-like CuO nanoparticles by using cathode glow discharge electrolysis plasma. Mater Lett 264:127316. https://doi.org/10.1016/j.matlet.2020.127316

[44] Saito G, Hosokai S, Tsubota M, Akiyama T (2011) Synthesis of copper/copper oxide nanoparticles by solution plasma. J Appl Phys 110:023302. https://doi.org/10.1063/1.3610496

[45] Jin S, Kim S-M, Lee S, Kim J (2014) Synthesis and characterization of silver nanoparticles using a solution plasma process. J Nanosci Nanotechnol 14: 8094–8097. https://doi.org/10.1166/jnn.2014.9428

[46] Goree J (1994) Charging of particles in a plasma. Plasma Sources Sci Technol 3:400–406. https://doi.org/10.1088/0963-0252/3/3/025

[47] Kang J, Li OL, Saito N (2013) Synthesis of structure-controlled carbon nano spheres by solution plasma process. Carbon N Y 60:292–298. https://doi.org/10.1016/j.carbon.2013.04.040

[48] Goia D V., Matijević E (1999) Tailoring the particle size of monodispersed colloidal gold. Colloids Surfaces A Physicochem Eng Asp 146: 139–152. https://doi.org/10.1016/S0927-7757(98)00790-0

[49] Kim SM, Kim GS, Lee SY (2008) Effects of PVP and KCl concentrations on the synthesis of gold nanoparticles using a solution plasma processing. Mater Lett 62:4354–4356. https://doi.org/10.1016/j.matlet.2008.07.025

[50] Lee H, Chung M, Ahn HG, et al (2015) Effect of the surfactant on size of nickel nanoparticles generated by liquid-phase plasma method. Int J Precis Eng Manuf 16:1305–1310. https://doi.org/10.1007/s12541-015-0170-6

[51] Xie SY, Ma ZJ, Wang CF, et al (2004) Preparation and self-assembly of copper nanoparticles via discharge of copper rod electrodes in a surfactant solution: A combination of physical and chemical processes. J Solid State Chem 177:3743–3747. https://doi.org/10.1016/j.jssc.2004.07.012

[52] Yao WT, Yu SH, Zhou Y, et al (2005) Formation of uniform CuO nanorods by spontaneous aggregation: Selective synthesis of CuO, Cu 2 O, and Cu nanoparticles by a solid-liquid phase arc discharge process. J Phys Chem B 109:14011–14016. https://doi.org/10.1021/jp0517605

Section 2

Transport Developments

Chapter 4

Polarized Thermal Conductivity of Two-Dimensional Dusty Plasmas

Aamir Shahzad, Madiha Naheed, Aadil Mahboob, Muhammad Kashif, Alina Manzoor and H.E. Maogang

Abstract

The computation of thermalt properties of dusty plasmas is substantial task in the area of science and technology. The thermal conductivity (λ) has been computed by applying polarization effect through molecular dynamics (MD) simulations of two dimensional (2D) strongly coupled complex dusty plasmas (SCCDPs). The effects of polarization on thermal conductivity have been measured for a wide range of Coulomb coupling (Γ) and Debye screening (κ) parameters using homogeneous non-equilibrium molecular dynamics (HNEMD) method for suitable system sizes. The HNEMD simulation method is employed at constant external force field strength (F^*) and varying polarization effects. The algorithm provides precise results with rapid convergence and minute dimension effects. The outcomes have been compared with earlier available simulation results of molecular dynamics, theoretical predictions and experimental results of complex dusty plasma liquids. The calculations show that the kinetic energy of $SCCDP_S$ depends upon the system temperature (≡ 1/Γ) and it is independent of higher screening parameter. Furthermore, it has shown that the presented HNEMD method has more reliable results than those obtained through earlier known numerical methods.

Keywords: Plasma thermal conductivity, complex dusty plasma, Homogenous non-equilibrium molecular dynamics, force field strength, system size, plasma parameters etc.

1. Introduction

Recently, thermophysical properties of complex materials are a major concern in the field of science and engineering. The term thermophysical properties used to pass on both thermodynamic and transport properties. Experimental or theoretical methods to study properties of fluids depend on microscopic and macroscopic categories [1–4]. The conventional macroscopic measurements depend on the state of stress, temperature, and density. Thermodynamic properties are defined by the equilibrium conditions of the system which consist of temperature, heat capacity, entropy, pressure, internal energy, enthalpy, and density, whereas the transport properties comprise thermal conductivity, diffusion viscosity, and waves with their instabilities. For further explanation of the process in detail for these systems, data that is applicable to thermodynamics, transport, optics, transmission, light, and other features are required for non-ideal plasma [5–9]. In this regard, various opinions regarding computer research methods including theoretical and numerical performance have greatly improved for non-ideal Plasma [10]. Determination for some reason, thermal conductivity is also a big problem for thermophysical researchers. Developmental aspects of heat transport in micron and nanoscale materials have shifted to the domain of technical issues as there are other areas, such as phonon heat transfer in semiconductor superlattices, which have received widespread attention from researchers. To study the internal energy of particles, their momentum, and heat transfer thus remains a crucial task. Therefore, thermal management, strategies sustainable high performance, reliability, and service life are main purposes. One such strategy is to develop new therapeutic materials based on dusty plasma that are more effective. Regulation with approval became a significant issue in modern technology [11]. Yet similar interests are present in plasma fusion, and it can be productive radiotoxic dust in plasma-wall reactions. In many ways, this chapter provides an update literature survey on thermal transport as well as heat flow strategies to determine thermal behaviors in two-dimensional (2D) complex liquids. The coefficients were computed through the Green Kubo (GK) equilibrium molecular dynamics (EMD) simulations by Salin and Caillol [12] and variance procedure (VP) estimation used by the Faussurier and Murillo [13]. Donkó and and Hartmann employed the inhomogenous non equilibrium MD (InHNEMD) method to investigate the transport and thermal conductivity [14]. Very recently, a homogeneous NEMD (HNEMD) and homogenous perturbed MD (HPMD) schemes are introduced by Shahzad and He (current authors) for strongly coupled complex dusty plasmas (SCCDPs) to compute the thermal transport and behaviors of SCCDPs [15–17]. For the computation of transport properties, in particular, numerical models are proposed in interest to investigate thermal behavior over a suitable range of system temperature and density values (Γ, κ). Complex fluids (dusty plasma fluids) have been used for many purposes, like power generation, semiconductors industry, cosmetics, paper industry, etc. [18].

1.1 Plasma

As we all know that 99% of matter exist in space is plasma and it is called forth state of matter. Basically plasma occurs in electrified gas form, where atoms dissociated into electrons and positive ions. It is form of matter in different areas of physics such as technical plasma, terrestrial plasma and in astrophysics. Plasma is produced artificially in laboratory used in many technical purposes likely in fluorescent lights, display, fusion energy research and other more. Term "Plasma" first time used by Irving Langmuir [19], who is an American physicist and defined plasma as "plasma is quasi-neutral gas of charged particles which exhibit collective behavior". Quasi-neutral means that gas becomes electrically neutral when number of ions equal to number of electrons ($ni \approx ne \approx n$).Where, ni is ion density, ne is electron density and n is number density. Collective behavior means that charged particles collide with each other due to coulomb potential and electric field. Plasma is extensively used in the field of science and technology. It plays a very significant role in over daily life. Plasma is used in over daily life fields such as laser, sterilizing of medical instruments, lightening, intense power beams, water purification planet and many more.

In 1922, American scientist Irving Langmuir was the only one person who defined plasma for the first time. In 1930, the study of plasma physics was started by some scholars; they are inspired by some particle problems. In 1940, hydromagnetic waves were advanced by Hanes Alfven [19] and these waves are called Alfven waves. Furthermore, he described that these waves would be used for the study of astrophysical plasma. At the start of 1950, the research on magnetic fusion energy was started at the same time in Soviet, Britain and USA. In 1958, the research on magnetic fusion energy was considered the branch of thermonuclear power. Primarily, this research was carried out as confidential but after the realization that controlled fusion research was not liked by military and therefore this research was publicized by above said three countries. Due to the reason, other countries may participate in fusion research based on plasma physics. At the end of 1960, plasma is created with different plasma parameters by Russian Tokomak configuration. In 1970 and 1980, various advanced tokomaks were built and approved the performance of tokomak. Moreover, fusion break almost achieved in tokomak and in 1990, the research on dusty plasma physics had begun. The dusty plasma is defined as "when charged particles absorbed in plasma, becomes four components plasma containing electrons, ions, neutral and dust particles" and dust particles alter the properties of plasma which is called "Dusty Plasma" [19].

1.2 Types of plasma

Plasma has complex characteristics and properties, characterized through temperature of electron and ion, density and degree of ionization. (i) **Hot plasma**: If plasma fulfills $T_e \cong T_i$ this condition then plasma is considered as hot plasma because hot plasma has very high temperature and also thermal equilibrium obtains due to frequent interactions between particles. Hot plasma is also called thermal plasma. It approaches to local thermodynamics equilibrium (LTE) and is created with high gas pressure in discharge tube in the laboratory. Hot plasma is produced by sparks, flames and atmospheric arcs. (ii) **Cold plasma**: When plasma satisfies $T_e > T_i > T_g$ this condition, plasma is called cold plasma. Where T_e, T_i, and T_g represent the temperature of electrons, ions and gas molecules. Cold plasma is created in laboratory with the positive column glow discharge tube. Motion of gas molecules is considered ignore because electron energy is very high as compared with gas molecules. Moreover, nonthermal equilibrium does not exist because collision between gas molecules and electrons is considers as low due to low gas pressure. On this regime, magnetic field is very weak and considered as ignore, only electric field is acted on charged particle. Application of cold plasma is self-decontaminating filter, food processing and sterilizing of tooth. (iii) **Ultracold plasma**: When the temperature of electrons and ions become low as 100mk and $10\mu k$ with density 2×10^9 cm^{-3}, then, plasma is called ultracold plasma. The behavior of ultracold plasma is obtained when Debye screening length becomes smaller than the sample size due to positive ions clouds trapped electrons. Ultacold plasma is considered as strongly coupled plasma because the coulomb interaction energy between the neighbor particles is more than thermal energy of charged particles. Such type of plasma is created in laboratory through pulsed laser and photoionizing laser cooled atoms [20].

1.3 Classification of dusty plasmas

Dusty plasma is characterized by an important parameter, coulomb coupling parameter Γ. The Coulomb coupling parameter is explained as, consider there are two dust particles, having same charge and separated by distance 'a' from each other. The coulomb potential energy of dust particle is $\varepsilon_c = \frac{q_d^2}{a} \exp.\left(-\frac{a}{\lambda_d}\right)$, where, q_d is the charge on dust particle, a is the distance between dust particles and λ_d is Debye screening length of dust particle. The thermal energy of dust particle is $K_B T_d$. Coulomb coupling parameter is defined as "ratio of coulomb potential energy to thermal energy". On the basis of coulomb coupling parameter, the dusty plasma is classified in ideal plasma (weakly coupled dusty plasma) and non-ideal plasma (strongly coupled dusty plasma) and is represented as Γ_c. (i) **Ideal plasma:** Ideal plasma is defined by plasma parameter called coulomb coupling and denoted as $\Gamma = \frac{P.E}{K.E}$ and is defined as "when kinetic energy of plasma is much larger than potential energy at low temperature and low density". Ideal plasma is also called weakly coupled dusty plasma and is known by $\Gamma > 1$ this condition. Ideal plasma does not have definite structure due to less collision between particles and low density. Moreover, weakly coupled plasma is defined by plasma parameter called coulomb coupling parameter Γ. When the value of coupling parameter becomes negligible then plasma is called weakly coupled plasma. Weakly coupled plasma is also called hot plasma. When the temperature of electron becomes equal to temperature of ion ($T_e \cong T_i$) then plasma is called hot plasma or ideal plasma. Hot plasma is generated in laboratory in the discharge tube with high gas pressure. Examples of hot plasma are flame, sparks and atmospheric arcs. Weakly coupled dusty plasma has not specific shape because at low density and high temperature and the interaction between interacting particles becomes very low. (ii) **Non-ideal plasma:** Dusty plasma will be strongly coupled when it satisfies this condition $\Gamma \geq 1$. Strongly coupled dusty plasma is also called nonideal plasma. Dust particle in several laboratory plasma systems is strongly coupled due to their small interparticle distance, low temperature and huge electric charge. Moreover, dusty plasma will be nonideal or strongly coupled, if average thermal energy of charged dust particle is much lesser than average potential energy. Examples of non-ideal plasma are laser generated plasma, brown dwarfs, exploding wires, high power electrical fuses, etc. Furthermore, the Yukawa potential or coulomb coupling potential Γ is used to define strongly coupled plasma. The ratio of potential energy to kinetic energy is called coulomb coupling potential. When kinetic energy becomes lower than potential energy i.e., $\Gamma > 1$. Its mean strongly coupled dusty plasma is also called cold plasma because of inter-particle kinetic energy decreases from potential energy and particles in plasma turn into crystalline shape. Crystalline shapes of particles in plasma have examined in many laboratory experiments [1–10]. Food processing and sterilization of tooth are the application of cold plasma. In strongly coupled plasma charge particles are affected by electric field but magnetic field affect is neglected for such type of cold plasma.

1.4 Complex (dusty) plasma and applications

Dusty plasma is generally electron ion plasma containing additional charged particulates. This charged component is sometimes termed as dust particle with size of micron. The properties of dusty plasma become more complex when charged particle immersed in plasma. Due to this reason such plasma is called dusty plasma and dusty plasma is also called complex plasma. Dust particles may be made of ice particles or it may be metallic. Dust particles are heavier than ions and their size ranging from few millimeters to nanometer. When dust particle coexists with plasma (electron, ions, neutral and dust particle) it becomes dusty plasma.
Dust particle exists in different shapes and size and it presents in entire universe and also in atmosphere. Usually it is solid form but also exists in liquid and gaseous form. Dust particle can be charge by the flow of electrons and ions. Charged dust particle is affected by electric and magnetic field and their electric potential varies from 1 to 10 V. Dust particle can be grown in laboratory. Dusty plasma has attracted attention of many researchers Transport properties of dusty plasma has played a very important role in the field of science and technology. Mostly the plasma exists in universe is dusty plasma. Dusty plasmas exist in atmosphere of stars, solar wind, sun, galaxies, planetary rings, cosmic radiation, magneto and ionosphere of earth.

Human life is influenced by plasma science. It plays a very significant role in laser developments of fusion energy, sterilizing of medical instruments, plasma processing, intense particle beam, high power energy sources, lightening, high power radiation sources and development of fusion energy controlling. Plasma governs diverse important devices and technological applications. Plasma processing technologies are one of the most important technologies. Plasma processing technologies are playing important role in advance modern technologies of superconductor film growth and diamond film. In addition, the practical application of plasma physics involves the treatment of materials by means plasma technologies. The ionization of system are used to produced particular physical characteristics of plasma, which involve three types of processes, Creation of new materials, Destruction of toxic materials and Superficial modification of existing materials. For industrial process, plasma technology uses two different types of plasma, the cold plasma and thermal plasma. The first type of plasma is cold plasma. Properties of cold plasma are described by electron temperature because electron temperature is greater than ion temperature. The surfaces modification is produced due to plasma particles interact with material, as a result different functional properties of materials are achieved. Cold plasma is produced in vacuum with microwave, dc source or low power rf. The second type of plasma is thermal plasma which is produced at high pressure by radio frequency, microwave source or direct or alternating current. Mostly, thermal plasma is used to devastate toxic materials. Furthermore, plasma has become one of the fast growing research fields which have attracted many researchers. Plasma has advanced applications in the field of industry, textile, plasma chemistry, fusion devices, environmental safety and printing technology and as well as in medical field. In the past decade, plasma physics has become fast developing research in medical field due to increase the atmospheric pressure of plasma sources. Plasma used in medicine has considered the latest developing novel research field with the connection of life science and plasma physics. Moreover, in past ten to fifteen years, World wild research group has set their attention on biological materials with cold atmospheric plasma interactions. Plasma used the field of life sciences in decontamination, in therapeutic medicine and in medical implant technology. The atmospheric pressure of plasma has used to reduce the efficiency of contaminations of food containers and food products. Feed gas humidity is used to adjust the level of contamination. Furthermore, plasma created in polymer tube, used in endoscopes [15–19]. Operating tools such as bone saw blade, neurosurgical and endoscope are sterilized before starting the surgery or dental treatments. Plasma plays a very dominant role in diagnostic system, treatments and in medical instruments. For decontamination of germs and sterilization operation tools, the non-equilibrium discharge plasma is used, is not dangerous for environment and patient as well [21].

2. Molecular dynamics simulations

MD simulation is a powerful technique that can be used to solve many physical problems in atomic material research. MD simulation is handled normally all microscopic information and molecular methods have proven to be the product of applied research. Plasmas and complex liquids have various uses, ranging from semiconductor chips, colloids, thin films, and electrochemistry to biochemical films and other important areas where structures play an important role. MD simulation plays an important role in all the advanced sectors, such as textile science, engineering, physics, plasma physics, astronomy, life sciences or organic sciences, and the chemical industry. Computer simulation has become increasingly important in detecting complex motion systems. Using faster and more sophisticated computer systems, it can be studied the habitat, composition, and behavior of large complex systems. In the 1950s, Alder [19], Wainwright and Rahman [19] used the first MD simulations for liquid argon [1–10], their references herein]. MD simulation has two basic kinds which rely on the properties so far which we can be going to calculate: one is EMD simulations (EMDS) and the other one is NEMD simulations (NEMDS). In the present work, NEMDS is applied to investigate the thermal conductivity of complex plasma at different dusty plasma parameters.

2.1 Numerical model and algorithm

NEMD simulation is used to detect the dust trajectory of an interacting system using Yukawa forces between dust particles. In present case, the HNEMD simulation (HNEMDS) method is used to calculate the thermal flow of complex (dust) plasma formed using Yukawa interaction taking in to account the charged particles with polarization effects and it is given in the form [22]:

$$\phi_{ij}(|\mathbf{r}|) = \frac{Q_d^2}{4\pi\varepsilon_0}\frac{e^{-|\mathbf{r}|/\lambda_D}}{|\mathbf{r}|} + \frac{d^2}{r^3}\left(1+\frac{r}{\lambda_D}\right)e^{-|\mathbf{r}|/\lambda_D}, \quad (1)$$

The first term in Eq. (1) provides the screened charge–charge interaction (form of Yukawa interaction) and the second term gives the screened dipole–dipole interaction. Yukawa potential model of dust particles interaction can be established to take into consideration the polarization effect, the temperature and the screening effects. Here '*r*' is the magnitude of interparticle distance, *Q* is the charge of dust particles, and λ_d is the Debye screening length. We have three normalized (dimensionless) parameters to characterize the Yukawa interaction model $\phi_Y(|\mathbf{r}|)$, the Coulomb coupling parameter $\Gamma = (Q^2/4\pi\varepsilon_0).(1/a_{ws}k_BT)$, where a_{ws} is Wigner Seitz radius and it is equal to $(n\pi)^{-1/2}$, here *n* is the number of particles per unit area (*N*/*A*). The k_B and *T* are Boltzmann constant and absolute temperature of the system, and *A* is the system area. The second is the screening strength (dimensionless inverse) $\kappa = {}^{aws}/_{\lambda_D}$, and additional normalized external force field strength, $F^* = (F_Z).(a_{ws}/\mathbf{J}_Q)$. GK relations ($GKR_S$) for the hydrodynamic transport coefficients of uncharged particles of pure liquids have applied to calculate the thermal conductivity of 2D complex plasma. Here, $\mathbf{J_Q}$ is the current heat vector at time *t* of 2D case [1–4].

$$\mathbf{J}_Q(t)A = \sum_{i=1}^{N} E_i\frac{\mathbf{p}i}{m} - \frac{1}{2}\sum_{i\neq j}(\mathbf{r}i - \mathbf{r}j).\left(\frac{\mathbf{p}i}{m}.\mathbf{F}ij\right) \quad (2)$$

where $\mathbf{F}_{ij}$ is the total interparticle force at time *t*, on particle *i* due to *j*, $\mathbf{r}_{ij} = \mathbf{r}_i - \mathbf{r}_j$ are the position vectors (interparticle separation), and $\mathbf{P}_i$ is the momentum vector of the *i*th particle. Where, E_i is the total energy of particle *i*.

$$E_i = \frac{\mathbf{p}^2{}_i}{2m} + \frac{1}{2}\sum_{i\neq j}\phi_{ij} \quad (3)$$

$$\dot{\mathbf{r}}_i = \frac{\mathbf{p}_i}{m}, \dot{\mathbf{p}}_i = \sum_{j=1}^{N}\mathbf{F}_i + \mathbf{D}_i(\mathbf{r}_i, \mathbf{p}_i).\mathbf{F}_e(t) - \alpha\mathbf{p}_i. \quad (4)$$

In Eq. (4), $\mathbf{F}_i = -\frac{d\phi_Y(|\mathbf{r}|)}{dr_i}$ is the Yukawa interaction force acting on particle *i*, where $\phi_Y|\mathbf{r}|$ is given from Eq. (1), and $\mathbf{D}_i = \mathbf{D}_i\{(\mathbf{r}_i, \mathbf{p}_i), i = 1,2, ... ,N\}$ is the phase space distribution function with $\mathbf{r}_i$ and $\mathbf{p}_i$ being the coordinate and momentum vectors of the ith particle in an N-particle system. A Gaussian thermostat multiplier (α) has been used to maintain the system temperature at equilibrium position and it is given as

$$\alpha = \frac{\sum_{i=1}^{N} \left[\mathbf{F}_i + \mathbf{D}_i(\mathbf{r}_i, \mathbf{p}_i).\mathbf{F}_i(t)\right].\mathbf{p}_i}{\sum_{i=1}^{N} p_i^2/m_i} \tag{5}$$

When an external force field is selected parallel to the z-axis $\mathbf{F}_e(t) = (0, F_z)$, in the limit $t \to \infty$. Then the thermal conductivity is given as [8].

$$\lambda = \frac{1}{2k_B A T^2} \int_0^\infty \left\langle \mathbf{J}_{Q_z}(t) \mathbf{J}_{Q_z}(0) \right\rangle dt = \lim_{F_z \to 0} \lim_{t \to \infty} \frac{-\left\langle \mathbf{J}_{Q_z}(t) \right\rangle}{T F_z} \tag{6}$$

where $\mathbf{J}_{Qz}$ is the z-component of the heat flux vector (energy current). All time series data are recorded during HNEMD and used in Eq. (6) to calculate λ. The detail of present scheme with all parameters (Gaussian thermostat multiplier, external force, $\mathbf{D}_i$, F_i, etc) for Yukawa interaction has been reported in our earlier works [8]. The most time consuming part of the algorithm is the calculation of particle interactions (energy and interaction forces). It has been shown in our previous work that the proposed method has the advantage of calculating Yukawa interaction and its associated energy with the appropriate computational power at the right time of the computer simulation. The actual HNEMD simulations are performed between 1.5 x 10^5 $/\omega_p$ and 3.0 x 10^5 $/\omega_p$ time units in the series of data recording of thermal conductivity (λ). Here, ω_p is the plasma frequency and it is defined as $\omega_{pd} = (Q_d^2/2\pi\varepsilon_0 m_d a^3)^{1/2}$, where m_d is the mass of a dust particle.

3. Simulation results and discussion

In this section, we have discussed the preliminary results obtained through HNMED simulation for Coulomb coupling parameters of Γ (= 10, 100), polarization values Γ_d = (0, 1, 10, 20, 50 and 100) and Debye screening (κ = 1.4, 2 and 3) at constant external force field (F^* = 0.02) of 2D strongly coupled dusty plasmas. We have chosen suitable number of particles (N = 400) in the simulation box with edge length (Lx, Ly). Periodic boundary conditions (PBCs) are applied along with minimum image convection in a simulation box of length L.

There are different conditions to improve the efficient results of thermal conductivity under polarization effects. These conditions include the system size (N), Coulomb coupling (Γ), Debye screening length (κ), system total length (t), simulation time step (dt), and external force field (F^*) strength and polarization values (Γ_d). We have chosen suitable parameters for precise results of thermal conductivity with increasing Γ_d and κ. The simulation data are for a suitable system size (N = 400) with different Γ, which covers the values of strongly coupled plasma states (nonideal gases-liquid to crystalline) at constant force field strength (F^* = 0.02) with varying polarization values Γ_d (0 to 100).

The polarized thermal conductivity of SCCDPs stated here may be scaled as $\lambda_0 = \lambda/nm\omega_p a^2$ (by the plasma frequency). It is demonstrated that the ω_E decreases with increasing κ, $\omega_E \to \omega_p/\sqrt{3}$ as $\kappa \to 0$, for the 3D case [1–10]. Furthermore, for the assessment of appropriate equilibrium range (nearly-equilibrium of external field strength) of 2D HNEMD scheme, various sequences of the polarized thermal conductivity corresponding to an increasing order of an external force field $[\mathbf{F}_e(t) = (0, F_z) \equiv F^* = (F_z)\ (a/J_Q)]$ are planned to measure the nearly equilibrium values of the λ_0. This possible appropriate F^* value gives the steady state λ_0 investigations, which are appropriate for the whole range of plasma states of $\Gamma \equiv (10, 100)$ and $\kappa \equiv (1.4, 3)$.

Figures 1–3 illustrate the normalized polarized thermal conductivity (plasma frequency, ω_p), as a function of Coulomb coupling (system temperature $\equiv 1 / \Gamma$) for the cases of $\kappa = 1.4$, 2 and 3, respectively, at constant force F^* with varying six values of polarizations $\Gamma_d = (0, 1, 10, 20, 50$ and $100)$. For three cases, the simulations are performed with setting $N = 400$ for $\kappa = 1.4$, 2 and 3, respectively, at constant

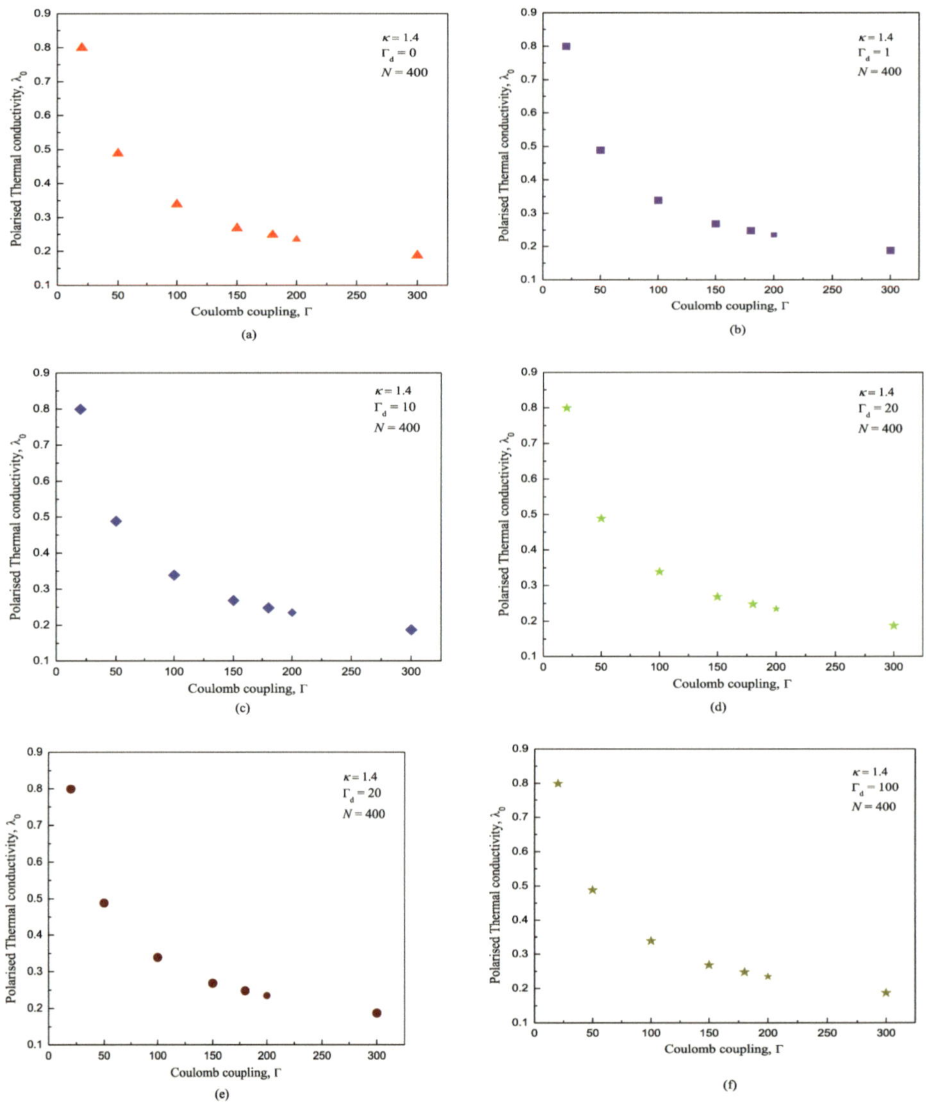

Figure 1.
Variations of thermal conductivity as a function of Coulomb coupling of strongly coupled complex plasma at $\kappa = 1.4$ *with* N = 400 *and (a)* $\Gamma_d = 0$, *(b)* $\Gamma_d = 10$, *(c)* $\Gamma_d = 20$, *(d)* $\Gamma_d = 50$, *(e)* $\Gamma_d = 50$ *and (f)* $\Gamma_d = 100$.

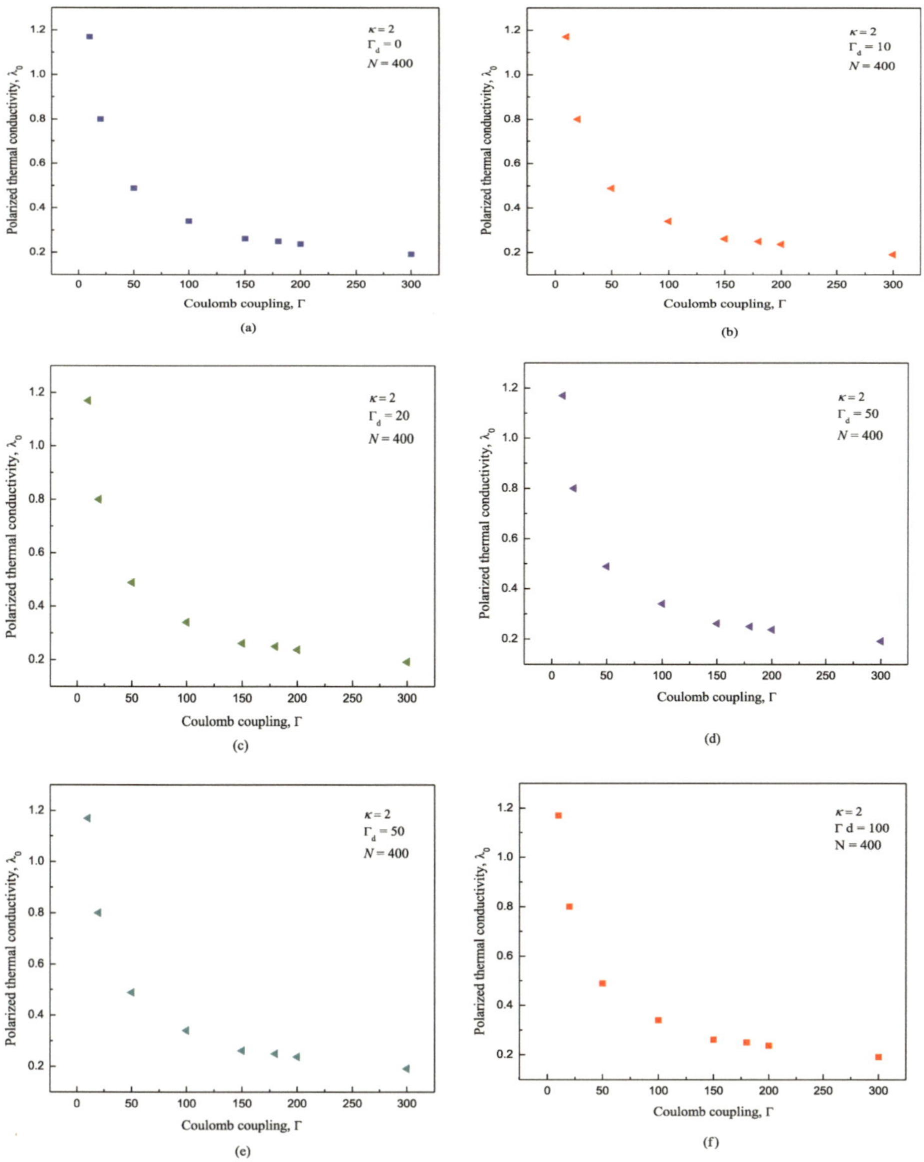

Figure 2.
Variations of thermal conductivity as a function of Coulomb coupling of strongly coupled complex plasma at $\kappa = 2.0$ *with* N = 400 *and (a)* $\Gamma_d = 0$, *(b)* $\Gamma_d = 10$, *(c)* $\Gamma_d = 20$, *(d)* $\Gamma_d = 50$, *(e)* $\Gamma_d = 50$ *and (f)* $\Gamma_d = 100$.

$F^* = 0.02$. Performing HENMD simulations with varying polarizations at constant F^* we examined the efficiency and reliability of the polarized λ_0 measurements. For three cases, we evaluate the six various simulation data sets covering from nonideal state ($\Gamma = 10$) to a strongly coupled liquid regime ($\Gamma = 100$). Figures show that the effects of polarization on the thermal conductivity have no significant changes and it is seen that the thermal conductivity remains constant under varying polarizations. However, the present results of thermal conductivity under varying polarizations are satisfactory agreement with earlier know available numerical data for a complete range of plasma parameters.

Figure 4 shows comparisons with earlier available 2D and 3D numerical data of thermal conductivity with setting $N = 1024$ and $\Gamma_d = 1$. The current results are

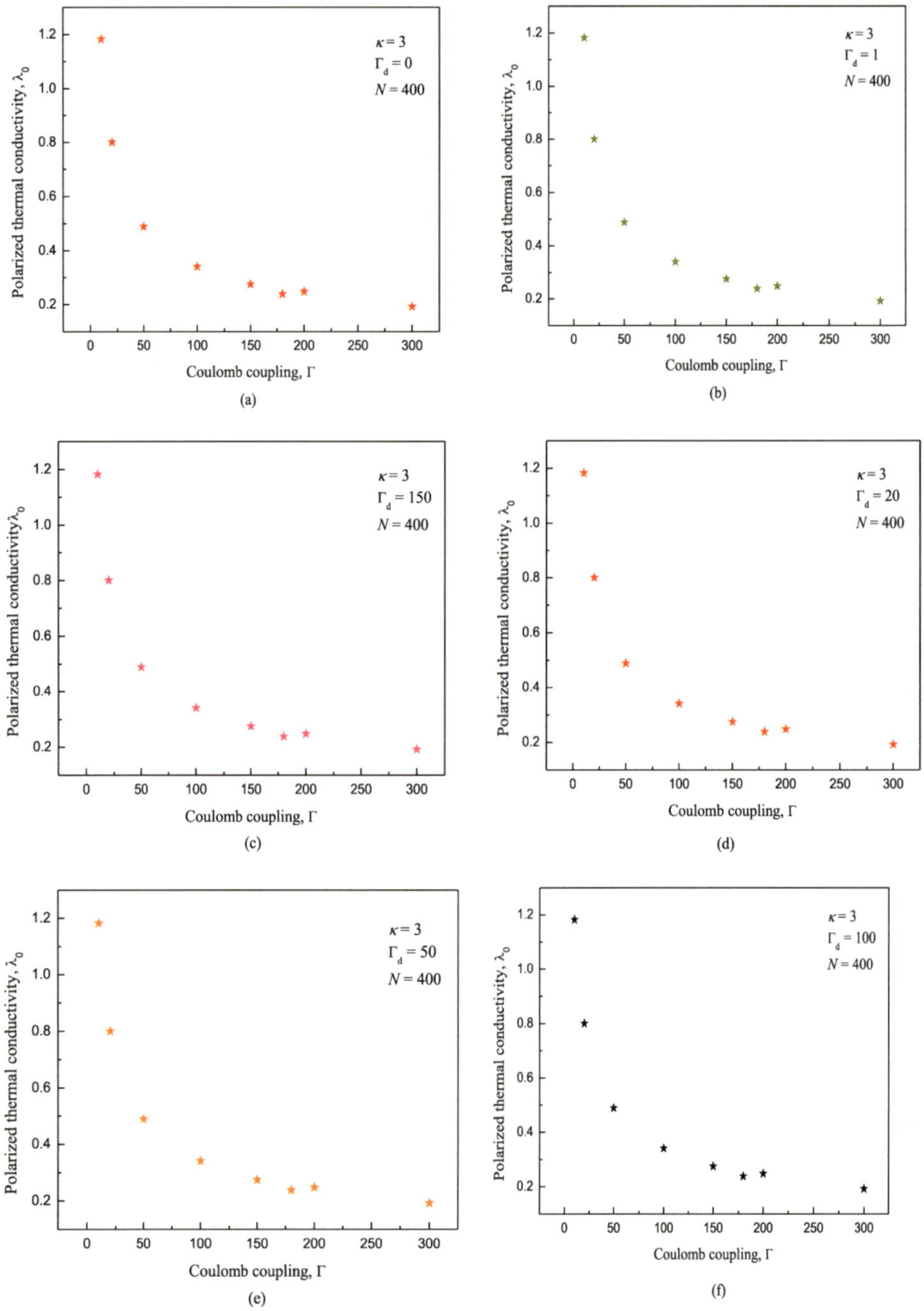

Figure 3.
Variations of thermal conductivity as a function of Coulomb coupling of strongly coupled complex plasma at $\kappa = 3.0$ *with* $N = 400$ *and (a)* $\Gamma_d = 0$, *(b)* $\Gamma_d = 10$, *(c)* $\Gamma_d = 20$, *(d)* $\Gamma_d = 50$, *(e)* $\Gamma_d = 50$ *and (f)* $\Gamma_d = 100$.

generally excellent agreement for the whole Coulomb coupling range and plot show overall the same behaviors as in the earlier simulation method s of 2D plasma systems. Figure involve the earlier work of 3D HNEMD and HPMD by Shahzad and He [8, 16, 23], EMD investigations of Salin and Caillol [12] and 3D theoretical prediction of Faussurier and Murillo [13] as well as 2D GKR-EMD of Khrustalyov and Vaulina [24].

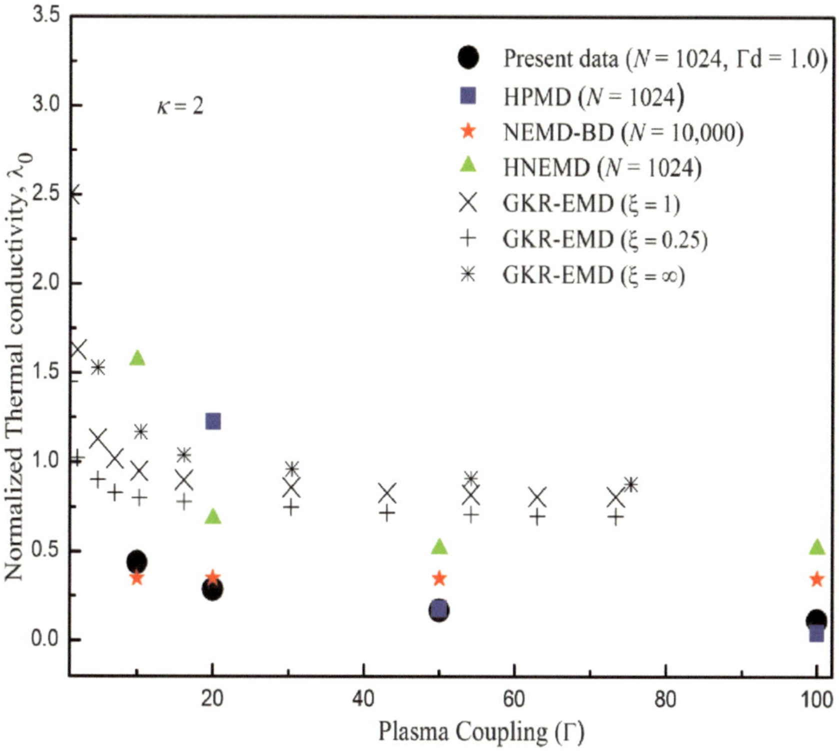

Figure 4.
Comparison of thermal conductivity as a function of Coulomb coupling of strongly coupled complex plasma at $\kappa = 2.0$ *with* $N = 1024$ *and* $\Gamma_d = 1$.

4. Conclusions

The HNEMD scheme is used for the investigation of thermal conductivity under the influence of varying polarization values for various screening lengths κ (= 1.4, 3) and Coulomb couplings Γ (= 10, 100) but constant force field strength. It has been shown that the current HNMED scheme with polarization and earlier HPMD and GKR-EMD techniques have comparable efficiency over the suitable range of plasma parameters, both generating reasonable data of polarized λ_0. New simulation results provide more reliable and ample results for the polarized thermal conductivity for a whole range of (Γ, κ) than previous known simulation data. Thermal conductivity results estimated from this newly developed HNEMD scheme are in well matched with available numerical results generally underpredicting within ±20% - 35%. The current HNEMD data with less statistical noise and best efficiency have been computed by taking different parameters (N, $1/\Gamma$, κ and Γd) for the current HNEMD scheme to measure the thermal conductivity of 2D DP systems, at constant $F^* = 0.02$. For future work, the HNEMD method introduced here may easily extend to other physical systems with varying force field with modifications.

Acknowledgements

We are grateful to the National High Performance Computing Center of Xian Jiaotong University and National Advanced Computing Center, National

Center of Physics (NCP), Pakistan for allocating computer time to test and run our MD code.

Abbreviations

Strongly coupled complex dusty plasmas	(SCCDP)
Homogeneous non-equilibrium molecular dynamics	(HNEMD)
Coulomb coupling	(Γ)
Debye screening length	(κ)
External force field strength	(F^*)
Homogenous non-equilibrium molecular dynamics	(HNEMD)
Non-equilibrium molecular dynamics	NEMD
Molecular dynamics	(MD)
Inhomogenous non-equilibrium molecular dynamics	(In HNEMD)
Strongly coupled plasma	(SCP)
Equilibrium molecular dynamics	(EMD)
Thermal conductivity	(λ)
Normalized thermal conductivity	(λ_0)
Number of Particles	(N)

Author details

Aamir Shahzad[1]*, Madiha Naheed[1], Aadil Mahboob[1], Muhammad Kashif[1], Alina Manzoor[1] and H.E. Maogang[2]

1 Department of Physics, Molecular Modeling and Simulation Laboratory, Government College University Faisalabad (GCUF), Faisalabad, Pakistan

2 Key Laboratory of Thermal Fluid Science and Engineering, Ministry of education (MOE), School of Energy and Power Engineering, Xi'an Jiaotong University, Xi'an, China

*Address all correspondence to: aamirshahzad_8@hotmail.com

References

[1] Shahzad A. Impact of Thermal Conductivity on Energy Technologies. London: IntechOpen; 2018. DOI: 10.5772/intechopen.72471

[2] Shahzad A, Haider SI, He MG, Yang F. Introductory Chapter: A Novel Approach to Compute Thermal Conductivity of Complex System, a Chapter from the Book of Impact of Thermal Conductivity on Energy Technologies. London: IntechOpen; 2018. 3p. DOI: 10.5772/intechopen.75367

[3] Shahzad A, He MG. Interaction contributions in thermal conductivity of three dimensional complex liquids. AIP Conference Proceedings. 2013; 1547: 172-180

[4] Shahzad A, Shakoori MA, He M-G, Bashir S. Sound waves in complex (Dusty) plasmas. In: Computational and Experimental Studies of Acoustic Waves. Rijeka, Croatia: InTech; 2018. DOI: doi.org/10.5772/intechopen.71203

[5] Killian T, Patard T, Pohl T, Rost J. Ultracold neutral plasmas. Physics Reports. 2007; 449: 77. DOI: 10.1016/j.physrep.2007.04.007

[6] Shukla PK, Mamun AA. Introduction to Dusty Plasma Physics. New York: CRC Press; 2015. ISBN 0 7503 0653

[7] Shahzad A, He MG. Thermal conductivity of three-dimensional Yukawa liquids (dusty plasmas). Contributions to Plasma Physics. 2012; 52(8):667. DOI: 10.1002/ctpp.201200002

[8] Shahzad A, He M-G. Numerical experiment of thermal conductivity in two-dimensional Yukawa liquids. Physics of Plasmas. 2015; 22(12):123707 htps://doi.org/10.1063/1.4938275

[9] Shahzad A, Aslam A, He M-G. Equilibrium molecular dynamics simulation of shear viscosity of two-dimensional complex (dusty) plasmas. Radiation Effects and Defects in Solids, 2014; 169(11):931-941 htps://doi.org/10.1080/10420150.2014.968852

[10] Shahzad A, Manzoor A, Wang W, Mahboob A., Kashif M, He M.-G, Dynamic Characteristics of Strongly Coupled Nonideal Plasmas. Arab J Sci Eng. Published: 05 July (2021). https://doi.org/10.1007/s13369-021-05954-4

[11] Shahzad A, Kashif M, Munir T, He M.-G and Tu X, Thermal conductivity analysis of two-dimensional complex plasma liquids and crystals. Phys. Plasmas, 2020; 27:103702. https://doi.org/10.1063/5.0018537

[12] Salin G, Caillol JM. Equilibrium molecular dynamics simulations of the transport coefficients of the Yukawa one component plasma. Physics of Plasmas. 2003; 10 (5):1220-1230

[13] Faussurier G, Murillo MS. Gibbs-Bogolyubov inequality and transport properties for strongly coupled Yukawa fluids. Physical Review E. 2003; 67 (4): 046404

[14] Donkó Z, Hartmann P. The thermal conductivity of strongly coupled Yukawa liquids. Physical Review E. 2004; 69 (1):016405

[15] Shahzad A, He MG. Thermal conductivity calculation of complex (dusty) plasmas. Physics of Plasmas. 2012;19(8):083707

[16] Shahzad A, He MG. Homogeneous nonequilibrium molecular dynamics evaluation of thermal conductivity in 2D Yukawa liquids. International Journal of Thermophysics. 2015, 36 (10 – 11):2565-2576

[17] Shahzad A, He MG, Irfan Haider S, Feng Y. Studies of force field effects on

thermal conductivity of complex plasmas. Physics of Plasmas, 2017; 24(9): 093701

[18] Shahzad A. Impact of Thermal Conductivity on Energy Technologies (IntechOpen, London, 2018), http://dx.doi.org/10.5772/intechopen.72471.

[19] Chen FF. Introduction to plasma physics and controlled fusion. New York: Sppring verlag; 2006, (2nd edition)

[20] Shahzad A, Shakoori MA, He MG, Yang F. Dynamical structure factor of complex plasmas for varying wave vectors. Physics of Plasmas. 2019; 26 023704

[21] Shahzad A, He M-G. Computer Simulation of Complex Plasmas: Molecular Modeling and Elementary Processes in Complex Plasmas. 1st ed. Saarbrücken, Germany: Scholar's Press; 2014. p.170

[22] Ramazanov T. S., Gabdulin A.Zh., Moldabekov Z.A., MD Simulation of Charged Dust Particles With Dipole Moments. IEEE Trans. Plasma Sci. 2015; 43: 4187–4189. 10.1109/TPS.2015.2490282

[23] Shahzad A, Haider S.I., Kashif M., Shifa M.S., Munir T. and He M.-G., Thermal Conductivity of Complex Plasmas Using Novel Evan-Gillan Approach. Commun. Theor. Phys. 2018; 69:704. https://doi.org/10.1088/0253-6102/69/6/704

[24] Khrustalyov YV, Vaulina OS. Numerical simulations of thermal conductivity in dissipative two-dimensional Yukawa systems. Physical Review E. 2012; 85 (4):046405

Chapter 5

Studies of Self Diffusion Coefficient in Electrorheological Complex Plasmas through Molecular Dynamics Simulations

Muhammad Asif Shakoori, Maogang He, Aamir Shahzad and Misbah Khan

Abstract

A molecular dynamics (MD) simulation method has been proposed for three-dimensional (3D) electrorheological complex (dusty) plasmas (ER-CDPs). The velocity autocorrelation function (VACF) and self-diffusion coefficient (D) have been investigated through Green-Kubo expressions by using equilibrium MD simulations. The effect of uniaxial electric field (M_T) on the VACF and D of dust particles has been computed along with different combinations of plasma Coulomb coupling (Γ) and Debye screening (κ) parameters. The new simulation results reflect diffusion motion for lower-intermediate to higher plasma coupling (Γ) for the sufficient strength of $0.0 < M \geq 1.5$. The simulation outcomes show that the M_T significantly affects VACF and D. It is observed that the strength of M_T increases with increasing the Γ and up to $\kappa = 2$. Furthermore, it is found that the increasing trend in D for the external applied M_T significantly depends on the combination of plasma parameters (Γ, κ). For the lower values of Γ, the proposed method works only for the low strength of M_T; at higher Γ, the simulation scheme works for lower to intermediate M_T, and D increased almost 160%. The present results are in fair agreement with parts of other MD data in the literature, with our values generally overpredicting the diffusion motion in ER-CDPs. The investigations show that the present algorithm more effective for the liquids-like and solid-like state of ER-CDPs. Thus, current equilibrium MD techniques can be employed to compute the thermophysical properties and also helps to understand the microscopic mechanism in ER-CDPs.

Keywords: Diffusion coefficient, electrorheological complex plasma, Molecular dynamics simulations, uniaxial electric field

1. Introduction

1.1 Electrorheological fluids

The dielectric fluids consisting of micro-sized (0.1-100 μm) solid particles, which display particular characteristics under the influence of the external applied electric field, are known as electrorheological fluids (ERFs). The dielectric fluids, such as olive oil, silicon oil hydrocarbon, etc., have low permittivity, conductivity and viscosity. The solid particles immersed in dielectric fluids are mostly polymers, metal oxide silica, and alumina silicates. These particles maintained the low viscosity of carrier fluids in the absence of external electric field strength. Without an external electric field, particles behave as a liquid. When the external electric field is turned on, these particles behave like solids due to changes in viscosity. The ERFs change their physical properties for the application of the external electric field. When the electric field is turned on the suspension of dielectric fluids, the solid particles are polarized and make a thin chain (string) along the direction of the applied electric field. The thickness of the particles depends on the intensity of the electric field. The viscosity of ERFs increased with increases an external applied electric field. If the electric field was turned off, the fluids reverse from solid to liquids within milliseconds. These fluids are also known as intelligent fluids. There are various types of rheological fluids, such as electrorheological fluids, magnetorheological fluids, positive electrorheological, negative electrorheological fluids etc. These fluids rapidly respond to the electric field and change their physical properties such as shear stress, elastic modulus and viscosity. These fluids are used in vehicle engineering, such as valves, clutches and breaks etc. The conventional ERFs have various industrial applications, such as vibration control in smart materials. Changes in the microstructure and physical properties are used in medical to control ultrasonic transmissions and sound transmission with low losses [1, 2].

1.2 Plasma

Plasma is a partially ionized gas that contains electrons, ions and neutral radicals. In our universe, 99% of physical matter is in the plasma state. In space, the most visible matter is in the plasma state; the sun and stars are the main examples of plasma in our universe. Plasma species show collective behaviors when any external perturbation is applied. The whole plasma is perturbed when an external force has applied this behavior of particles called collective behavior of plasma. Classification of plasma depends on the species temperatures such as hot plasma, cold plasma, ultracold plasma, ideal and non-ideal plasma etc. [3].

1.2.1 Complex (dusty) plasmas

Complex (dusty) plasmas (CDPs) contained micro to submicron-sized conductive, and dielectric particles called grain in addition to plasma species (neutral atom, electron, positive or negative ions). The conductive grain has a $3e^{-11}$ kg mass and about e^4 eV charge. Mostly having a negative charge but in the rear case also have positive charge depend on charging phenomena. Dust particles are naturally present in the plasma and can be manually inserted in a plasma medium through sputtering, etching and chemical reaction—the dust particles made by a single element or composition of different elements. The dust particles like a swimmer in the sea of electrons and ions and respond to electromagnetic forces. There are different mechanisms of charging dust particles. The charge amount depends on the charging phenomena. Dusty plasmas illustrate an astrophysical matter in white dwarfs, neutron stars, giant planetary interiors and supernova core. In laboratory ultra-cold plasma, charged stabilized collides and electrolytes, laser-cooled ions in cryogenic traps, and dusty plasma. The warm dense matter and strongly coupled complex (dusty) plasmas (SCCDPs) are relevant models for nuclear fusion devices [3–6].

The CDPs are classified according to the energies of interacting charged dust particles. When interacting particle's potential energy exceeds their kinetic energy, then CDPs are called as SCCDPs. This system appears in a wide variety of physical systems. The Weakly coupled complex (dusty) plasmas (WCCDPs) are inverts of SCCDPs. The SCCDPs also in high order structural form or exist in the crystalline state at low temperature with high density. The WCCDPs are mostly remained in the gaseous state of the plasma and having high temperature with low density. Various states of dusty plasma are easily observed through a video camera under laboratory conditions. SCCDPs are found in nature and also in laboratory experiments [7, 8]. The phase transition (condensation) can be observed by reducing the temperatures of CDPs. The structural order of dust particles is formed under some external conditions. The dusty plasmas are encountered in astrophysics and are extensively believed to play a significant role in cosmology to reveal the structure and origin of the universe and its galaxies. The transport and thermodynamic properties are well studied for CDPs through experiment, theoretical and computational methods. Thermophysical properties are well investigated in the recent decade, such as thermal conductivity [9–11], diffusion phenomena [12, 13], shear viscosity [13], thermodynamic properties [14], dynamical structure factor [4, 15] and propagation of different waves [3, 5].

1.3 Applications of complex dusty plasmas

Initially, CDPs originate in astrophysics and space physics; nowadays, it becomes a fascinating, applicable field in space physics as an analogue to unravel issues like the role of dust accumulation super high speed crashed in space and the formation of planets and many industries on different scales. The CDPs play an essential role as an analogue system for investigating multifaceted cross-disciplinary phenomena, such as an experimental study of non-linear dynamics and long-range interaction in strongly correlated systems. These systems often have challenging investigations because they generally require extreme temperature and pressure conditions, such as very low temperature and high density. These systems belong to different research institutes. The CDPs allow the study and formation of the strongly coupled systems under laboratory conditions at room temperature and for easily attainable pressure. The CDPs analogue was recently used to model crystallization dynamics in 2D and excitation of quantum dots, viscoelastic material, shock and non-linear waves and recently electrorheological fluids [16].

In industrial applications, CDPs are directly applicable in the processing of microelectronics devices. It is used for deposition integrated circuits, masking, and stripping. The capacitors and transistors make on the silicon wafer chip millions of transistors put on the Pentium chip with the help of plasma to safe from contaminations. Dust particles have both advantages and disadvantages in the technologies. It plays a vital role in the scientific research of various technologies and industries. It plays a significant role during the thin-film depositions, processing of ceramic, insulation, filtration processes, petroleum industry, biomechanics, paper industry, packed bed reactors etc. Dust particles are helped to enhance the efficiency and stability of solar cells, LED (light emitted diode), improving lighting source, display, and laser technology. Through plasma processing, the coating of particles has been produced. It has grown or modified existing materials in the semiconductor industry [17]. The scientific communities currently focus on controlling nuclear fusion reactions and developing devices such as tokamak to produce efficient and carbon-free energy.

In the field of medicine and healthcare, the CDPs have become an emerging field, which combines plasma physics, life science, and clinical medicine. In the area of life science, it is directly used in biological medicine such as double-helix molecular interaction, sterilizing surgical instruments and implants, wound healing, cancer therapy, break DNA damage for human prostate cancer [18]. The CDPs were found to help kill cancer cells without affecting the healthy tissues. It is also used to inactive the bacterial in the field of health care. It is also used for food storage technologies. Furthermore, the CDPs are helpful as a diagnostic technique for a precise calculation of plasma parameters such as non-linear laser spectroscopy, spectrally resolved nanosecond imaging, phase-resolved optical emission spectroscopy, and laser-induced fluorescence VUV, UV, VIS and IR, etc. [19].

1.4 Electrorheological complex dusty plasma

CDPs have electrorheological characteristics like conventional electrorheological fluids when an external ac electric field is applied. The dust particles respond quickly to the external applied electric field and make a chain (string) sheet-like as conventional ERFs so-called electrorheological CDPs (ER-CDPs). The first time ER-CDPs were observed by Ivlev A. V. et al., in 2008 during the microgravity experiments (PK-3 plus laboratory) and MD simulation. They show the phase transition of CDPs from isotropic to anisotropic with increasing the external applied electric field [20]. Under the influence of electric field dc mode, attractive attractions between charged dust particles are introduced by ions streaming. In this way, wake potential behind the dust grains produced, which make particles strings (sheet), in the dc discharge plasma sheath, where charged dust particles levitated against the electric field's gravity force. Such a type of system is non-Hamiltonian, and wake potential is asymmetric. The ac electric fields have a much smaller frequency than ion plasma frequency, but larger than dust plasma frequency was applied to the complex plasma. In this way, the wake potential is symmetric, and a system known as Hamiltonian due to the electric field, dipole–dipole attractions increased in ER-CDPs, particles arrange themselves in sheet, string, or crystalline structure same as conventional ERFs. The phase transition in ER-CDPs has studied experiments and simulations method [20–23].

After discovering ER-CDPs, it opens up new dimensions of research for plasma science and technologies communities. There is little literature available to understand the macroscopic phenomena of ER-CDPs. Ivlev A. V. et al., done a PK-3 plus experiments under microgravity conditions and molecular dynamics simulations, observed phase transition from an isotropic to string with increasing external ac electric field [24]. Later they have done *PK*-4 dc discharge experiment and observed an anisotropic structure under the influence of an external ac electric field [22]. Yaroshenko V. V. theoretically studied the propagations of dust lattice waves along the electric field in a one-dimensional string. He found the instability leads to spontaneous excitations of compressional waves [25]. Rosenberg *M*. theoretically studied the formation and excitation of waves in one-dimensional ER-CDPs under the ac electric field [26]. Kana *et al*. explored the phase transition in ER-CDPs using MD simulation and observed the anisotropic structure under the influence of ac electric field and did not find the anisotropic structure for the dc electric field mode [21]. Sukhinin *et al*. used a Monte Carlo (MC) simulation of plasma polarization around dust particles in an external applied electric field. They mentioned that due to induced dipole potential, the formation of dust particles alignment, chain (string) multi-layered structure, and coagulation of charged dust particles [27]. A self-consistent model was developed for plasma anisotropy (string) of charged dust particles under the external electric field's action by Sukhinin *et al*. [28]. No evidence has been found of thermophysical properties in ER-CDPs. In the future, for precise tailoring, new materials may be modeled with the help of ER-CDPs. The ER-CDPs can play a significant role in modeled new smart materials. It is also be used to generate negative dipolar interparticle interactions [24].

1.5 Diffusion coefficient

The diffusion coefficient is one of the transport properties of gas, liquids. During the diffusion process, the particles migrate from high concentrations to low concentrations. The diffusion rate can be controlled by variations of external parameters such as temperature, concentration, external electromagnetic forces (electric and magnetic). Investigations of diffusion have different purposes in liquids, such as exploring the dynamical properties and understanding the microscopic phenomena. For the CDPs research direction, the investigations of diffusion motion for applying external fields such as electric and magnetic become a hot topic in current research. Different types of diffusion motion exist in dusty plasma regimes depends on temperature and forces. Diffusion also plays an essential role in exploring the dynamical properties (structure, waves, and instabilities) of many biological, physical, and chemical systems. The diffusion motion of dust particles in CDPs continues as one of the active research topics, an essential consequence of understanding the transport and dynamical properties [12, 29–33]. It is one of the primary sources to lose energy (stopping power) in CDPs. Therefore, we can easily understand the microscopic phenomena of particles in the different states [34]. An extensive amount of previous studies have been made to understand the diffusion motion in CDPs. Molecular dynamics simulations were performed and investigate the velocity autocorrelation function for 2D WCCDPs [12], Langevin dynamics simulation for 2D SCCDPs [35] and diffusion coefficient [36]. For the investigations of self-diffusion coefficient MD simulations were performed for 3D CDPs [30, 33, 37], one-component plasma [31] and ionic mixtures [32].

The main objective of the present book chapter is to give an overview of electrorheological (ER) fluids and ER-CDPs. We have computed the effect of the external applied uniaxial electric field on the velocity autocorrelation function and self-diffusion coefficient of 3D ER-CDPs using MD simulations over a wide range of input CDPs parameters.

2. MD simulation algorithm and parameters

The Computer simulation provides a linkage between theoretical and experimental research work. In thermal fluids, science and engineering, the installations of experimental setups are very complex and high cost. In the age of modern technology, the cost-effective and low time consumption high computational power is most prominent. Nowadays, it is a trend that before starting the experimentations, first test with computational tools than verified with experiments. There are various computer algorithms and techniques are used for the calculation of various properties in various materials. We perform computer simulations to test a theoretical model, verify experimental data, and also for comparison purposes [38]. Different computational techniques can be designed for extreme conditions, such as for very low temperature and high density. It also acts as a bridge between microscopic length, time scale, and macroscopic worlds. MD and Monte Carlo (MC), and Langevin dynamics simulations are the main methods used to compute CDPs' physical properties. Different software's are also available to compute various physical properties and build a new model by following one simulation scheme.

MD simulations have become a prominent computational tool to investigate various properties such as thermophysical and dynamical properties in different types of fluids and materials. The MD simulations consist of the numerical solution of the Newton equation of motion for the spherical particles system [9]. This section explained the MD simulation algorithm to calculate the self-diffusion coefficient (D). Equilibrium MD simulations are used to investigate the D in ER-CDPs. Yukawa potential (screened Coulomb) is the most commonly used potential for CDPs systems, while other physical systems such as physics of chemical and polymer, medicine and biology systems, astrophysics, and environmental research are adopted. Most scholars used Yukawa potential to screen charged dust particles in the dusty plasma results in isotropic interactions. In the present research used applied a uniaxial ac electric field for additional dipole–dipole-like interactions. This idea was taken from Pk-3 plus experiment under the microgravity conditions [20], theoretical approach [39], MD simulation study [24], and *Pk*-4 experiments [23]. For ER-CDPs, the charged dust particles interact with each other through Yukawa potential and Quadrupole interactions due to external applied ac electric field: the equation for ER-CDPs becomes as

$$W(r,\theta) = Q_d^2\left[\frac{1}{4\pi\varepsilon_o}\frac{\exp^{-r/\lambda_D}}{r} - 0.43\frac{M_T^2\lambda_D^2}{r^3}\left(3\cos^2\theta - 1\right)\right] \quad (1)$$

First-term in the above equation presents pair-wise Yukawa potential in the absence of an electric field, and the second term shows the Quadrupole interaction between particles due to uniaxial ac external electric field [22]. Where λ_D is Debye length, Q_d is the charge on dust particles, r distance between interacting particles, and ϵ_o permittivity of space. The θ is the angle between the electric field ($\boldsymbol{E}$) and interacting dust particles. In the present study, we take $\theta = 0^\circ$ for uniaxial ac electric field for anisotropic interactions. The $\boldsymbol{r}$ is the distance between the interacting particles, $\boldsymbol{r} = \boldsymbol{r}_j - \boldsymbol{r}_i$. The M_T is thermal Mach number normalized with the thermal velocity of charged dust particles $M_T^2 = \langle u_d^2\rangle/v_T^2$ where $v_T^2 = T_d/m_d$ the drift velocity is proportional to an electric field ($u_d \propto E$), an electric field can be measured in the unit of M_T. For small values of M_T, in ER-CDPs, the interactions of particles are the same as dipolar interaction in conventional ERFs. In the prior study, strings (chain) of dust particles were observed in typical conditions when an external ac electric field was applied.

There are two central (dimensionless) parameters, which are fully-characterized complex plasma systems. The first parameter is the plasma coupling parameter (define same as Coulomb systems) $\Gamma = Q_d^2/4\pi\epsilon_o a_{ws} k_B T$, a_{ws} Wigner-Seitz radius defined as $(3/4\pi n)^{\frac{1}{3}}\pi n^{\frac{-1}{2}}$ Where n is the equilibrium dust number density, k_B the Boltzmann constant, and T is the system's absolute temperature. The plasma coupling (Γ) parameter is the ratio of average potential energy to the average kinetic energy of interacting dust particles. The SCCDPs associated with $\Gamma > 1$ inversely account as weakly coupled dusty plasma ($\Gamma < 1$). Another essential parameter is the Debye screening parameter, $\kappa = a_{ws}/\lambda_D$. The plasma dust frequency $\omega_{pd} = \left(Q_d^2/3\pi\epsilon_0 m a^3\right)^2$ describes the time scale of a complex plasma system; here, m is the dust particle's mass.

2.1 Green-Kubo relation of diffusion coefficient

Green-Kubo integral formula were used to calculate the self-diffusion coefficient (D) for 3D CDPs [30, 33], one component plasma [31] and ionic mixtures [32]. Here we have used the same numerical models for the investigations of D in 3D ER-CDPs; the Green-Kubo relation is given as following.

$$D = \frac{1}{3N}\int_0^\infty Z(t)dt \quad (2)$$

In Eq. (2), $Z(t)$ is known as the velocity autocorrelation function (VACF), 3 shows the 3D system, and N represents the number of simulated dust particles. The D is the integral of VACF calculated over all the particle velocity product's ensemble average segments at a time (t) and initial time (t_0). If $Z(t)$ decays too slowly, Yukawa particles' motion is described as anomalous diffusion for the integral equation one converges. The transport coefficient's existence, autocorrelation function must rapidly decay enough for integral to convergence. Different types of diffusion motion were analyzed through VACF [12], defined as

$$Z(t) = \langle v_j(t).v_j(0) \rangle \tag{3}$$

Where $v_j(t)$ denote the jth particle's velocity at simulation time (t) and (0), the brackets $\langle . ... \rangle$ represent the canonical ensemble average all over particles. $Z(t)$ demonstrates the decay that is characterized by plasma frequency. The existence of the transport coefficient, autocorrelation function must rapidly decay enough for integral convergence [29, 33, 34].

In this method, classical molecular dynamics simulations are used to map the trajectories of N = 500 and calculated the diffusion coefficient. Periodic boundary conditions (PBCs) were used to be constrained enforced during the N-particles simulation under statistical microcanonical ensemble (NVE). The edge length of the 3D cubic simulation box is L with 12.79a. Particles are placed randomly in the simulation cubic cell for the beginning of the simulations [15]. An appropriate time step (dt = 0.001ω_{pd}) was selected to calculate particle position and velocity at each step. The leapfrog integration method is used to integrate the equation of motion and calculate each particle's force, acceleration, and position for every time step. The whole scheme in MD dynamically simulates the equation of motion of N-dust particles with interaction through Yukawa potential, and Quadrupole interaction was given in Eq. (1). The Ewald summation method was used for calculations of Yukawa and dipolar potential energies and forces. Here we select the Ewald convergence parameter (γ = 5.6/L) for the Yukawa and anisotropic interactions. Suitable input parameters are selected to get the accuracy and consistency of the model. The appropriate system temperature (($T = 1/\Gamma$) Yukawa Ewald-summation method, screening length, Mach number, time step, number of particles, simulation run time, etc., are selected to provide 3D ER-CDPs diffusion motion. Negative derivatives of Eq. (1) calculate forces on each interacting charged dusty particle in the Yukawa system. MD simulations are performed 2 x $10^5/\omega_{pd}$ time unit to record the SDC (Γ, κ, M_T). The 12-core-processor takes about three hours to perform each simulation. Total 156 simulations are performed and calculated the VACF and D for a wide range of plasma parameters and external applied uniaxial electric field. For the present chapter, the EMD method is used of ER-CDPs and calculated the D over a wide range of Γ = 2–500, $1 < \kappa \geq 3$, and N = 500. In addition, the conservation of total energies and momentum is checked and verify that system has self-consistency.

3. Simulation results and discussions

This book chapter used the equilibrium MD simulation method using Yukawa (screened Coulomb) potential and dipolar interactions. Without external forces (electric and magnetic), Yukawa potential was used and calculated thermophysical properties. In this work, we add dipolar interactions with the same numerical schemes and analyses the diffusion motion through VACF and D. First of all; we compute the VACF for Screening parameters ($1 < \kappa \geq 3$), plasma coupling parameters ($1 < \Gamma \geq 500$), uniaxial electric field strength ($0.0 \leq M_T \geq 1.50$) for the same number of particles (N = 500). After that, D is calculated for nearly the same parameters used for the VACF.

The single-particle motions are computed for 3D ER-CDPs through the equilibrium MD simulation using VACF for diffusion by employing Eq. (3). VACF are analyzed and discussed for plasma screening κ = 1.4, 2 and 3, plasma coupling (Γ = 2–500), number of spherical charged dust particles (N = 500) with the variation of the external uniaxial electric field. The VACF has been widely used for 3D CDPs over a wide combination range of plasma parameters (κ, Γ) [12, 30, 36]. The VACF was computed at M_T = 0.0 for comparison purposes with earlier MD results. It has been shown that MD outcomes have comparable fair agreements with previous results. Here we defined the critical strength of the applied external electric field (M_{CT}). The M_{CT} is directly proportional to plasma coupling strengths or inversely proportional to system temperature. We have been found that above the M_{CT} system goes out of thermal equilibration. We have also got noisy results above the M_{CT} strength of the external applied uniaxial electric field. The critical strength of the uniaxial electric field is different for the different combinations of CDPs parameters.

The simulation outcomes of VACF in 3D ER-CDPs as a function of the simulation time scale for four different plasma coupling values ($\Gamma \equiv$ 20, 50, 100, and 250) at constant κ = 1.40 and N = 500 are displayed in **Figure 1**. The effects of variation uniaxial electric field ($M_T \equiv$ 0.01–1.30) on VACF are computed to analyze the diffusion motion of charged dust particles. It has been shown that M_T significantly depends on plasma temperature and the screening of charged particles. At Higher plasma temperature (Γ = 20), one can easily observe the rapid decay of VACF with very weak particle oscillations. This regime M_T does not have significant effects on the dynamics of dust particles. **Figure 1(b)** shows the oscillatory damped motion and slightly decreasing amplitude of VACF up to uniaxial electric field strength (M_T = 0.90). For higher plasma coupling strength ($\Gamma \equiv$ 100,250), results of VACF show the higher oscillations with slightly damping phenomena. **Figure 2** demonstrated the VACF at κ = 2.0 for four different regimes of strongly coupled CDPs (Γ = 20, 50, 200, 400) N = 500 with a variation of uniaxial electric field ($0.0 < M_T \geq 1.50$). It is observed that the effect of M_T on VACF is nearly the same as was observed in **Figure 1**. It has also been shown that the M_{CT} increases with increases in the screening strength. Furthermore, we increased the screening strength κ = 3.0 and Γ = 30, 75, 250, and 500 are shown in **Figure 6**. The trend of damping and oscillations of dust particles was observed the same as in **Figures 1** and **2**. Therefore, the VACF below the M_{CT} simulation outcomes is founded in good agreement with previous data [13, 30, 35, 36].

It is observed that at higher temperatures (low coupling, Γ), the M_T does not affect the dynamics and oscillation of the particles. So what we can say that at higher system temperature, the radius between interacting charged dust particles is large. Due to the large distance, Eq. (1) remains invalid for higher uniaxial electric field strength. The motion of a particle in this regime indicates only thermal motion. Complete damped oscillations of dust particles were observed for a long time at intermediate values of Γ. The magnitude of oscillations slightly decreased with increasing M_T, increasing the diffusion at a low rate of thermal motion. In ER-CDPs, the effect of M_T on VACF can be easily observed from intermediate to higher plasma coupling (Γ) strength. The simulation time length scales, maximum at intermediate values of Γ and further increasing Γ, the length scale slightly decreased. It was observed that the oscillation was not fully damped at higher Γ. It is concluded that a uniaxial electric field can affect diffusion motion in ER-CDPs for the sufficient strength of M_T., so what can say the presented numerical method gives us precise and reliable data for comparative study.

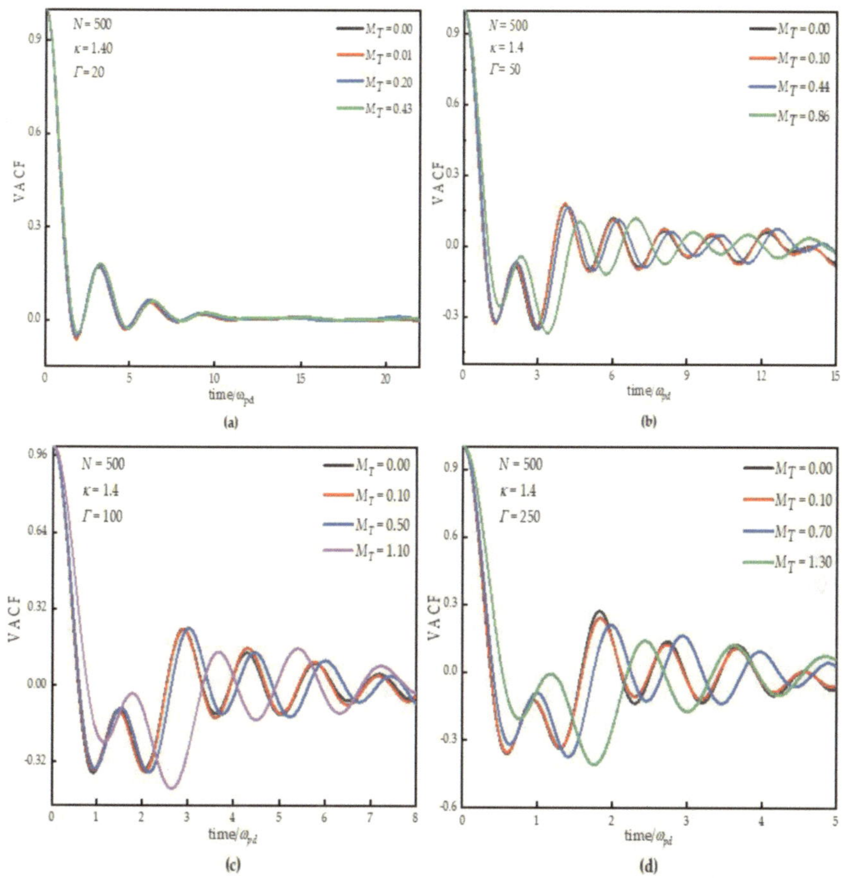

Figure 1.
VACF as a function of molecular dynamics simulation time scale for constant plasma screening parameter (κ = 1.40). MD simulations results obtained for N = 500 *simulated dust particles with the variation of uniaxial external electric field (0.0 < $M_T \leq$ 1.30) and four different plasma regimes (a) Γ = 20, (b) Γ = 50, (c) Γ = 100 and (d) Γ = 250.*

Now we focused on the primary outcomes through the equilibrium MD method. The simulation data of this work are that the self-diffusion coefficient (D) of ER-CDPs can be obtained with minimum statistical error at M_T = 0.0 by equilibrium MD. Moreover, it is shown that the MD with GKR has a fair good agreement with previous simulation outcomes of Ohta and Hamaguchi [30], Daligault [31, 32], and Begum and Das [33, 37] at M_T = 0.0. **Figures 3–5** display the primary outcomes of D measured from the equilibrium MD method for ER-CDPs at κ = 1.4, 2, and 3, respectively. D of ER-CDPs was calculated for different uniaxial electric field strengths ($0 < M_T < 1.5$). Here we have explained the D for a wide range of combinations of plasma parameters (Γ, κ) over acceptable M_T ranges below the M_{CT}. This section focused on the variation of plasma coupling ($1/T$), electric field, Debye screening length, and constant N = 500 simulated particles.

The effect of M_T on D in 3D ER-CDPs is calculated for plasma coupling ($\Gamma \equiv 20$, 50, 100, and 200), Debye screening length ($\kappa \equiv 1.4$), uniaxial electric field ($0.0 \leq M_T \leq 1.30$), and N = 500 number of particles shown in **Figure 3**. We have performed several MD simulations at constant κ = 1.4 and N = 500 to analyze the effect of M_T on D. The reported data of D shows good agreements for small M_T values with previous results [30–32]. With respective of Γ different regimes are under consideration. It is found that the M_{CT} values are the same as in **Figures 1, 2, 6**

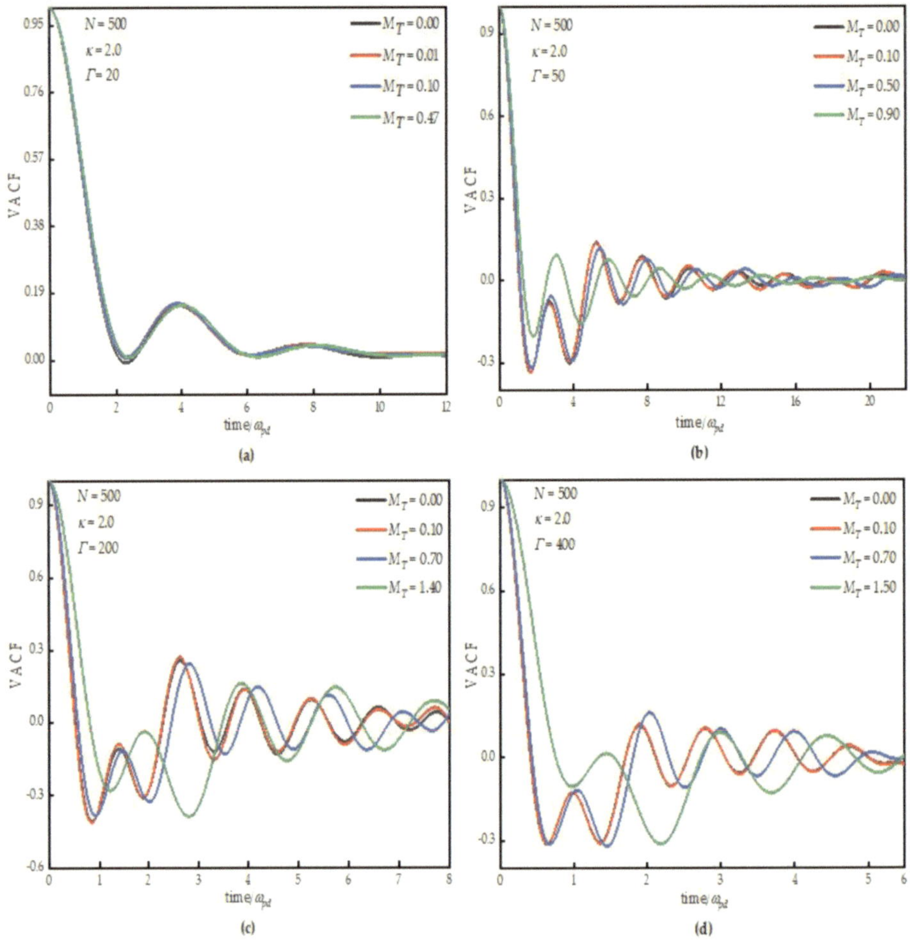

Figure 2.
VACF as a function of molecular dynamics simulation time scale for constant plasma screening parameter (κ = 2.0). MD simulations results obtained for N = 500 *simulated dust particles with the variation of uniaxial external electric field (0.0 <* $M_T \leq 1.30$*) and four different plasma regimes (a)* Γ = 20*, (b)* Γ = 50*, (c)* Γ = 200 *and (d)* Γ = 400*.*

for the same combinations of plasma parameters. It is noted that for plasma coupling (Γ = 20), the critical values ($M_{CT} > 0.42$) for κ = 1.4 display in **Figure 3**. We have found the increasing trend of M_T with increases the plasma coupling values or decreasing the system temperature such as Γ = 50, 100 and 200 critical $M_{CT} \geq 0.88$, 1.12, 1.25 respectively. The possible reason for the low strength of M_T in high system temperature is the large interparticle distance show with the term ($1/\mathbf{r}^3$) in Eq. (1). The increasing values of D with respective of Γ at κ = 1.4 can be easily observed from four panels of **Figure 3**. Upon further increasing M_T, the MD simulation gives an error in the form of out of thermal equilibrium. At lower plasma coupling strength (Γ = 20), the D does not significantly increase under the applied external electric field's influence; only 0.03% observed the integral values of VACF. The increasing of diffusion was observed for Γ = 50 are 20%, Γ = 100 are 75% and Γ = 200 are 160%. The electric field effect on D in ER-CDPs can be found in the same as conventional electrorheological fluids and ionic liquids [40].

The simulation outcomes of D in 3D ER-CDPs at constant N = 500, different values of Γ as a function of the uniaxial electric field (M_T) for κ = 2 and 3 are shown in **Figures 4** and **5**, respectively. The effect of M_T was computed in four different

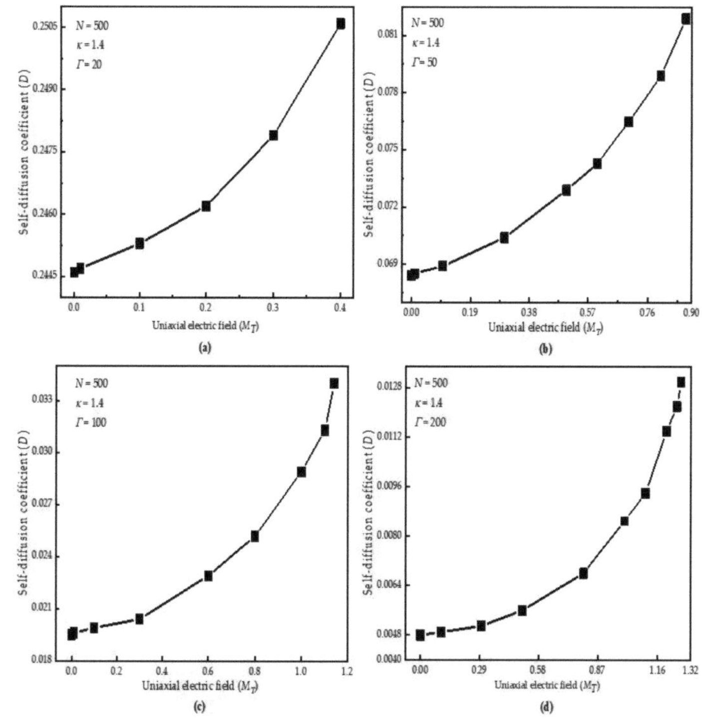

Figure 3.
Self-diffusion coefficient as a function of external applied uniaxial electric field ($0.0 < M_T < 1.30$), for constant plasma screening ($\kappa = 1.4$) and a number of particles ($N = 500$), considered four different ER-CDPs states (a) $\Gamma = 20$, (b) $\Gamma = 50$, (c) $\Gamma = 100$ and (d) $\Gamma = 200$.

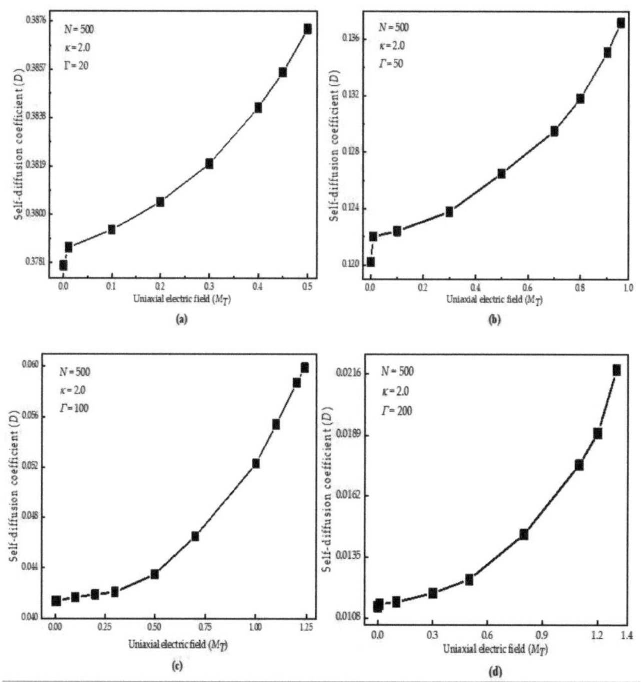

Figure 4.
Self-diffusion coefficient as a function of external applied uniaxial electric field ($0.0 < M_T < 1.40$), for constant plasma screening ($\kappa = 2$) and a number of particles ($N = 500$), considered four different ER-CDPs states (a) $\Gamma = 20$, (b) $\Gamma = 50$, (c) $\Gamma = 100$ and (d) $\Gamma = 200$.

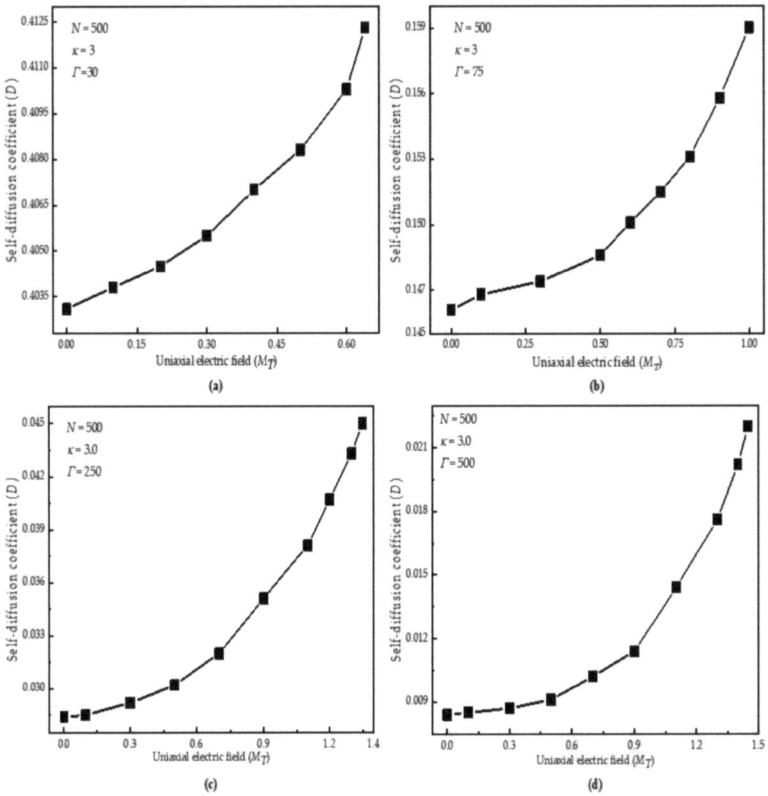

Figure 5.
Self-diffusion coefficient as a function of external applied uniaxial electric field ($0.0 < M_T < 1.50$), for constant plasma screening ($\kappa = 3$) and a number of particles ($N = 500$), considered four different ER-CDPs states (a) $\Gamma = 30$, (b) $\Gamma = 75$, (c) $\Gamma = 250$ and (d) $\Gamma = 500$.

states of ER-CDPs by performed almost thirty-three simulations. The M_{CT} values for $\kappa = 2.0$ at $\Gamma = 20$, $M_T > 0.45$, $\Gamma = 50$, $M_{CT} > 0.90$, $\Gamma = 100$, $M_{CT} > 1.23$, and $\Gamma = 200$, $M_{CT} > 1.35$. The increase in D under the action of external applied uniaxial electric field for $\Gamma = 20$, 50, 100, and 200 are 1.7, 11, 42, and 115%. The increases in the D for $\kappa = 3.0$ same N were observed nearly the same as said above for $\kappa = 2.0$. The effect of M_T on the D is prominent at higher values of Γ parameters. It was also noted that M_T does not significantly affect D lower to intermediate values of Γ. From **Figures 3–5**, we can conclude that the proposed MD simulation method worked for the limited strength of M_T. It was noted that the system's critical values above the system go out of thermal equilibrations, and kinetic energy increased very largely up to 23 orders of magnitude, but the potential energy does not change higher order of magnitude. Below the critical strength, the potential energy of interacting Yukawa dust particles increases with M_T.

4. Conclusion

An equilibrium MD simulation has been performed to report the velocity autocorrelation function (VACF) and self-diffusion coefficient (D) of three-dimensional (3D) electrorheological complex (dusty) plasmas (ER-CDPs) for the analysis of diffusion motion of dust particles. The interactions and forces between dust particles were modeled by Yukawa potential and Quadrupole interactions of charged dust particles. This paper highlights the outcomes of VACF and D for 3D

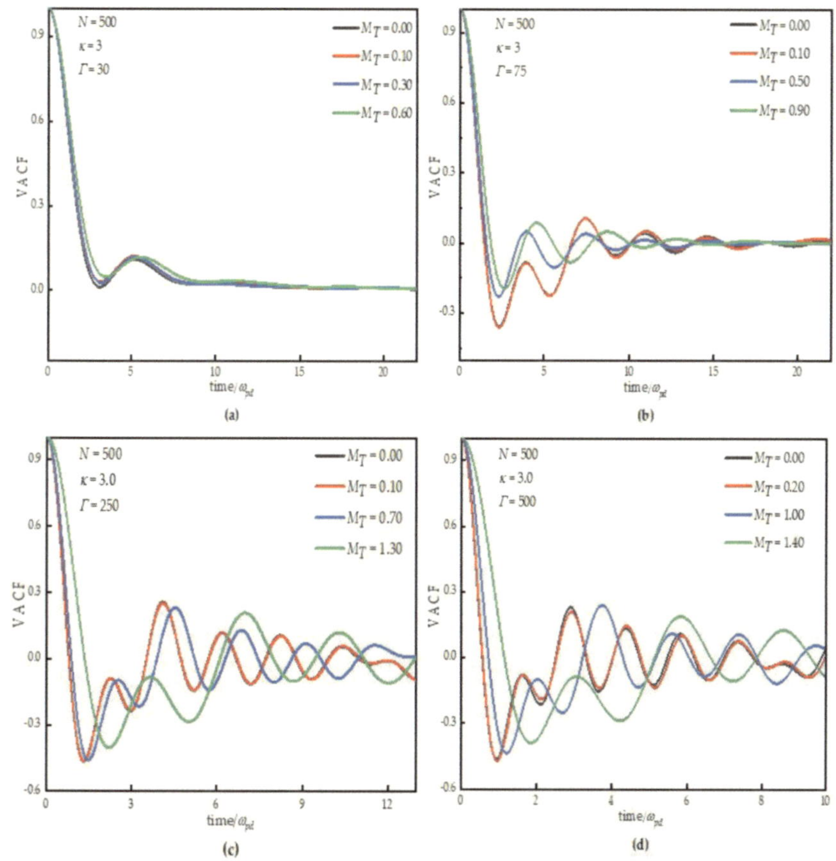

Figure 6.
VACF as a function of molecular dynamics simulation time scale for constant plasma screening parameter (κ = 3.0). MD simulations results obtained for N = 500 simulated dust particles with the variation of uniaxial external electric field (0.0 < $M_T \le$ 1.30) and four different plasma regimes (a) Γ = 30, (b) Γ = 75, (c) Γ = 250 and (d) Γ = 500.

ER-CDPs in the presence of a uniaxial electric field (M_T). We used Green-Kubo expression to calculate VACF and D over a wide range of CDPs Coulomb coupling (Γ) and Debye screening (κ) parameters. The VACF and D are significantly dependent on the plasma parameters (κ, Γ) and M_T. The calculated results are highly consistent with other MD data in the absence of M_T. It was observed that D decreased with increasing the Γ. The M_T significantly affects the diffusion motion in ER-CDPs from intermediate to higher values of Γ and does not affect low Γ. The D increases with increasing the M_T and κ. A new investigation gives more comprehensive and reliable data for VACF and D over given plasma parameters. It has been demonstrated that the presented numerical results for a given range of plasma parameters (Γ, κ) and the number of particles are good performances. The proposed numerical model is suitable for liquid-like and solid-like states for 3D CDPs. We can be concluded that the developed MD simulations approach will be employed to investigate the thermophysical properties of ER-CDPs.

Acknowledgements

The support provided by the National Science Fund for Distinguished Young Scholars of China (No. 51525604), the Foundation for Innovative Research Groups of the National Natural Science Foundation of China (No. 51721004) for the completion of the present work is gratefully acknowledged. The authors would also like to thank Dr. X. D. Zhang at the Network Information Center of Xi'an Jiaotong University for supporting the HPC platform and the National Centre for Physics (NCP) Islamabad allocation computational power to check and run the MD code.

Abbreviation and symbol

EMD	Equilibrium molecular dynamic
ERFs	Electrorheological fluids
κ	Screening strength
Γ	Coulomb coupling
M_T	Thermal mach number
N	Number of particles
CDP	Complex (dusty) plasma
SCCDPs	Strongly coupled complex (dusty) plasmas
WCCDPs	Weakly coupled complex (dusty) plasmas
3D	Three dimensional
PK-3 plus	Plasma Kristall-3 plus
VACF	Velocity autocorrelation function
D	Self-diffusion coefficient
GKR	Green-Kubo relation
PBCs	Periodic boundary conditions
k_B	Boltzmann constant
ω_{pd}	Plasma frequency
T	Plasma temperature
a_{ws}	Wigner Seitz radius
θ	The angle between the electric field and particles vector

Author details

Muhammad Asif Shakoori[1], Maogang He[1]*, Aamir Shahzad[2] and Misbah Khan[3]

1 Key laboratory of Thermal Fluid Science and Engineering, Ministry of Education (MOE), School of Energy and Power Engineering, Xi'an Jiaotong University, Xi'an, China

2 Molecular Modeling and Simulation Laboratory, Department of Physics, Government College University Faisalabad (GCUF), Faisalabad, Pakistan

3 Department of Refrigeration and Cryogenics Engineering, School of Energy and Power Engineering, Xi'an Jiaotong University, Xi'an, China

*Address all correspondence to: mghe@xjtu.edu.cn

References

[1] Szary ML, Noras M. Experimental study of sound transmission loss in electrorheological liquids. ASME Int. Mech. Eng. Congr. Expo. Proc. 2002; **240**:137–142.DOI:10.1115/ IMECE2002-33349

[2] Stanway R, Sproston JL, El-Wahed AK. Applications of electrorheological fluids in vibration control: A survey. Smart Mater. Struct. 1996;5:464–482. DOI:10.1088/0964-1726/5/4/011

[3] Shahzad A, Shakoori MA, He M-G, Bashir S. Sound Waves in Complex (Dusty) Plasmas. *Comput. Exp. Stud. Acoust. Waves, Intech open*.2018. DOI: 10.5772/intechopen.71203

[4] Shahzad A, Shakoori, MA, He M-G, Feng Y. Numerical Approach to Dynamical Structure Factor of Dusty Plasmas. Book entitled "Plasma Science and Technology Basic Fundamentals and Modern Applications". *Intech Open.* 2020; DOI:org/10.5772/intechopen.78334

[5] Shahzad A, Shakoori MA, He M-G. Wave Spectra in Dusty Plasmas of Nuclear Fusion Devices. Fusion energy. *Intech Open*. 2020.DOI:org/10.5772/ intechopen.91371

[6] Shahzad A, He M-G, shakoori MA. Thermal Transport and Non-Newtonian behaviors of 3D Complex Liquids Using Molecular Simulations. Proceeding of the Proceedings of 14th International Bhurban Conference on Applied Sciences & Technology (IBCAST); 10th - 14th January 2017; Islamabad, Pakistan, 2017.p.472–474.

[7] Shahzad A, Khan MQ, Shakoori MA, He M-G Feng Y. Thermal Conductivity of Dusty Plasmas through Molecular Dynamics Simulations. *Intech Open. 2020;* DOI:http://dx.doi.org/10.5772/57353

[8] Shahzad A, He M-G. Calculations of thermal conductivity of complex (dusty) plasmas using homogenous nonequilibrium molecular simulations. Radiat. Eff. Defects Solids.*2015;***170**:758–770.DOI:10.1080/10420150.2015.1108316

[9] Shahzad A, Kashif M, Munir T, He M-G, Tu X. Thermal conductivity analysis of two-dimensional complex plasma liquids and crystals. Phys. Plasmas. *2020*;**27**:103702.DOI:10.1063/ 5.0018537

[10] Shahzad A, He M-G. Thermal conductivity calculation of complex (dusty) plasmas. Phys. Plasmas.2012;**19**: 083707.DOI:10.1063/1.4748526

[11] Shahzad A, He M-G, Haider SI, Feng Y. Studies of force field effects on thermal conductivity of complex plasmas. Phys. Plasmas. 2017;**24**:093701. DOI: 10.1063/1.4993992

[12] Shahzad A, He M-G, He K. Diffusion motion of two-dimensional weakly coupled complex (dusty) plasmas. Phys. Scr. 2013;**87**:035501.DOI: 10.1088/0031-8949/87/03/035501

[13] Shahzad A, He M-G. Shear viscosity and diffusion motion of two-dimensional dusty plasma liquids. Phys. Scr. 2012;**86**:015502.DOI:10.1088/ 0031-8949/86/01/015502

[14] Shahzad A, He M-G. Thermodynamic characteristics of dusty plasma studied by using molecular dynamics simulation. Plasma Sci. Technol. 2012;**14**:771–777.DOI: 10.1088/ 1009-0630/14/9/01

[15] Shahzad A, Shakoori MA, He M-G, Feng Y. Dynamical structure factor of complex plasmas for varying wave vectors. Phys. Plasmas. 2019; **26**: 023704.DOI: 10.1063/1.5056261

[16] Donkó I, Hartmann P, Donkó Z. Molecular dynamics simulation of a two-dimensional dusty plasma. Am. J.

Phys. 2019;**87**:986–993.DOI: 10.1088/0031-8949/86/01/015502

[17] Adamovich I. *et al.* The 2017 Plasma Roadmap: Low temperature plasma science and technology. J. Phys. D. Appl. Phys. 2017;**50**:323001 20.DOI: 10.1088/1361-6463/aa76f5

[18] Hirst AM, *et al.* Low-temperature plasma treatment induces DNA damage leading to necrotic cell death in primary prostate epithelial cells. Br. J. Cancer. 2015**112**:1536–1545.DOI:10.1038/bjc.2015.113

[19] Laroussi M. Cold Plasma in Medicine and Healthcare: The New Frontier in Low Temperature Plasma Applications. Front. Phys. 2020;**8**:1–7. DOI: 10.3389/fphy.2020.00074

[20] Ivlev AV, *et al.* First observation of electrorheological plasmas. Phys. Rev. Lett. 2008;**100**:1–4.DOI:10.1103/PhysRevLett.100.095003

[21] Kana D, Dietz C, Thoma MH. Simulation of electrorheological plasmas with superthermal ion drift. Phys. Plasmas. *2020;*27:103703.DOI:10.1063/5.0010021

[22] Ivlev AV, Thoma MH, Räth C, Joyce G, Morfill GE. Complex plasmas in external fields: The role of non-hamiltonian interactions. Phys. Rev. Lett. 2011;**106**:155001.DOI:10.1103/PhysRevLett.106.155001

[23] Schwabe M *et al.* Slowing of acoustic waves in electrorheological and string-fluid complex plasmas. *New* J. Phys. 2020;**22**:083079.DOI: 10.1080/10420150.2015.1108316

[24] Ivlev AV *et al.* Electrorheological complex plasmas. IEEE Trans. Plasma Sci. 2010;**38**:733–740.DOI: 10.1109/TPS.2009.2037716

[25] Yaroshenko VV. Charge-gradient instability of compressional dust lattice waves in electrorheological plasmas. Phys. Plasmas. 2019;**26**:083701.DOI: 10.1063/1.5115346

[26] Rosenberg M. Waves in a 1D electrorheological dusty plasma lattice. J. Plasma Phys. 2015;**81**:905810407.DOI: 10.1017/S0022377815000422

[27] Sukhinin GI, Fedoseev AV, Khokhlov RO, Suslov SYU. Plasma Polarization Around Dust Particle in an External Electric Field. Contrib. to Plasma Phys. 2012;**52**:62–65.

[28] Sukhinin GI *et al.* Plasma anisotropy around a dust particle placed in an external electric field. Phys. Rev. E. *2017;*95:063207.DOI:10.1103/PhysRevE.95.063207

[29] Hartmann P, *et al.* Self-diffusion in two-dimensional quasimagnetized rotating dusty plasmas. Phys. Rev. E. *2019;*99:013203.DOI:10.1103/PhysRevE.99.013203

[30] Ohta H, Hamaguchi S. Molecular dynamics evaluation of self-diffusion in Yukawa systems. Phys. Plasmas. *2000;*7:4506–4514.DOI:10.1063/1.1316084

[31] Daligault J. Practical model for the self-diffusion coefficient in Yukawa one-component plasmas. Phys. Rev. E - Stat. Nonlinear, Soft Matter Phys. 2012; **86**:1–2.DOI:10.1103/PhysRevE.86.047401

[32] Daligault J. Diffusion in ionic mixtures across coupling regimes. Phys. Rev. Lett. 2012; **108**:1–5.DOI:10.1103/PhysRevLett.108.225004

[33] Begum M, Das N. Self-Diffusion of Dust Grains in Strongly Coupled Dusty Plasma Using Molecular Dynamics Simulation. 2016;**8**:49–54.DOI:10.9790/4861-0806024954

[34] Ott T, Bonitz M, Donkó Z, Hartmann P. Superdiffusion in quasi-two-dimensional Yukawa liquids. Phys.

Rev. E - Stat. Nonlinear, Soft Matter Phys. 2008;**78**:1–6.DOI:10.1103/PhysRevE.78.026409

[35] Dzhumagulova KN, Masheeva RU, Ramazanov TS, Donkó Z. Effect of magnetic field on the velocity autocorrelation and the caging of particles in two-dimensional Yukawa liquids. *Phys. Rev. E - Stat. Nonlinear, Soft Matter Phys.* 2014;**89**:1–7.DOI: 10.1103/PhysRevE.89.033104

[36] Dzhumagulova KN, Ramazanov T S, Masheeva RU. Velocity Autocorrelation Functions and Diffusion Coefficient of Dusty Component in Complex Plasmas. Contrib. to Plasma Phys. 2012;**52**:182–185.DOI:10.1002/ctpp.201100070

[37] Begum M, Das N. Self-diffusion as a criterion for melting of dust crystal in the presence of magnetic field. Eur. Phys. J. Plus. 2016;**131**:1–12.DOI: 10.1140/epjp/i2016-16046-2

[38] Hollingsworth SA, Dror RO. Molecular Dynamics Simulation for All. Neuron. 2018; **99**:1129–1143.DOI: 10.1016/j.neuron.2018.08.011

[39] Kompaneets R, Morfill GE, Ivlev AV. Design of new binary interaction classes in complex plasmas. Phys. Plasmas. 2009;**16**:043705.DOI: 10.1063/1.3112703

[40] Clark R. *et al.* Effect of an external electric field on the dynamics and intramolecular structures of ions in an ionic liquid. J. Chem. Phys. 2019;**151**: 164503.DOI: 10.1063/1.5129367

Chapter 6

Transverse Thermal Instability of Radiative Plasma with FLR Corrections for Star Formation in ISM

Sachin Kaothekar

Abstract

Impact of porosity, rotation and finite ion Larmor radius (FLR) corrections on thermal instability of immeasurable homogeneous plasma has been discovered incorporating the effects of radiative heat-loss function and thermal conductivity. The general dispersion relation is carried out with the help of the normal mode analysis scheme taking the suitable linearized perturbation equations of the difficulty. This general dispersion relations is further reduces for rotation axis parallel and perpendicular to the magnetic field. Thermal instability criterion establishes the stability of the medium. Mathematical calculations have been performed to represent the impact of different limitations on the growth rate of thermal instability. It is found that rotation, FLR corrections and medium porosity stabilize the growth rate of the medium in the transverse mode of propagation. Our outcome of the problem explains that the rotation, porosity and FLR corrections affect the dens molecular clouds arrangement and star configuration in interstellar medium.

Keywords: Thermal instability, Rotation, FLR Correction, Radiative heat-loss functions, ISM

1. Introduction

In different research fields of science, now a day's plasma physics have been one of the most important growing areas of research. Also the plasma instabilities are studied from several decades to understand the process of formation of small and big structures in interstellar medium in astronomy and astrophysics. The learning of thermal unsteadiness is suiting fashionable because this is the major development that agreements through outside warming and radiative codling in cosmological plasma environment and in the interstellar medium. The formation of a high amount of cosmological constructions given as interstellar clouds, solar eminences, concentrated organizations in planetary nebulae, etc., can be clarified by resource of thermal unsteadiness. Thermal unsteadiness occurs in an environment that can be developed into colder because of radiation and fluid reduction. Along with this, a reduction of temperature builds the environment unbalanced and directs to the configuration of novel arrangements because of density concentration (Parker [1] & Field [2]). In the unsteadiness, the serious length scale is lesser than that of the other dynamical unsteadiness's such as the Jeans gravitational unsteadiness; i.e., an environment can be develop into thermally unbalanced still if the environment is steady beside the gravitational unsteadiness. So, it is clear that the substantial foundation of slighter-range configurations is because of thermal unsteadiness in spite of the dynamical unsteadiness. While the masses of these contained dense entities are lower than those needed for gravitational reduction, the situation of thermal unsteadiness gets gratify. Certainly, when the gravitational power of originally uniform plasma is comparatively tiny the solar eminences (Tandberg-Hanssen [3]; Priest [4]) and numerous kinds of interstellar clouds (Spitzer [5]; Hollenbach & Thronson [6]; & Burton et al. [7]) are structured because of the procedure of thermal strengthening. Thermal unsteadiness has been studied by several investigators for more than seven decades in astrophysical objects and plasma physics applications (Aggarwal and Talwar [8]; Bora and Talwar [9]; Fukue & Kamaya [10]; Prajapati et al. [11]; Kaothekar et al. [12]; Sharma and Jain [13]; Prajapati et al. [14]). More recently Kaothekar [15] has discussed the thermal instability of partially ionized viscous plasma with Hall effect FLR corrections flowing through porous medium. Thus, we find that a large number of studies are done for magneto-thermal and radiative plasma with different parameters under various assumptions.

In addition to this, the problem of thermal instability of plasma flowing through porous medium has much significance in the learning of large and small astrophysical entities, such as comets, meteorites and interplanetary dust. More over the learning of flow via porous media is of considerable attention because of its diversity of applications in geophysical circumstances, magneto-hydrodynamics (MHD) flows, laboratories, industries and in petroleum and chemical engineering. A large amount of the pioneer work in the field of plasma flow via porous medium is analyzed by Nield and Bejan [16] and Vafai [17]. Many investigators have discussed the importance of porosity in thermal instability with different physical parameters in different plasma environments (Somerton & Catton [18]; Poulikakos [19]; Nield & Kuznetsov [20]; Shue [21]; Kumar [22]; Kaothekar & Chhajlani [23]; Nield & Kuznetsov [24]; Kaothekar [25]. More recently Nguyen-Thoi et al. [26] have discussed the magneto-hydrodynamic nano-fluid radiative thermal behavior by means of Darcy law inside a porous media. Thus we see that, porosity of the medium plays a crucial role in instability and stability examinations of the thermally magnetized plasma flowing through porous medium.

Along with this, in current days the significance of FLR in thermal instability and gravitational instability of plasma is important owing to its huge relevance in astrophysics. Many researchers (Jukes [27]; Roberts & Taylor [28]; Rosenbluth et al. [29]; Singh & Hans [30]; Herrnegger [31]; Sharma [32]; Chhonkar & Bhatia [33]; Devlen & Pekunlu [34]; Kaothekar & Chhajlani [35]; Kaothekar et al. [36]; Kaothekar [37]) have discussed the importance of FLR corrections in thermal instability with different parameters. More recently Kaothekar [38] has investigated the problem of Jeans instability of finitely conducting radiative rotating plasma with FLR corrections f lowing through porous medium. Thus it is clear that FLR is a significant restriction in argument of thermal instability and Jeans-gravitational instability.

From the above study we discover that combined influence of the rotation, FLR corrections, radiative heat-loss function, thermal conductivity and porosity on the thermal instability is not taken. Thus remaining in brains the importance of rotation and FLR corrections in formation of astrophysical small and big structures, we attempt to argue the outcomes of rotation, porosity and FLR corrections on thermal instability of plasma with thermal conductivity and radiative heat-loss function.

This paper is organized in following ways. In Section 2 linearized equations are presented. Section 3 contains the dispersion relation which is derived by linearized perturbation equations and discussed mathematically for transverse wave propagation. Section 4 represents the linear growth rate of the dispersion relation in transverse wave propagation and finally Section 5 contains discussion of the presented problem and result.

2. Linearized perturbation equations of the problem

We consider an infinite homogeneous, thermally conducting, radiating, porous plasma with FLR corrections in the presence of magnetic field $\boldsymbol{H}\,(0, 0, H)$. The perturbation in fluid pressure, density, temperature, velocity, magnetic field and heat-loss function are given as δp, $\delta\rho$, δT, $\boldsymbol{u}(\delta u_x, \delta u_y, \delta u_z)$, $\delta\boldsymbol{H}(\delta H_x, \delta H_y, \delta H_z)$, and L respectively. The perturbation circumstances is given as.

$$p = p_0 + \delta p, \rho = \rho_0 + \delta\rho, T = T_0 + \delta T, \boldsymbol{u} = \boldsymbol{u_0} + \delta\boldsymbol{u}, \boldsymbol{H} = \boldsymbol{H_0} + \delta\boldsymbol{H}, \text{and } L = L_0 + L. \tag{1}$$

Suffix '0' represents the initial equilibrium state, which is independent of space and time. L (ρ, T) is the heat-loss function of the material limited of thermal conduction and is in general a cause of the local values of density and temperature Field [2]. The operator (d/dt) is the substantial derivative given as $(d/dt) = (\partial/\partial t + (1/\varepsilon)\ \boldsymbol{u}.\nabla)$. With these effects the linearized perturbation equations of the problem are

$$\left(\frac{1}{\varepsilon}\right)\partial_t\delta\boldsymbol{u} = -\left(\frac{\nabla\delta p}{\rho}\right) - \left(\frac{\nabla\cdot\mathbf{P}}{\rho}\right) + \left(\frac{1}{4\pi\rho}\right)(\nabla\times\delta\boldsymbol{H})\times\boldsymbol{H} + 2(\boldsymbol{u}\times\boldsymbol{\Omega}), \tag{2}$$

$$\varepsilon\ \partial_t\delta\rho + \rho\nabla.\delta\boldsymbol{u} = \mathbf{0}, \tag{3}$$

$$\left(\frac{1}{\gamma-1}\right)\partial_t\delta p - \left(\frac{\gamma}{\gamma-1}\right)\left(\frac{p}{\rho}\right)\partial_t\delta\rho + \ \rho\left[\delta\rho\left(\frac{\partial L}{\partial\rho}\right)_T + \delta T\left(\frac{\partial L}{\partial T}\right)_\rho\right] - \ \lambda\nabla^2\delta T = 0, \tag{4}$$

$$\left(\frac{\delta p}{p}\right) = \left(\frac{\delta T}{T}\right) + \left(\frac{\delta\rho}{\rho}\right), \tag{5}$$

$$\partial_t\delta\boldsymbol{H} = \left(\frac{1}{\varepsilon}\right)\nabla\times(\boldsymbol{u}\times\boldsymbol{H}), \tag{6}$$

$$\nabla.\delta\boldsymbol{H} = \ 0, \tag{7}$$

where $(\partial L/\partial T)_\rho$, $(\partial L/\partial\rho)_T$ are the partial derivatives of temperature dependent heat-loss function L_T and density dependent heat-loss function L_ρ respectively. The components of pressure tensor **P**, considering the finite ion gyration radius for the magnetic field along z-axis as given by Roberts and Taylor [28] are

$$\begin{aligned} &P_{xx} = -\rho\upsilon_0\left[(\partial\delta u_y/\partial x) + (\partial\delta u_x/\partial y)\right], P_{yy} = \rho\upsilon_0\left[(\partial\delta u_y/\partial x) + (\partial\delta u_x/\partial y)\right],\\ &P_{xy} = P_{yx} = \rho\upsilon_0\left[(\partial\delta u_y/\partial x) - (\partial\delta u_y/\partial y)\right],\\ &P_{xz} = P_{zx} = -2\rho\upsilon_0\left[(\partial\delta u_y/\partial z) + (\partial\delta u_z/\partial y)\right],\\ &P_{yz} = P_{zy} = 2\rho\upsilon_0[(\partial\delta u_z/\partial x) + (\partial\delta u_x/\partial z)],\ \ P_{zz} = 0. \end{aligned} \tag{8}$$

The parameter υ_0 has the dimensions of the kinematics viscosity and called as magnetic viscosity defined as $\upsilon_0 = \Omega_L R_L^2/4$, where R_L is the ion-Larmor radius and Ω_L is the ion gyration frequency.

We seek plain wave solution of the form

$$\exp\left(i\sigma t + ik_x x + ik_z z\right), \tag{9}$$

where σ is the frequency of harmonic disturbance, k_x and k_z are the wave numbers of the perturbations along x and z axes. Such that

$$k^2 = k_x^2 + k_z^2 \tag{10}$$

The components of Eq. (6) may be given

$$\delta H_x = (iH/\varepsilon\omega)k_z\delta u_x, \quad \delta H_y = (iH/\varepsilon\omega)k_z\delta u_y, \quad \delta H_z = -(iH/\varepsilon\omega)k_x\delta u_x. \tag{11}$$

where $i\sigma = \omega$.

Using Eqs. (4), (5) and (9) we write

$$\delta p = \frac{\left\{(\gamma-1)\left[TL_T - \rho L_\rho + \left(\frac{\lambda k^2 T}{\rho}\right)\right] + \omega c^2\right\}}{\left\{(\gamma-1)\left[\left(\frac{T\rho}{p}\right)L_T + \left(\frac{\lambda k^2 T}{p}\right)\right] + \omega\right\}\delta\rho}, \tag{12}$$

Using Eqs. (3)–(11) in Eq. (2), we may engrave the subsequent algebraic equations for the constituents of Eq. (2)

$$\delta u_x\left[\omega + \left(V^2k^2/\omega\right)\right] + \delta u_y\left[\varepsilon\upsilon_0\left(k_x^2 + 2k_z^2\right) - 2\varepsilon\Omega_z\right] + \varepsilon\left(ik_x/k^2\right)\Omega_T^2\,\mathrm{s} = 0, \tag{13}$$

$$-\delta u_x\left[\varepsilon\upsilon_0\left(k_x^2 + 2k_z^2\right) - 2\Omega_z\right] + \delta u_y\left[\omega + \left(V^2k_z^2/\omega\right)\right] - \delta u_z[2\varepsilon(\upsilon_0 k_x k_z + \Omega_x)] = 0, \tag{14}$$

$$\delta u_y[2\varepsilon(\upsilon_0 k_x k_z + \Omega_x)] + \delta u_z\omega + \varepsilon\left(ik_z/k^2\right)\Omega_T^2\,\mathrm{s} = 0. \tag{15}$$

Taking divergence of Eq. (2) and using Eqs. (3)–(11), we obtain as

$$\delta u_x\left[ik_x\left(\frac{V^2k^2}{\varepsilon\,\omega}\right)\right] + \delta u_y\left[i\upsilon_0 k_x\left(k_x^2 + 4k_z^2\right) + 2i(k_z\Omega_x - k_x\Omega_z)\right] - \mathrm{s}\left[\omega^2 + \Omega_T^2\right] = 0, \tag{16}$$

we have made following substitutions $\alpha = (\gamma-1)\left[TL_T - \rho L_\rho + \left(\frac{\lambda k^2 T}{\rho}\right)\right]$, $\beta = (\gamma-1)\left[\left(\frac{T\rho L_T}{p}\right) + \left(\frac{\lambda k^2 T}{p}\right)\right]$, $s = \frac{\delta\rho}{\rho}$, $\Omega_T^2 = \left[\frac{\left(\Omega_I^2 + \omega\Omega_J^2\right)}{(\omega+\beta)}\right]$, $\Omega_J^2 = c^2k^2$, $\Omega_I^2 = k^2\alpha$, $V^2 = \left(\frac{H^2}{4\pi\rho}\right)$, $c = (\gamma p/\rho)^{1/2}$ is the adiabatic velocity of sound in the medium.

3. Dispersion relation

The nontrivial solution of the determinant of the matrix gained from Eqs. (13)–(16) with $\delta u_x, \delta u_y, \delta u_z, s$ having various coefficients that should disappear is to give the subsequent dispersion relation

$$
\begin{aligned}
&\left(\omega^2+\Omega_T^2\right)\left[\omega+\left(V^2k^2/\omega\right)\right]\left\{\omega\left[\omega+\left(V^2k_z^2/\omega\right)\right]+4\varepsilon^2\left(v_0k_xk_z+\Omega_x\right)^2\right\}-\left(2\varepsilon^2\Omega_T^2/k^2\right)\\
&\times\left(v_0k_xk_z+\Omega_x\right)\left[\omega+\left(V^2k^2/\omega\right)\right]\left[v_0k_xk_z\left(k_x^2+4k_z^2\right)+2\left(k_z^2\Omega_x-k_xk_z\Omega_z\right)\right]\\
&+\omega\left[\varepsilon v_0\left(k_x^2+2k_z^2\right)-2\varepsilon\Omega_z\right]^2\left(\omega^2+\Omega_T^2\right)+\left(2\varepsilon k_xk_zV^2/\omega\right)\Omega_T^2\left(v_0k_xk_z+\Omega_x\right)\\
&\times\left[\varepsilon v_0\left(k_x^2+2k_z^2\right)-2\varepsilon\Omega_z\right]-\omega\left(\varepsilon\Omega_T^2/k^2\right)\left[\varepsilon v_0\left(k_x^2+2k_z^2\right)-2\varepsilon\Omega_z\right]\\
&\times\left[v_0k_x^2\left(k_x^2+4k_z^2\right)+2\left(k_xk_z\Omega_x-k_x^2\Omega_z\right)\right]-\omega\left[\omega+\left(V^2k_z^2/\omega\right)\right]\left(V^2k_x^2/\omega\right)\Omega_T^2\\
&-\left(4\varepsilon^2k_x^2V^2/\omega\right)\Omega_T^2\left(v_0k_xk_z+\Omega_x\right)^2=0.
\end{aligned}
\tag{17}
$$

The dispersion relation (17) demonstrates the jointed influence of rotation, FLR corrections, radiative heat-loss function, thermal conductivity and porosity on the thermal instability of homogeneous plasma flowing through porous medium. The above dispersion relation is long and to learn the consequence of all parameter we now diminish the dispersion relation (17) for transverse mode of transmission.

4. Conversation of the dispersion relation

4.1 Transverse mode of transmission (K⊥B)

In this situation the perturbations are in use to be vertical to the path of the magnetic field (*i.e.* $k_x=k,\ \ k_z=0$). The dispersion relation (17) reduces to

$$
\begin{aligned}
&\left(\omega^2+\Omega_T^2\right)\left\{\left[\omega+\left(V^2k^2/\omega\right)\right]\left(\omega^2+4\varepsilon^2\Omega_x^2\right)+\omega\left(\varepsilon v_0k^2-2\varepsilon\Omega_z\right)^2\right\}\\
&-\Omega_T^2\left\{\omega\left[\left(V^2k^2/\omega\right)+\varepsilon^2\left(v_0k^2-2\Omega_z\right)^2\right]+4\varepsilon^2\Omega_x^2\left(V^2k^2/\omega\right)\right\}\\
&=0.
\end{aligned}
\tag{18}
$$

This dispersion relation (18) provides the control of rotation, FLR corrections, radiative heat-loss function thermal conductivity and porosity on thermal unsteadiness of plasma for transverse mode of transmission. Now we discuss the dispersion relation (18) for rotation axis parallel and vertical to the magnetic field.

4.1.1 *Axis of rotation along the magnetic field (Ω//B)*

For the case of axis of rotation along the magnetic field, we put $\Omega_x=0$ and $\Omega_z=\Omega$ in dispersion relation (18) which reduces to

$$
\omega^3\left\{\omega\left[\omega+\left(V^2k^2/\omega\right)\right]+\varepsilon^2\left(v_0k^2-2\Omega\right)^2+\left(\Omega_I^2+\omega\Omega_J^2\right)/(\omega+\beta)\right\}=0. \tag{19}
$$

Eq. (19) has two independent factors. The first factor of Eq. (19) gives $\omega^3=0$, which is a marginal stable mode. The second factor of Eq. (19) gives the following dispersion relation on alternating the values of $\Omega_I^2,\ \Omega_J^2,\alpha$ and β.

$$
\begin{aligned}
&\omega^3+\left\{\left[(\gamma-1)\left((T\rho L_T/p)+\left(\lambda k^2T/p\right)\right)\right]\right\}\omega^2+\left[4\varepsilon^2\Omega^2+\varepsilon^2v_0^2k^4\right.\\
&\left.+V^2k^2+c^2k^2-4\varepsilon^2\Omega v_0k^2\right]\omega+(\gamma-1)\left[(T\rho L_T/p)+\left(\lambda k^2T/p\right)\right]\\
&\times\left(4\varepsilon^2\Omega^2+\varepsilon^2v_0^2k^4+V^2k^2-4\varepsilon^2\Omega v_0k^2\right)+k^2(\gamma-1)\\
&\quad\times\left[TL_T-\rho L_\rho+\left(\lambda k^2T/\rho\right)\right]\}\\
&\quad=0.
\end{aligned}
\tag{20}
$$

This dispersion relation symbolizes the consequence of direct addition of rotation, FLR corrections, radiative heat-loss function, thermal conductivity and porosity on the thermal unsteadiness of the organization. When constant term of Eq. (20) is less than zero this allows at least one positive real root which converses to the unsteadiness of the organization. The situation of unsteadiness obtained from steady term of Eq. (20) is specified as

$$\left\{k^2\left[TL_T-\rho L_\rho+\left(\frac{\lambda k^2 T}{\rho}\right)\right]+\left[\left(\frac{T\rho L_T}{p}\right)+\left(\frac{\lambda k^2 T}{p}\right)\right]\left(4\varepsilon^2\Omega^2+\varepsilon^2 v_0^2 k^4+V^2k^2-4\varepsilon^2\Omega v_0 k^2\right)\right\}<0, \tag{21}$$

Eq. (21) symbolizes the modified form of thermal instability criterion by enclosure of rotation, FLR corrections, radiative heat-loss function and thermal conductivity. From Eq. (21) we conclude that rotation and FLR corrections stabilize the radiative instability.

In nonappearance of FLR corrections ($v_0 = 0$) Eq. (20) grows to be

$$\begin{aligned}&\omega^3+\{(\gamma-1)[(T\rho L_T/p)+(\lambda k^2 T/p)]\}\,\omega^2+\left[4\varepsilon^2\Omega^2+V^2k^2+c^2k^2\right]\omega\\&+\{k^2(\gamma-1)[TL_T-\rho L_\rho+(\lambda k^2T/\rho)]+(\gamma-1)[(T\rho L_T/p)+(\lambda k^2 T/p)]\\&\times\left[V^2k^2+4\varepsilon^2\Omega^2\right]\}=0.\end{aligned} \tag{22}$$

When constant term of Eq. (22) is less than zero this permits at least one positive real root which communicates to the instability of the system. The condition of instability attained from constant term of Eq. (22) is given as

$$\left\{k^2(\gamma-1)\left[TL_T-\rho L_\rho+\left(\frac{\lambda k^2T}{\rho}\right)\right]+(\gamma-1)\left[\left(\frac{T\rho L_T}{p}\right)+\left(\frac{\lambda k^2T}{p}\right)\right]\left[V^2k^2+4\varepsilon^2\Omega^2\right]\right\}<0, \tag{23}$$

The above situation of instability is the changed form of thermal condition by addition of rotation and magnetic field strength. From Eq. (23) we bring to a close that rotation and magnetic field stabilize the radiative instability.

Now Eq. (20) can be written in the following form

$$\begin{aligned}&\omega^3+c_s\left\{k_T+\left(\frac{k^2}{k_\lambda}\right)\right\}\omega^2+c_s^2\left[\frac{4\varepsilon^2\Omega^2}{c_s}+\frac{\varepsilon^2v_0^2k^4}{c_s}+\frac{V^2k^2}{c_s}+k^2-\frac{4\varepsilon^2\Omega v_0k^2}{c_s}\right]\omega+c_s^3\left[k_T+\left(\frac{k^2}{k_\lambda}\right)\right]\\&\times\left[\frac{4\varepsilon^2\Omega^2}{c_s}+\frac{\varepsilon^2v_0^2k^4}{c_s}+\frac{V^2k^2}{c_s}+k^2-\frac{4\varepsilon^2\Omega v_0k^2}{c_s}\right]+\left(\frac{c_s^3k^2}{\gamma}\right)\left[k_T-k_\rho+\left(\frac{k^2}{k_\lambda}\right)\right]=0.\end{aligned} \tag{24}$$

We have used

$$k_\rho=\left[(\gamma-1)\rho L_\rho\right]/(Rc_sT),\ k_T=[(\gamma-1)L_T]/(Rc_s),\ k_\lambda=(Rc_s\rho)/[(\gamma-1)\lambda], \tag{25}$$

To investigate the effect of viscosity, porosity, rotation and radiative heat-loss functions on the growth rate of thermal instability, we solve Eq. (24) numerically. Therefore Eq. (24) can be written in non-dimensional form with the help of following dimension-less quantities as given in Field [2]

$$\omega^*=\omega/k_\rho c_s,\ \Omega^*=\Omega k_\rho/c_s,\ k^*=k/k_\rho,\ k_\lambda^*=k_\rho/k_\lambda,\ k_T^*=k_T/k_\rho, \tag{26}$$

In astrophysical circumstances, instability of the organization is one of the most significant reasons of arrangement of entities. So we learn the consequences of medium porosity ε, rotation Ω^*, and FLR corrections v_0^* on the growth rate of unstable mode. Using Eq. (26), we write Eq. (24) in non-dimensional form as

$$\omega^{*3} + \left(k_T^* + k_\lambda^{*2}\right)\omega^{*2} + \left[4\varepsilon^2\Omega^{*2} + \varepsilon^2 v_0^{*2}k^{*4} + V^{*2}k^{*2} + k^{*2} - 4\varepsilon^2\Omega^* v_0^* k^{*2}\right]\omega^* \\ + \left(4\varepsilon^2\Omega^{*2} + \varepsilon^2 v_0^{*2}k^{*4} + V^{*2}k^{*2} - 4\varepsilon^2\Omega^* v_0^* k^{*2}\right)\left(k_T^* + k_\lambda^{*2}\right) \\ + \left(k^{*2}/\gamma\right)\left[k_T^* - 1 + k_\lambda^* k^{*2}\right] = 0. \tag{27}$$

Mathematical computations were executed to decide the roots of (ω^*) as a function of wave number (k^*) for moderately a few values of dissimilar parameters occupied captivating $\gamma = 5/3$. Out of three modes, only one mode is unstable for which the computations are at presented in **Figures 1–5**, where the growth rate ω^* has been sketched versus the wave number k^* to display the reliance of the growth rate on the dissimilar substantial limitations such as porosity, rotation and FLR corrections. It is clear from **Figure 1** that the max out rate of the growth rate reduces with augment in the rate of medium porosity. Thus the consequence of medium porosity is stabilizing on the growth rate of the environment. From **Figure 2** we see that the growth rate diminishes with augment in the value of rotation. Thus it is bring to a close that rotation stabilizes the growth rate of the environment. One can examine from **Figure 3** that the growth rate diminishes with rising FLR corrections. Thus the effect of FLR corrections is stabilizing on the growth rate of the environment. From **Figure 4** it is clear that growth rate diminishes on raising the value of

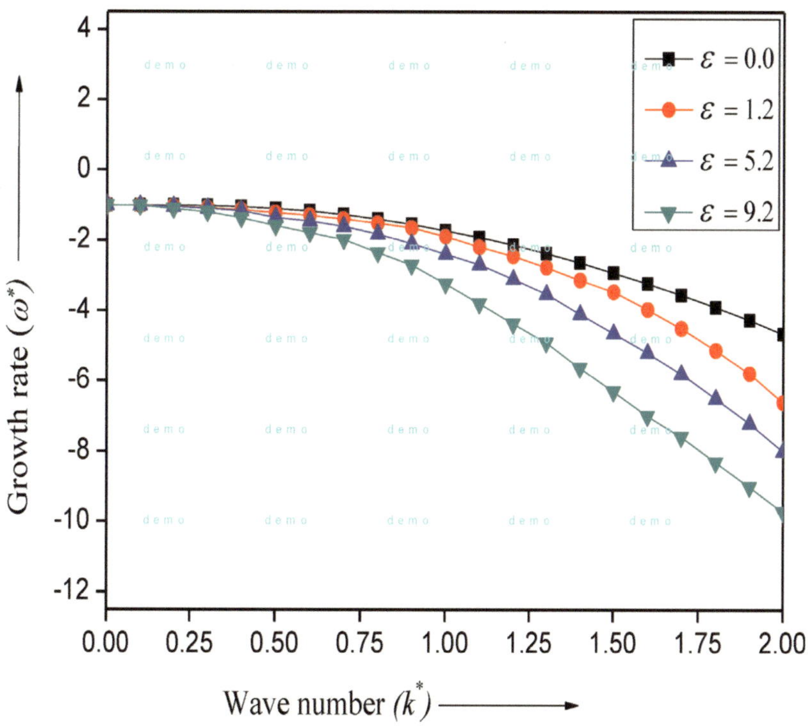

Figure 1.
Growth rate (ω^*) against wave number k^* for four values of parameter ε keeping the other parameters fixed $K_T{}^* = 1$, $K_\lambda{}^* = 0$, $V^* = 1$, $\nu_o{}^* = 1$, $\Omega^* = 1.0$.

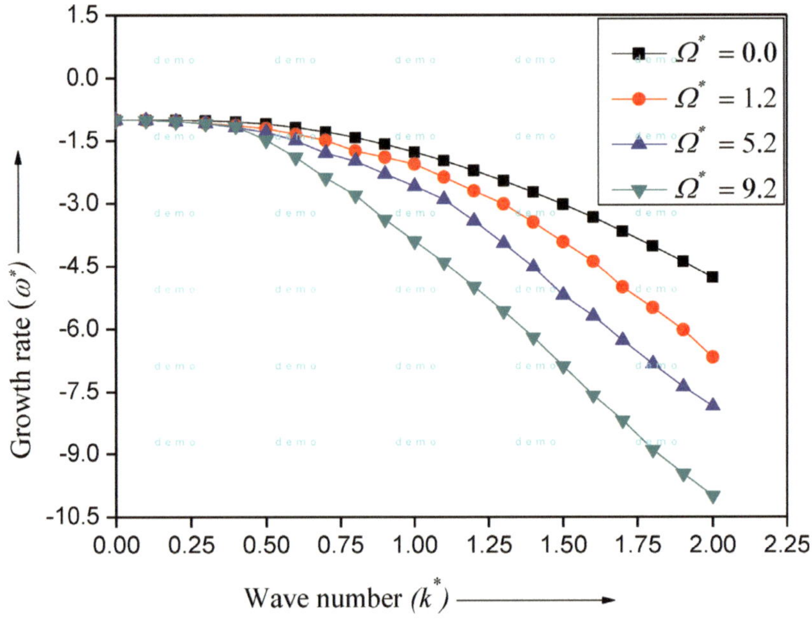

Figure 2.
Growth rate (ω^*) against wave number k^* for four values of parameter Ω^* keeping the other parameters fixed $K_T^* = 1.0$, $K_\lambda{}^* = 0$, $V^* = 1$, $\nu_0{}^* = 1$, $\varepsilon = 1.0$.

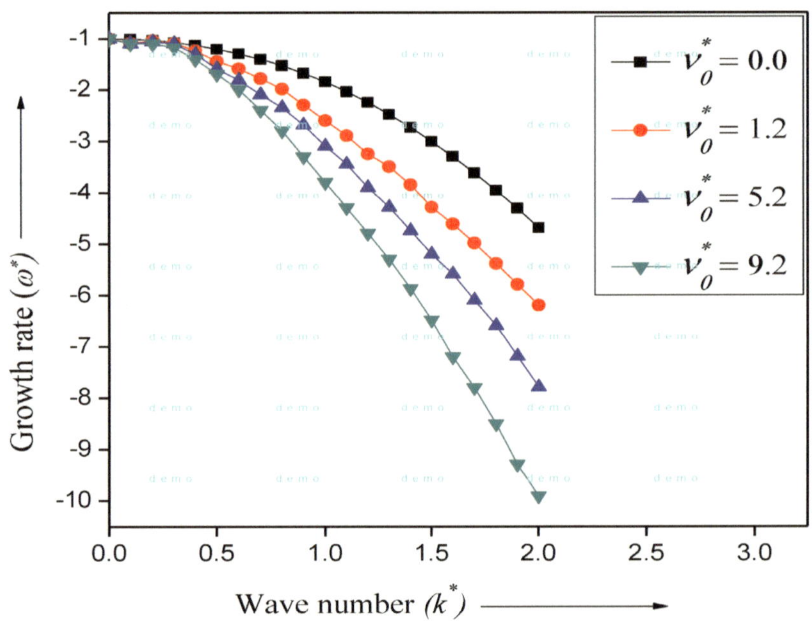

Figure 3.
Growth rate (ω^*) against wave number k^* for four values of parameter $\nu_0{}^*$ keeping the other parameters fixed $K_T^* = 1.0$, $K_\lambda{}^* = 0$, $V^* = 1$, $\Omega^* = 1$, $\varepsilon = 1.0$.

K_T^*. So K_T^* shows stabilizing effect on the growth rate of the environment. One can observe from **Figure 5** that as the value of K_λ^* increases the growth rate of the environment decreases. So it is clear that K_λ^* stabilize the growth rate of the environment.

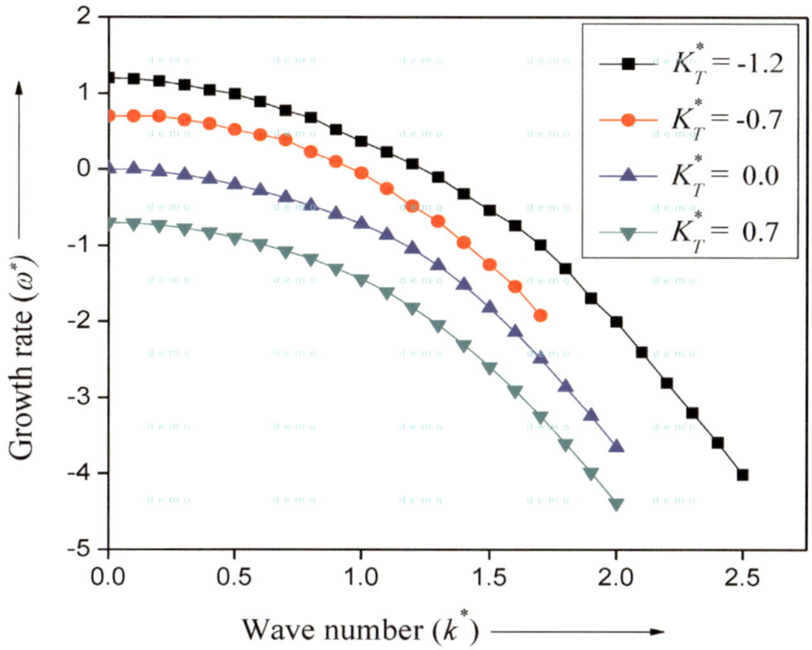

Figure 4.
Growth rate (ω^*) against wave number k^* for four values of parameter K_T^* keeping the other parameters fixed $K_\lambda^* = 1, V^* = 1, \ \Omega^* = 1, \varepsilon = 1.0$.

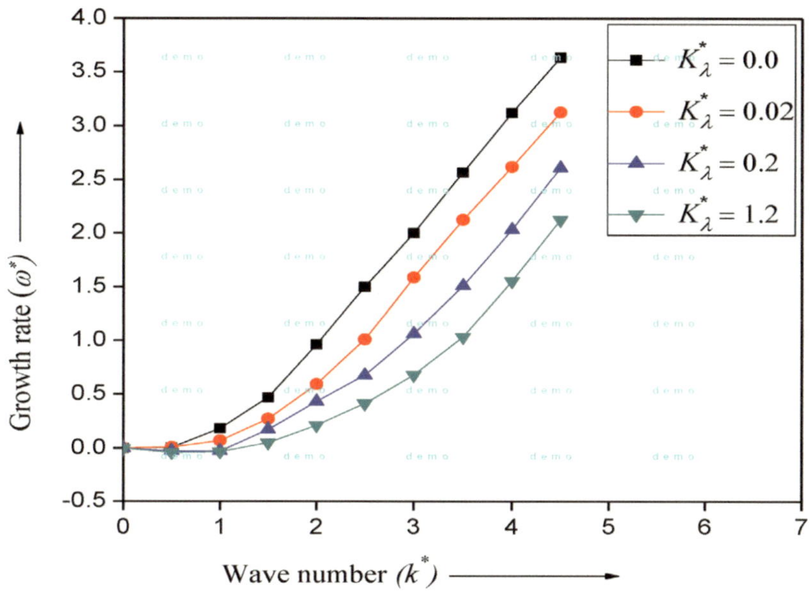

Figure 5.
Growth rate (ω^*) against wave number for four values of parameter K_λ^* keeping the other parameters fixed $K_T^* = 1, V^* = 1, \Omega^* = 1, \varepsilon = 1.0$.

4.1.2 Axis of rotation vertical to the magnetic field ($\Omega \perp B$)

In the case of axis of rotation perpendicular to the magnetic field, we put $\Omega_x = \Omega$ and $\Omega_z = 0$ in the dispersion relation (18) which reduces to

$$\omega\left\{\omega^2(\omega^2+4\varepsilon^2\Omega^2+\varepsilon^2 v_0^2 k^4)+(\omega^2+4\varepsilon^2\Omega^2)\left[V^2k^2+\frac{\left(\Omega_I^2+\omega\Omega_J^2\right)}{(\omega+\beta)}\right]\right\}=0. \tag{28}$$

This dispersion relation symbolizes the joint influence of FLR corrections, rotation, porosity magnetic field, radiative heat-loss function and thermal conductivity on the thermal instability of the considered organization. Eq. (28) has two independent factors. The first factor of Eq. (28) gives $\omega = 0$, which is a marginal stable mode. The second factor of Eq. (28) gives the following dispersion relation on replacing the values of Ω_I^2, Ω_J^2, and β.

In nonattendance of rotation ($\Omega = 0$) Eq. (28) becomes

$$\begin{aligned}&\omega^3+\left\{(\gamma-1)\left[\left(\frac{T\rho L_T}{p}\right)+\left(\frac{\lambda k^2 T}{p}\right)\right]\right\}\omega^2\\&+\left[\varepsilon^2v_0^2k^4+V^2k^2+c^2k^2\right]\omega+\left\{k^2(\gamma-1)\times\left[TL_T-\rho L_\rho+\left(\frac{\lambda k^2T}{\rho}\right)\right]\right.\\&\left.+(\gamma-1)\left[\left(\frac{T\rho L_T}{p}\right)+\left(\frac{\lambda k^2T}{p}\right)\right]\left[\varepsilon^2v_0^2k^4+V^2k^2\right]\right\}\\&=0.\end{aligned} \tag{29}$$

The situation of instability acquired from constant term of Eq. (29) is given as

$$\left\{k^2(\gamma-1)\left[TL_T-\rho L_\rho+\left(\frac{\lambda k^2T}{\rho}\right)\right]+(\gamma-1)\left[\left(\frac{T\rho L_T}{p}\right)+\left(\frac{\lambda k^2T}{p}\right)\right]\left[\varepsilon^2v_0^2k^4+V^2k^2\right]\right\}<0. \tag{30}$$

In present case situation of instability and growth rate of instability both depend on FLR corrections and porosity.

5. Conclusions

In the above present problem we have approved out the consequence of rotation, porosity and FLR corrections on the thermal instability of plasma counting the effects of radiative heat-loss function and thermal conductivity. The general dispersion relation is attained, which is customized due to the attendance of calculated physical limitations. This dispersion relation is condensed for transverse wave propagation to the route of magnetic field, which is additional argued for rotation axis parallel and vertical to the route of magnetic field.

In the situation of transverse wave propagation to the direction of magnetic field with axis of rotation along magnetic field we gained two modes. The first one is a marginal stable mode. The second one represents the thermal mode amended by rotation, porosity, FLR corrections and radiative heat-loss function. It is concluded that the condition of thermal unsteadiness is modified due to the attendance of porosity, rotation, FLR corrections, radiative heat-loss function, and thermal conductivity. For the case of non-FLR medium, it is found that the condition of radiative unsteadiness and expression of critical thermal wave number both are amended due to the occurrence of porosity, rotation, and it explains the stabilizing influence. It is found that for non-FLR the condition of radiative unsteadiness and expression of critical thermal wave number both are amended due to the occurrence of porosity, rotation and magnetic field. It is self-governing of FLR corrections.

In the case of axis of rotation vertical to the magnetic field for transverse wave propagation, we obtained two modes. The first one is a marginal stable mode. The second one symbolizes the impact of porosity, rotation, FLR corrections, radiative heat-loss function and thermal conductivity on thermal unsteadiness of plasma. It is concluded that the condition of unsteadiness is sovereign of porosity, rotation and FLR corrections and it depends on radiative heat-loss function and thermal conductivity. But the growth rate of the organization is exaggerated by the attendance of rotation, porosity and FLR corrections. For the case of non-rotating medium, it is found that condition of radiative unsteadiness is amended by the presence of FLR corrections, porosity and magnetic field, and it demonstrates stabilizing authority.

Acknowledgements

Author (S.K.) is grateful to CA Avnish Gupta, Vice Chairman Prashanti Institute of Technology & Science Ujjain, for continuous support.

Author details

Sachin Kaothekar
Department of Physics, Prashanti Institute of Technology and Science, Ujjain, M.P., India

*Address all correspondence to: sachinmgi007@gmail.com; sackaothekar@gmail.com

References

[1] Parker EN. Astrophy. J. 1953;**117**: 413-436. DOI: 10.1086/146579

[2] Field GB. Astrophys. J. 1965;**142**: 531-567. DOI: 10.1086/148317

[3] Tandberg-Hanssen E. Solar Prominences (Dordrecht: Reidel)1974. DOI: 10.1007/978-94-010-2136-4

[4] Priest ER. Dynamics and Structures of Quiescent Solar Prominences (Dordrecht, Kulwer)1989. DOI: 10.1007/978-94-009-3077-3

[5] Spitzer L. Physical Processes in the Interstellar Medium (New York: Wiley) 1978. DOI: 10.1002/9783527617722

[6] Hollenbach D. J., & Thronson H. A., 1987, Interstellar Processes (Dordrecht: Reidel). DOI:10.1002/asna.2113100329

[7] Burton W. B., Elmegreen B. G. & Genzel R., 1992, The Galactic Interstellar Medium (Berlin: Springer). DOI: 10.1007/3-540-31629-9

[8] Aggarwal M, Talwar SP. Mon. Not. R. Astro. Soc. 1969;**146**:235-242. DOI: 10.1093/mnras/146.3.235

[9] Bora MP, Talwar SP. Phys. Fluids B. 1993;**5**:950-955. DOI: 10.1063/1.860944

[10] Fukue T, Kamaya H. Astrophys. J. 2007;**669**:363-377. DOI: 10.1086/521268

[11] Prajapati R. P., Pensia R. K., Kaothekar S., & Chhajlani R. K., Astrophys. SpaceSci., 2010, **327**, pp. 139–154. DOI: 10.1007/s10509-010-0273-6

[12] Kaothekar S., Soni G. D. & Chhajlani R. K., AIP Adv., 2012, **2**, pp. 1-18 (042191). DOI: 10.1063/1.4773348

[13] Sharma P. & Jain S., Phys. Scr., 2016, **23**, pp. 1-8 (015602). DOI: 10.1088/0031- 8949/91/1/015602

[14] Prajapati RP, Bhakta S, Chhajlani RK. 1-8 (053703). Phys. Plasmas. 2016:**23** DOI.10.1063/1.4950821

[15] Kaothekar S. J. Porous Media. 2018; **21**:679-699. DOI: 10.1615/JPorMedia.2018017559

[16] Nield D. A., & Bejan, A., Convection in Porous Media, 2nd ed., Berlin: Springer, 1999. DOI: 10.1007/978-1-4614-5541-7

[17] Vafai K. Handbook of Porous Media. New York: Marcel Dekker; 2000. DOI: 10.1201/b18614

[18] Somerton C. W. & Catton I., J., Heat Transfer, 1982, **104**, pp. 160-165. DOI: 10.1115/1.3245044

[19] Poulikakos D. Numerical Heat Transfer. 1987;**12**:83-99. DOI: 10.1080/10407788708913575

[20] Nield DA, Kuznetsov AV. Int. J. Heat Mass Trans. 2009;**52**:5796-5801. DOI: 10.1016/j.ijheatmasstransfer.2009.07.023

[21] Shue LJ. Transport in Porous Media. 2011;**88**:461-477. DOI: 10.1007/s11242-011-9749-2

[22] Kumar P. Heat Transfer Research. 2012;**43**:167-185. DOI: 10.1615/HeatTransRes.2012003422

[23] Kaothekar S, Chhajlani RK. J. Porous Media. 2013;**16**:709-724. DOI: 10.1615/JPorMedia.v16.i8.30

[24] Nield DA, Kuznetsov AV. Int. J. Heat Mass Trans. 2014;**68**:211-214. DOI: 10.1016/j.ijheatmasstransfer.2013.09.026

[25] Kaothekar S. Astrophys. Space Sci. 2017;**362**:107. DOI: 10.1007/s10509-017-3085-0

[26] Nguyen-Thoi T., Sheikholeslam M., Shah Z., Kumam P., & Shafee A., Sci. Rep.2019, **9**, pp. 12765. DOI: 10.1109/ACCESS.2019

[27] Jukes JD. Phys. Fluids. 1964;7:52-58. DOI: 10.1063/1.1711401

[28] Roberts KV, Taylor JB. Phys. Rev. Lett. 1962;**8**:197-198. DOI: 10.1103/PhysRevLett.8.197

[29] Rosenbluth MN, Krall N, Rostoker N. Nucl. Fusion Suppl. 1962;**1**:143-150. DOI: 10.1088/0029-5515/2/1-2/012

[30] Singh S, Hans HK. Zeit. Astrophys. 1965;**62**:12. DOI: 10.1017/S0022377800021061

[31] Herrnegger F. J. Plasma Phys. 1972; **8**:393-400. DOI: 10.1017/S0022377800007248

[32] Sharma RC. Astrophys. Space Sci. 1974;**29**:L1-L4. DOI: 10.1007/BF00642728

[33] Chhonkar RPS, Bhatia PK. J. Plasma Phys. 1977;**18**:273-286. DOI: 10.1017/S0022377800021061

[34] Devlen E., & Pekunlu E. R., Mon. Not. R. Astron. Soc., 2010, **404**, pp. 830-836.DOI: 10.1111 /j.1365-2966.2010.16299.x

[35] Kaothekar S. & Chhajlani R. K., ISRN Atron. Astrophys., 2012, **2012**, pp. 1-14 (420938). DOI: 10.5402/2012/420938

[36] Kaothekar S., Soni G. D., Prajapati R. P. & Chhajlani R. K., Astrophys. Space Sci., 2016, **361**, pp. 204. DOI: 10.1007/s10509-016-2796-y

[37] Kaothekar S., Astrophys. Space Sci., vol. 2020, **365**, pp. 80. DOI: 10.1007/s10509-020-03792-8

[38] Kaothekar S., Radiation Effects & Defects in Solids, (2020) **176**, pp. DOI: 10.1080/10420150.2020.1812073.

Section 3

Wave and Instabilities

Chapter 7

Parametric Interaction of VLF and ELF Waves in the Ionosphere

Vladimir I. Sotnikov

Abstract

In this Chapter we analyze a non-linear parametric interaction between Very Low Frequency (VLF) and Extremely Low Frequency (ELF) waves in the ionosphere. We demonstrate that nonlinear parametric coupling between quasi-electrostatic Lower Oblique Resonance (LOR) and ELF waves significantly contributes to the VLF electromagnetic whistler wave spectrum. Analytical and numerical results are compared with experimental data obtained during active space experiments and satellite data. These data clearly show that presence of VLF waves in the region of plasmasphere boundary layer, where there are no injected due to substorm/storm activity energetic electrons with energies of tens keV can strongly affect the radiation belt boundary.

Keywords: Ionosphere, wave interaction, whistlers

1. Introduction

The generation of VLF sideband emissions due to parametric interaction of LOR and ELF waves was first suggested in [1, 2] in an attempt to explain an experimental results observed in the ionosphere by the Aureol 3 satellite [3, 4] and during the CHARGE 2B ionospheric rocket experiment [5]. Sideband VLF wave emissions can be explained as secondary peaks above and below the primary peak. They results from parametric interaction of excited VLF and ELF waves. Next, nonlinear parametric interactions between quasi-electrostatic LOR and ELF waves was proposed as possible generation mechanisms of VLF whistler waves in the Turbulent Plasmosphere Boundary Layer (TPBL). Excitation of these waves was analyzed through an assessment of observations from the Cluster spacecraft and Van Allen Probes [6]. To further validate a model developed in [1, 2] and adapted in [6] to explain the observations of whistler waves in the plasmasphere. In [7] a numerical solution of a system of nonlinear equations describing parametric interactions between LOR and ELF pump waves excited in the TPBL by the diamagnetic ion currents and hot ion ring instabilities [8, 9] was analyzed. Obtained results show that nonlinear coalescence of the LOR and ELF waves leads to oblique electromagnetic VLF (whistler) emissions at frequencies much greater than the LH resonance frequency, in agreement with the observations. Finally Particle-In-Cell (PIC) simulation of parametric generation of electromagnetic whistler waves will be discussed. This simulation will be initiated by excitation of the forced wave electric field at VLF and ELF frequencies. Such initiation is possible due to the ability of the PIC code known as Large Scale Plasma (LSP) code [10] to excite traveling plane waves in a simulation box. Wave vectors and frequencies of excited in this way

IntechOpen

modes are chosen to satisfy the ELF and VLF dispersion relations. It was demonstrated that quasi-electrostatic VLF and electromagnetic ELF waves in the process of nonlinear interaction were able to excite electromagnetic whistler waves. These results was obtained implementing a Lagrangian fluid model – part of the LSP package. Simulation results also reveal generation of multiple sideband emissions around the pump VLF wave. These simulation results strongly support analytical model presented in [1, 2] and used in [6, 7] to explain the observations of whistler waves in the plasmasphere boundary layer [11].

2. VLF waves in the ionosphere

We analyze excitation of waves with frequencies ω several times above the lower hybrid resonance frequency, but below the one half of electron cyclotron frequency *i.e.*:

$$\omega_{LH} < \omega < \frac{1}{2}\omega_{ce}, \tag{1}$$

where the lower hybrid frequency ω_{LH} in the case when $\omega_{ce}^2 << \omega_{pe}^2$ is given by:

$$\omega_{LH}^2 = \frac{\omega_{pi}^2}{1+\omega_{pe}^2/\omega_{ce}^2}. \tag{2}$$

and ω_{pe} is an electron plasma frequency. It is well known that in this case in a cold plasma only one mode can be excited. The character features of excited wave field at large distances from the source region can be explained using a plot presented at **Figure 1**. This plot is similar to the commonly used wave refractive index surface plot. The plot at **Figure 1** was obtained using the dispersion relation of VLF waves presented below:

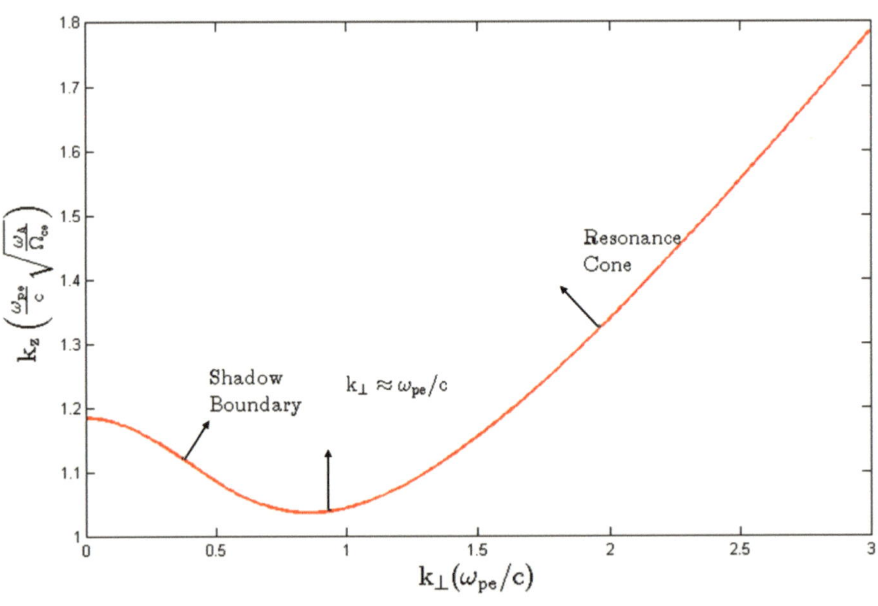

Figure 1.
Wave number surface for a constant $\omega_{LH} < \omega < \frac{1}{2}\omega_{ce}$ with three critical points.

$$\omega^2 = \frac{m_i}{m_e}\frac{k_z^2}{k^2}\frac{\omega_{LH}^2}{\left(1+\frac{\omega_{pe}^2}{k^2c^2}\right)^2} \tag{3}$$

In (3) the wave vector k is defined as $k^2 = k_\perp^2 + k_z^2$ where $k_\perp$ and k_z are the wave vector components perpendicular and along an external magnetic field. In **Figure 1** the wave vector component k_z along the magnetic field is plotted versus $k_\perp$ assuming a constant ω. Most of the radiated by an antenna power can be found in a region in k space occupied by the quasi-electrostatic whistler waves with the parameter $\omega_{pe}^2/k^2c^2 \le 1$. Another part of the wave spectrum in k space which satisfies the condition $\omega_{pe}^2/k^2c^2 >> 1$ belongs to the electromagnetic whistler waves and is radiated up to an angle 19.5° in oblique direction. This is the shadow boundary determined by the long wavelength inflection point and radiated power of these waves is small compared to the power radiated into the quasi-electrostatic part of the wave spectrum.

In [12] it was shown that most of the wave power is radiated perpendicular to the curve presented in **Figure 1** and depends from the distance as R^{-1} except three critical points which define three directions. Two of them are inflection points $d^2k_z/dk_\perp^2$. In these points a wave field dependence is given as $R^{-5/6}$. The third critical point is defined from the equation $d^2k_z/dk_\perp^2 = 0$ and provides wave field dependence in the form $R^{-1/2}$ and corresponds to the wave power radiated along the direction of magnetic field.

3. Parametric excitation of VLF waves in the ionosphere

Nearly monochromatic signals injected from ground-based VLF transmitters are known to experience bandwidth expansion as they traverse the ionosphere [13–17] and magnetosphere [18]. Several mechanisms have been proposed to explain this phenomenon based upon linear and nonlinear scattering assuming existence of magnetic-field-aligned plasma density irregularities. In the absence of ionosphere irregularities a mechanism based on a parametric instability was proposed in [19–21].

Reports on sideband signals associated with VLF transmitter signals are rather scarce. Spectral peaks have been identified near the magnetic equatorial plane on the ISEE satellite at approximately ±55 Hz of the carrier frequency (13.1 and 13.6 kHz) of Omega pulses [22]. Similar peaks seem to be observed on the COSMOS 1809 satellite and generated in the ionosphere by the carrier frequency 19 kHz [17]. Sidebands at approximately ±500 Hz of the carrier frequency (11.9 and 12.65 kHz) of Alpha pulses have been observed in the ionosphere by the AUREOL 3 satellite [3, 4].

At first sight, the 50-Hz sidebands observed on AUREOL 3 seem to correspond the Riggin and Kelly [19] prediction in which the transmitted wave decays into a lower hybrid wave and an ion-acoustic type of oscillation. To account for the existence of two symmetric spectral peaks, one may replace the three-wave parametric instability considered by these authors by a four-wave parametric instability (or modulation instability) as suggested in [21]. According to this scheme, the ELF branch is due to a purely growing electrostatic mode with wave vector ***k*** large enough to provide sidebands $\pm|kV_s|$ off the transmitter frequency.

(V_s is the satellite velocity). This mode is excited in course of a four-wave process by the incident VLF transmitter wave. In our case, the ELF wave branch is

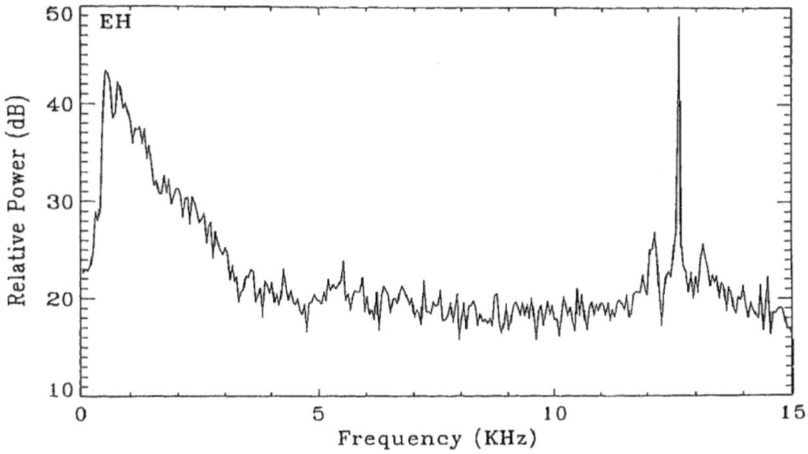

Figure 2.
Averaged power spectral density of electric field. ELF natural emission at 500 Hz, VLF transmitted emission at 12.65 kHz, sidebands at frequencies (12.65 + 0.5) kHz and (12.65–0.5) kHz. This data were observed in ionosphere on AUREOL 3 satellite during experiments in framework of ARCAD project [3].

clearly electromagnetic and as such is of natural origin. Therefore, another explanation in accord with this experimental data has to be found.

Sotnikov et al. [1] proposed another mechanism for the production of 500-Hz sidebands. It is based on nonlinear coupling between the transmitted wave and the ELF emission above the local proton gyrofrequency. The sidebands are shown to be forced oscillations, excited only where the coupling take place. We consider the nonlinear coupling model described in the articles [1, 2].

Next, analysis of the parametrically generated VLF turbulence has been developed in the articles by Sotnikov et al., [1, 2] as attempt to explain appearance of symmetric sidebands in frequency. Such parametrically generated waves were observed in multiple ionospheric experiments. In these experiments two types of waves were present, waves excited by a VLF transmitter and ELF waves excited due to natural processes in the ionosphere. It was demonstrated that beat wave excitation mechanism can be responsible for appearance of observed sidebands with comparable wave amplitudes. Sidebands were excited at combination frequencies given by:

$$\omega_{\pm} = \omega_{k_1} \pm \omega_{k_2}, \tag{4}$$

which can result in sideband emissions. The sideband wave numbers are matched according to

$$k_{\pm} = k_1 \pm k_2 \tag{5}$$

Note that sidebands are not plasma eigenmodes but forced oscillations excited only where VLF to ELF wave coupling take place (**Figures 2** and **3**).

Using the cold plasma approximation the equations for the perpendicular to magnetic field sideband electric field components can be derived in the form [2]:

$$\begin{aligned} E_{\perp k_+} &= \frac{e}{2m} \frac{k_+}{\Omega_e \delta\omega_+} [E_{k_1} \times E_{k_2}]_z \\ E_{\perp k_-} &= \frac{e}{2m} \frac{k_-}{\Omega_e \delta\omega_-} [E_{k_1} \times E_{k_2}]_z \end{aligned}, \tag{6}$$

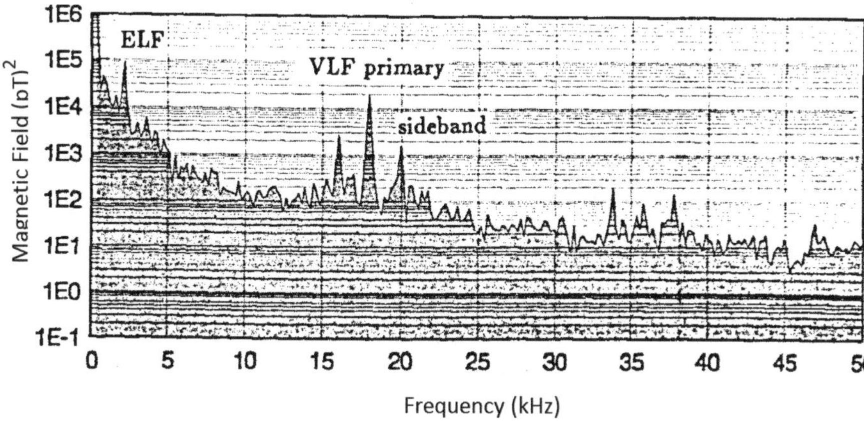

Figure 3.
ELF natural emission at 2 kHz, VLF transmitted emission at 17.95 kHz, sidebands at frequencies (17.95 + 2) kHz and (17.95–2) kHz. This data were observed during cooperative high-altitude rocket gun experiment (CHARGE 2B) carried out in march 1992.

where

$$\frac{\delta\omega_{\pm}}{\Omega_e} = \frac{k_{z1}/k_1}{1+\omega_{pe}^2/\left(k_1^2c^2\right)} \pm \frac{k_{z2}/k_2}{1+\omega_{pe}^2/\left(k_2^2c^2\right)} - \frac{(k_{z1} \pm k_{z2})/k_{3\pm}}{1+\omega_{pe}^2/\left(k_{3\pm}^2c^2\right)}, \tag{7}$$

$k_{3\pm}^2 = k_1^2 + k_2^2 \pm 2k_1k_2\cos(\theta)$ and θ is the angle between **k** vectors.

Sidebands may be a result of nonlinear coupling of the VLF transmitter wave and the natural ELF emission above the local proton gyrofrequency. The VLF wave propagate through the ionosphere as a whistler mode.

$$\omega_{k_1} = \Omega_e \frac{k_{1z}/k_1}{1+\omega_{pe}^2/\left(k_1^2c^2\right)} \tag{8}$$

For the transmitted frequency $\omega/2\pi = 12$ kHZ, the corresponding wave number is $k_1 \approx 2 \cdot 10^{-4}\text{cm}^{-1}$. For the known parameters whistler propagates with $\omega_{pe}^2/k_1^2c^2 = 1$ at large angle to the magnetic field.

The characteristic frequency ω_{k_z} of the ELF wave is slightly above the ion gyrofrequency and $k_2 \approx 3 \cdot 10^{-5}$ cm^{-1}. These waves generally propagate at large angle to the geomagnetic field. As $\omega_{pe}^2/k_2^2c^2 >> 1$, it is described by

$$\omega_{k_2} = \Omega_e \frac{k_{2z}}{k_2}\frac{k_1^2c^2}{\omega_{pe}^2}, \tag{9}$$

where Ω_e is the electron cyclotron frequency.

4. Excitation of whistler waves in a turbulent plasmapause boundary layer

In this section we will discuss parametric interaction of quasi-electrostatic lower oblique resonance waves excited by electron and ion diamagnetic currents and hot anisotropic ion distributions [6–9, 23, 24] with ELF waves in the turbulent plasmasphere boundary layer (TPBL). It is demonstrated below that this nonlinear mechanism can be responsible for generation of broadband, oblique Very Low Frequency (VLF) whistler (W) waves at frequencies much greater than the LH resonance frequency. It is important because the well known whistler generation mechanism by energetic electrons is unavailable in the TPB. Below we present the results of numerical solution of a system of nonlinear equations describing parametric interactions between LOR and ELF pump waves excited in the TPBL by the diamagnetic ion currents and hot ion ring instabilities [8, 9]. Due to instabilities the LOR and ELF waves are generated. Results of simulation confirm that due to nonlinear interaction electromagnetic whistler waves propagating in oblique direction are excited. The frequency of excited waves is well above the Lower Hybrid frequency what is in agreement with experimental results.

In general, parametric interaction of two waves, ω_{k_1} and Ω_{k_2}, produces sidebands at the combination frequencies, $\omega_{\pm}$, that satisfy the matching conditions (4) and (5). In the case in question, we have $\omega_{\pm} \quad \omega_{k_1} >> \Omega_{k_2}$, that is, the high-frequency (VLF) and low-frequency (ELF) counterparts, with $|k_{\pm}| << |k_1| \quad |k_2|$. A general approach for solution of this problem was developed to explain symmetric sidebands, observed during active experiments with injection of a high-power VLF pump whistler wave [1] and modulated electron beam [2] into the ionosphere. It was demonstrated that beat wave interaction between the artificially excited VLF wave and natural ELF emissions can produce observed VLF sidebands. This can be viewed as a first step in the process of a broad VLF spectrum formation because subsequent interaction produces secondary sideband waves and this process continues until the broad range of wavenumbers in k-space is excited. This leads to the requirement that analytical description of the problem should be capable to correctly capture nonlinear interaction in the broad range of wavenumbers and wave frequencies.

In [1, 2, 6, 25] equations written in the Fourier space were used and it was sufficient for obtaining the estimate for sideband amplitudes. To study a nonlinear stage of excited wave turbulence we will switch from the Fourier analysis to description in time and space. To do so Maxwell's equations together with equations of motion of magnetized electrons and unmagnetized ions in hydrodynamic approximation will be used. It is well known that to describe correctly nonlinear evolution of VLF turbulence it is necessary to use 3D description [26–28]. We will use two different systems of equations for description of ELF and VLF waves. They are connected through nonlinear terms containing vector nonlinearities. As a result of cumbersome but straightforward manipulations as in [1, 2, 25], we can obtain nonlinear set of equations for parametric interaction of the VLF and ELF waves, which can be found in [7]. The resulting system of nonlinear equations which describes evolution of VLF turbulence and appearance of electromagnetic whistler waves was solved numerically. Using the developed FORTRAN code which employs the predictor–corrector quasi-spectral numerical scheme detailed analysis of nonlinear mechanism of electromagnetic whistler wave generation from the quasi-electrostatic LOR wave spectra was demonstrated. Main results of this analysis can be found in [25, 29].

Numerical analysis was carried out in a simulation box with the grid size 256 x 32 x 128. In the x direction it includes 16 VLF wavelengths, in the y direction 2 ELF wavelengths and in the z direction 1 ELF wavelength, what corresponds to 4 VLF wavelengths. In all directions periodic boundary conditions were applied.

An initial value problem was solved with VLF and ELF pump waves turn-on all the time. The adaptive time stepping was implemented with initial dimensionless time step $\Delta\bar{t} = 2\pi \cdot 10^{-2} (\Delta t \approx 10^{-5}\ \text{s})$. The computation takes a few days on a standard PC. The input conditions are taken close to the observed values in the plasmasphere [23]:

$B_0 = 0.003$ G, $n_0 = 10^2\ cm^{-3}$, $\omega_{ce} = 5.3 \cdot 10^4\ s^{-1}$, $\omega_{pe} = 5.6 \cdot 10^5\ s^{-1}$ and $\omega_{LH} \approx 1.2 \cdot 10^3\ s^{-1}$. The input VLF pump wave is a monochromatic quasi-electrostatic LOR wave at $\omega_1 \approx 5\omega_{LH}$, 3-D wavevector $k_1 = \frac{\omega_{pe}}{c}(8, 0, 0.94\mu^{-1/2})$, and the amplitude $E_1 = 2\ mV/m$. The input ELF wave is a monochromatic MS wave wit $h\Omega_2 \approx 0.77\ \omega_{LH}$, $k_2 = \frac{\omega_{pe}}{c}(0, 0.05, 0.13\mu^{-1/2})$, and $E_2 = 2$ mV/m. The values of ω_1 and k_1, as well as Ω_2 and k_2, satisfy the dispersion equation for Fast Magnetosonic (FMS) waves:

$$\omega_k^2 = \frac{\omega_{LH}^2}{1 + \frac{\omega_{pe}^2}{k^2c^2}}\left[1 + \frac{M}{m}\frac{k_z^2}{k^2}\frac{1}{1 + \frac{\omega_{pe}^2}{k^2c^2}}\right]$$

Note that the pump wave parameters are chosen specifically so that they are close to but not exactly satisfy the resonance conditions (Eq. (3)) required to get the maximal efficiency of parametric interaction, as described in [1, 2]. However, in the resonance case, the collisionless system of nonlinear equations crashes after only a few time steps because of singularities that cannot be avoided, unless collisional terms are included. Spatial spectra in 2D of the electrostatic potential $\delta\Phi$ of nonlinearly excited VLF waves (frames b, c, and d) and a pump wave Φ_0 (frame a) are presented in **Figure 4**. They were taken in the middle of the computational box (y = 16) at 10^{-3}, 0.1, and 0.18 sec from the beginning of the computational run. Presented in **Figure 4** results clearly demonstrate that the spectral density of electromagnetic modes with $kc << \omega_{pe}$ absent initially starts to grow with time due to the wave cascade towards smaller wavenumbers. Eventually we find that electromagnetic VLF whistlers with frequencies from the range $3 < \omega/\omega_{LH} < 7.5$ produce noticeable part of the excited wave spectrum. The wavenumbers of these waves are inside the rectangles in **Figure 4b–d**. These waves represent the long wavelength part of the dispersion relation for the FMS waves, which corresponds to an electromagnetic VLF whistler wave. Calculations were initiated with $k_\perp \approx k_x >> k_y$. Time evolution of oblique electromagnetic VLF whistler waves with the wavenumbers

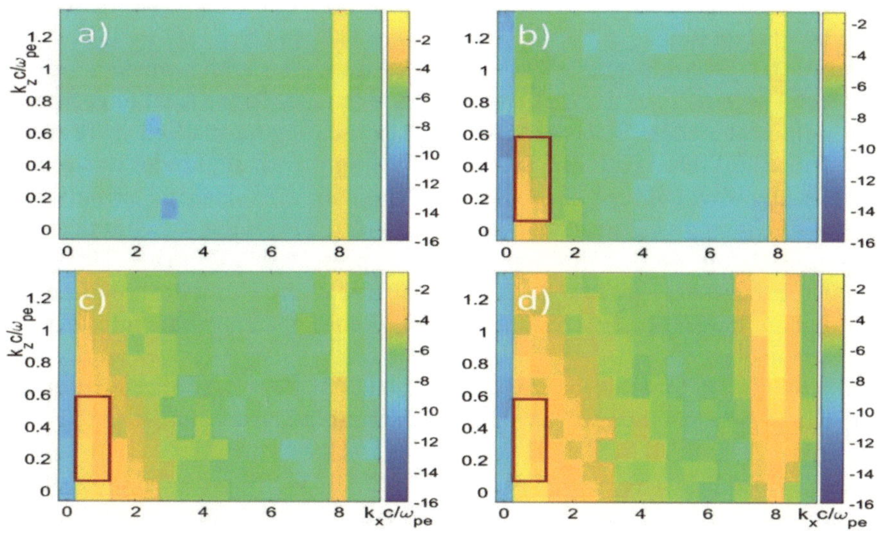

Figure 4.
(a) – 2D representation of the VLF pump wave in Fourier space. (b) – 2D Fourier spectra of VLF density perturbations at time T_1. (c) - 2D Fourier spectra of VLF density perturbations at time $96 \cdot T_1$. (d) -2D Fourier spectra of VLF density perturbations at time $171 \cdot T_1$.

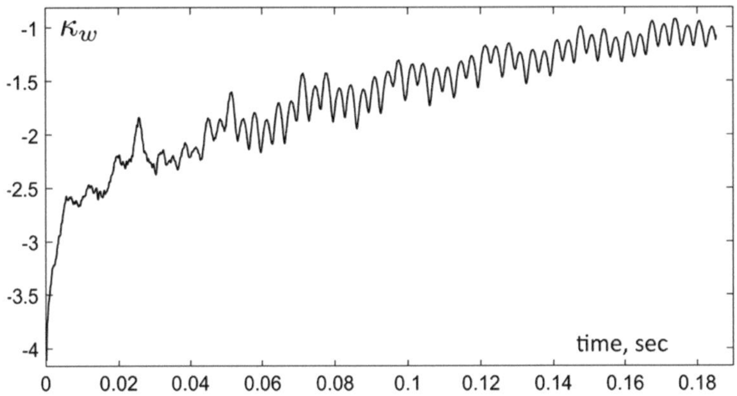

Figure 5.
The whistler generation efficiency, $\kappa_w \approx (\omega_{ce}/2\omega_w)^{1/2}E_w/E_{LH}$ *in logarithmic scale versus time.*

from the rectangles in **Figure 4** is presented in more details in **Figure 5**. It shows the change in time of efficiency of wave transformation from quasi-electrostatic to an electromagnetic part of the wave spectra $\kappa_w \approx (\omega_{ce}/2\omega_w)^{1/2}E_w/E_{LH}$. The root of the mean-square amplitude of the quasi-electrostatic VLF wave field energy density is denoted as E_{LH} whereas E_w represents an electromagnetic wave energy density inside the rectangles. Numerical results show that the amplitude of electromagnetic whistler waves increases with time and eventually reaches the value $\sim 0.1E_{LH}$. It is worth mentioning that the analytical estimate presented in [6] provides similar values and is also consistent with the experimental data. Presented numerical results allow to implement the following scenario of electromagnetic VLF whistler waves generation based on beat wave excitation mechanism. A VLF pump wave in the process of nonlinear interaction with an ELF pump wave generates sideband waves, which in turn generate another sidebands and spread of the wave spectrum in k-space. This in turn leads to appearance of long wavelength waves corresponding to an electromagnetic whistlers. Amplitudes of these waves rapidly grow and can achieve very large values, up to the 30% of the quasi-electrostatic pump wave amplitudes.

In (a) – (d) electromagnetic VLF density perturbations with frequencies from the interval $3 < \omega/\omega_{LH} < 7.5$ are placed inside a rectangle. To the right of each panel one can find color codes in logarithmic scale representing normalized wave amplitudes. Numerical setup with constant in time pump waves amplitudes was used. Chosen pump waves did not obey resonance conditions for sideband excitation and this resulted in relatively small sideband amplitudes. This is the reason it takes so much time to form a broad VLF wave spectrum presented in **Figure 4**. This in turn leads to relatively slow growth of VLF type perturbations. Numerical results also show that nonlinearly excited ELF perturbations does not contribute much to VLF perturbations.

This conclusion follows from the comparison with the results of simulations with exactly the same input parameters but without taking account of the ELF disturbance.

These waves have also been detected in the TPBL, which is devoid of substorm-injected kiloelectronvolt electrons [11, 22]. These emissions represent a distinctive subset of the substorm/storm-related VLF whistler activity and provide the rate of pitch angle diffusion of the radiation belt (RB) electrons that can explain the plasmapause-radiation belt boundary correlation [11]. As the "standard" whistler generation mechanism by energetic electrons is unavailable in the TPBL, [6] suggested nonlinear interactions between quasi-electrostatic LH oblique resonance (LOR) and ELF waves to be the source.

Free energy for enhanced waves comes from electron diamagnetic currents in the entry layer near the TPBL's outer boundary [6, 22], while diamagnetic ion currents and anisotropic (nearly ring like) hot ion distributions are the main contributors near the inner boundary [8, 9]. It is worth mentioning that electromagnetic VLF whistler waves with frequencies far exceeding the Lower Hybrid frequency were produced as a result of numerical solution of nonlinear equations describing interaction of quasi-electrostatic lower oblique resonance (LOR) waves and externally excited ELF waves. This result supports the suggestion that experimentally detected in the TPBL electromagnetic VLF whistler waves with frequencies well above the Lower Hybrid frequency can be produced in the process of nonlinear interaction between the LOR and ELF waves. Experimental results also show that one of the possible mechanisms for changes in the outer radiation belt boundary is connected with the presence of electromagnetic VLF whistler waves. Taking into account that due to the absence of substorm-injected kiloelectronvolt electrons the well known whistler generation mechanism is not applicable to the plasma sheet inner boundary, we can conclude that described above nonlinear generation mechanism can play an important role in this region. This statement is also supported by observations.

5. Parametric excitation of whistler waves: LSP simulation results

A well-developed particle-in-cell plasma simulation code called Large Scale Plasma (LSP) [10] was used to perform 3D simulations of VLF field excitation. We have used the Large Scale Plasma (LSP) simulation code to force the VLF and ELF modes in a cold, magnetized plasma. One of the built in LSP models is the Lagrangian fluid model for both the ion and the electron species. This model was used to obtain presented results. Fluid particles in the model carry the fluid velocity. It is updated every time step with the help of the momentum equation. The fluid particles characteristics such as velocity and position are weighted on the simulation grid. This allows to involve the source terms which define excited electromagnetic fields through supplied density and current density. In addition, using the known density and fluid velocity an equation for temperature can also be solved on the grid. In this model plasma pressure can also be found on the grid assuming an ideal gas approximation. Next, the pressure gradient on the grid can be used in the momentum equation to update the particle velocities and fields. To allow for larger spatial grids and simulation time steps in LSP an implicit energy conservation scheme is used. In this way the scheme provides the Lorentz force push. Electric and magnetic fields are solved self-consistently. This approach allows to substantially reduce simulation time in comparison with implementation of the explicit field solvers. The initial distribution function is assumed to be Maxwellian to allow plasma to behave as an ideal gas.

Simulations to compare the Lagrangian model with the fully PIC results with application to the nonlinear interaction of VLF and ELF waves restricted by 2D approximation were carried out. Obtained results were very close and this was the reason the Lagrangian approach was used. This approach due to the dramatic reduction of simulation time allowed to perform 3D simulations what is necessary to obtain correct nonlinear description of parametric interaction. There are several other advantages in using a Lagrangian fluid approach. Fewer particles per cell are needed and simulations are much quieter. In presented simulations only eight particles per cell were used. To obtain similar quality result using a fully kinetic approach it was needed to use 200 particles per cell. To carry out simulations a 3D Cartesian geometry was used with imposed externally magnetic field directed along a z axis. The amplitude of the magnetic field was chosen to be 0.3 G and plasma density $\sim 10^5$ cm^{-3} what corresponds to the ionospheric parameters. In simulations hydrogen ions with a mass ratio of 1836:1 were used. An outlet boundary conditions were used and the wave was allowed to propagate out of a simulation box minimizing reflections and wave return back into the simulation box. In a cold plasma used in simulations no thermal expansion was observed and no particles were leaving a simulation box.

In LSP we can impose a traveling plane wave inside a simulation box. Both VLF and ELF waves are excited simultaneously. We choose k_x, k_y and ω. We then use the VLF and ELF dispersion to solve for k_z for these waves. Therefore, unlike the waveguide approach which excites a seemingly random set of k-vectors consistent with the dispersion relation, we can target any mode we desire. Because we are directly exciting specific modes, we call this method the "Direct Excitation" (DE) method. The boundary conditions are also much simpler using the DE method. We have chosen to use outlet boundaries. The waves are free to leave the simulation domain and there is little reflection of the waves at the boundaries. Finally, the simulation domain is much smaller, which allows is to make the plasma region larger and therefore resolve smaller k-vectors. Lastly, because we can target specific k-vectors, using the DE method we can test the theory and compare results from LSP with the direct solution to the equations which describe the parametric interaction. The following k-vectors and frequencies are used in the simulation results shown below: $k_{VLF} = (3.5, 0, 1.1)\omega_{pe}/c$ and $k_{ELF} = (0.1, 0.1, 0.11)\omega_{pe}/c$. The ELF and VLF frequencies are $0.8\omega_{LH}$ and $12\omega_{LH}$. This set of parameters leads to $\delta\omega_+ \approx 0.013\omega_{LH}$ and $\delta\omega_- \approx 0.05\omega_{LH}$. Therefore, we expect the positive sideband to be a larger amplitude since $\delta\omega_+$ is smaller than $\delta\omega_-$ and therefore closer to resonance. Indeed, we have seen that the positive sideband is often a larger amplitude. The simulation was run for ~11 ELF periods. The VLF and ELF waves are exciting by specifying the y-component of the electric field only. This leads to the excitation of the other electric field components and the magnetic field. Note that we do not specifically excite B_y. Therefore all field components are excited self-consistently when only one field component is excited. The black curves in **Figure 6** represent the solutions to the VLF and ELF dispersion equations. The most prominent modes occur at the wavenumbers which are driven externally. The VLF dispersion curve

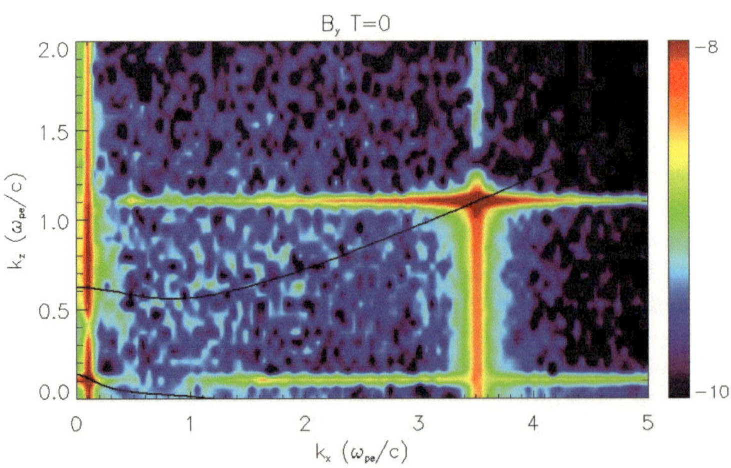

Figure 6.
VLF/ELF wave power spectra of magnetic field component by at t = 0.

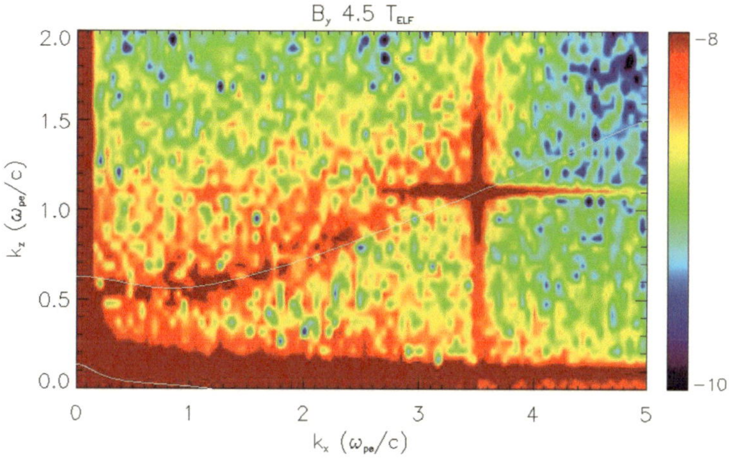

Figure 7.
VLF/ELF wave power spectra of the magnetic field component B_y at time $t = 4.5 \times T_{ELF}$. Wave power spectra presented at $t = 4.5 \times T_{ELF}$ are consistent with the VLF dispersion relation presented in ***Figure 1****. This simulation result confirms nonlinear transformation of a quasi-electrostatic LOR to an electromagnetic VLF whistler waves.*

crosses the externally driven VLF k-vector and ELF dispersion crosses the externally driven ELF k-vector. Therefore, we are confident that we are driving the correct modes (**Figure** 7).

6. Conclusion

In this Chapter using analytical methods and PIC simulation we analyzed efficiency of excitation of electromagnetic VLF whistler waves due to parametric interaction of quasi-electrostatic LOR and ELF waves in the ionospheric plasma.

$\delta\omega_{\pm} < < \Omega_{k_s}$ and values of the sideband amplitudes in agreement with experimental results. It is also possible to satisfy the condition for resonance excitation of VLF waves. If we take into account resonance broadening $\Delta\omega$ due to finite collisions then $\Delta\omega \sim \frac{\omega}{\omega_{ce}}\nu$, where ν is the collision frequency. This means that for nonresonant excitation of sidebands to occur $\delta\omega_{\pm} > \Delta\omega$ must be satisfied. In the opposite case when $\delta\omega_{\pm} < \Delta\omega$ resonant excitation mechanism takes place.

A numerical model describing nonlinear parametric coupling of LOR with ELF waves in cold collisionless plasma has been developed in order to explain the generation of electromagnetic VLF whistler waves in the TPBL in the absence of energetic electrons. These electrons are usually viewed as a source for generation of electromagnetic VLF whistler waves and absence of them in the satellite data was an unanswered question for understanding of a generation mechanism. The results of the 3D LSP simulation confirm that nonlinearly excited waves exhibit spectral features consistent with the observed electromagnetic VLF whistler waves.

Using PIC simulations we have directly tested the nonlinear mechanism suggested in [6] by forcing a quasi-electrostatic whistler wave (i.e., a LOR wave) and an ELF mode to allow parametric interaction. Obtained simulation results confirm that this generation mechanism is capable to explain observed electromagnetic VLF modes. Simulation results clearly show that the LOR mode has cascaded to lower wave number electromagnetic VLF whistler modes. Therefore, the model proposed in [6] that the observed whistler waves are due to a parametric interaction between the LOR and ELF waves is consistent with the findings from the simulation results.

Acknowledgements

It is a great pleasure to thank all my collaborators and co-authors of several papers related to this topic, especially E.Mishin, N. Gershenzon, D. Main.

Public release approval record: AFRL-2021-2303.

Author details

Vladimir I. Sotnikov
Air Force Research Laboratory, Sensors Directorate, Wright-Patterson AFB, OH, USA

*Address all correspondence to: vladimir.sotnikov.1@us.af.mil

References

[1] Sotnikov V. I., Fiala V., Lefeuvre F., Lagoutte D., and Mogilevskii M., Excitation of sidebands due to nonlinear coupling between VLF transmittesr signal and a natural ELF emission, J. Geophys. Res. 1991; 96, 11363-11374, DOI:10.1029 /91JA00695

[2] Sotnikov V.I., Schriver D., Ashour-Abdalla M., and Ernstmeyer J., Excitation of sideband emissions by a modulated electron beam during the CHARGE 2B mission, J. Geophys. Res. 1994; 99, 8917-8926, DOI:10.1029/ 93JA03024

[3] Tanaka, Y., Lagoutte, D., Hayakawa, M., Tajima, M.S., Geophys. Spectral broadening of VLF transmitter signals and sideband structure observed on Aureol 3 satellite at middle latitudes, J. Geophys. Res. 1987; 92,7551-7561

[4] Lagoutte D., Lefeuvre, F., Hanasz, J., Application of bicoherence analysis in study of wave interactions in space plasma, J. Geophys. Res. 1989; 94, 435-442

[5] Myers, N.B., and Ernstmeyer, J., Notes on the CHARGE 2B electron beam experiment platform, Tech. Memo. RL TM-92-28, Utah State University, Logan, Utah, 1992.

[6] Mishin, E.V., and Sotnikov, V.I., The turbulent plasmasphere boundary layer and the outer radiation belt boundary, Plasma Phys. Control. Fusion 2017; 59, 124003-124012

[7] Mishin E.V., Sotnikov V.I., Gershenzon N., and Amit Sharma, Whistler Waves in the Plasmasphere Boundary Layer: Nonlinear Parametric Excitation, Geophys. Res. Letters 2019; 10.1029/2019GL083432, 1-6

[8] LaBelle, J., Treumann, R., Current-driven lower hybrid waves at the inner edge of the ring current; Journal of Geophysical Research 1988; 93, 2591–2598.

[9] Mishin, E., Burke, W., Storm time coupling of the ring current, plasmasphere and topside ionosphere: Electromagnetic and plasma disturbances, Journal of Geophysical Research 2005; 110, A07209. DOI: 10.1029/2005JA011021

[10] Welch D.R., Rose V., Cueno M.E., Campbell, R.B. and Mehlhorn T.A., Integrated Simulation of the generation and transport of proton beams from laser-target interaction, Phys. Plasmas 2006; 14, 25-35

[11] Mishin, E., Albert, J., & Santolik, O., SAID/SAPS-related VLF waves and the outer radiation belt boundary. Geophysical Research Letters 2011; 38, 21101-21109, DOI: 10.1029/ 2011GL049613

[12] Fisher R.K. and Gould R.W., Resonances cones in the field pattern of a radio frequency probe in a warm anisotropic plasma, Phys. Fluids 1971; 857-865

[13] Bell T.F., James, H.G., Inan, U.S., Katsufrakis, J.P., The apparent spectral broadening of VLF transmitter signals during transionospheric propagation. J. Geophys. Res. 1983; 88, 4813 - 4821

[14] Titova E.E., Di, V.I., Yurov, V.E., Raspopov, O.M., Trakhtengertz, V.Yu., Jiricek, T., Triska P., Interaction between VLF waves and the turbulent ionosphere. Geopys. Res. Lett. 1984, 11, 323-331

[15] Inan U.S., Bell, T.F., Spectral broadening of VLF transmitter signals observed on DE 1: A quasi-electrostatic phenomenon, J. Geophys. Res. 1985; 90, 1771-1783

[16] Tanaka, Y., Lagoutte, D., Hayakawa, M., Tajima, M.S., Spectral

broadening of VLF transmitter signals and sideband structure observed on Aureol 3 satellite at middle latitudes, J. Geophys. Res. 1987; 92,7551-7562

[17] Chmyrev V.M., Parametric excitation of ELF waves and acceleration of ions during the injection of strong VLF waves into ionospher, Kosm. Issled. 1989; 27, 249-263

[18] Bell T.F. and Ngo, H.D., Electrostatic waves stimulated by coherent VLF signals propagating in and near the inner radiation belt. J. Geophys. Res. 1988; 93, 2599-261

[19] Riggin, D. and Kelley, M.C., The possible production of lower hybrid parametric instabilities by VLF ground transmitters and by natural emissions, J. Geophys. Res. 1982; 87, 2545-2553

[20] Lee M.C. and Kuo, S.P., Production of lower hybrid waves and field-aligned plasma density striations by whistlers, J. Geophys. Res. 1984; 89, 10873-10885

[21] Groves K.M., Lee, M.C., Kuo, S.P., Spectral broadening of VLF radio signals traversing the ionosphere, J. Geophys. Res. 1988; 93, 14, 683-697

[22] T. F. Bell, J. Geophys. Res. 1985; 90, 2792-2802, DOI: 10.1029/JA090iA03p02792

[23] Mishin, E., Interaction of substorm injections with the subauroral geospace: 1.Multispacecraft observations of SAID, Journal of Geophysical Research: Space Physics 2013; 118, 5782–5796, DOI: 10.1002/jgra.50548

[24] Mishin, E., Puhl-Quinn, P. A., & Santolik, O. (2010). SAID: A turbulent plasmaspheric boundary layer. Geophysical Research Letters 2010; 37, L07106, DOI: 10.1029/2010GL042929

[25] Sotnikov, V., Kim, T., Caplinger, J., Main, D., Mishin, E., Gershenzon, N., Parametric excitation of very low frequency (VLF) electromagnetic whistler waves and interaction with energetic electrons in radiation belt. Plasma Physics and Controlled Fusion 2018; 60, 7-14, DOI: 10.1088/1361-6587

[26] Ganguli, G., Rudakov, L., Scales, W., Wang, J., & Mithaiwala, Three dimensional character of whistler turbulence, Physics of Plasmas 2010; 17, 52310-52321, DOI: 10.1063/1.3420245

[27] Shapiro, V., Shevchenko, V., Solov'ev, G., Kalinin, V., Bingham, R., Sagdeev, R., Wave collapse at lower-hybrid resonance, Physics of Fluids, 1993; 5, 3148–3162.

[28] Sotnikov V., Shapiro V., Shevchenko V., Macroscopic consequences of collapse at the lower hybrid resonance. Soviet Journal of Plasma Physics 1978, 4, 252–257.

[29] Sotnikov V.I., Ivanov V.V., Presura R., Leboeuf J.N., Onishchenko O.G., Oliver B.V., Jones B., Mehlhorn T.A., Deeney C., Investigation of Compressible Electromagnetic Flute Mode Instability in Finite Beta Plasma in Support of Z-pinch and Laboratory Astrophysics Experiments, Commun. Comput. Phys. 2008; 4, 611–623

Chapter 8

Free-electron Driven Terahertz Wave Sources Based on Simth-Purcell Effect

Weihao Liu, Zijia Yu and Zhi Tao

Abstract

Terahertz electromagnetic wave is one of the hottest research topics in nowadays scientific world thanks to its broad applications in material characterization, medical imaging, wireless communication, and security checking etc. Using free-electron beams to interact with periodic structures via the famous Smith-Purcell effect is an efficient way of generating high-power terahertz radiation. In this chapter, we introduce the basic theory and latest developments of the terahertz radiation schemes using a free-electron beam (including continuous electron beam, a single electron bunch, and a train of electron bunches, etc.) to interact with periodic electromagnetic structures, including grating, surface plasmonics, and subwavelength hole arrays, via a special Smith-Purcell effect or Cherenkov-like effect. A kind of free-electron lasers based on the special Smith-Purcell radiation in the terahertz region is proposed and investigated, which can be developed as high-power terahertz wave sources for practical applications.

Keywords: Terahertz, free-electron beam, beam-wave interaction, diffraction radiation

1. Introduction

Terahertz electromagnetic wave is an attractive topic to researchers thanks to its broad application prospects in fields as diverse as biological imaging, materials science, and astrophysics [1]. However, the development of compact, high-power, broadly tunable terahertz sources is challenging [2, 3]. Compared with other kinds of terahertz sources, the free-electron driven sources, such as conventional vacuum electron devices (VEDs) [4] and free electron lasers (FELs) [5], can generate electromagnetic radiation with high power and desirable coherence. However, VEDs can hardly reach the frequency as high as 1 THz and FELs require tremendous costs and cumbersome peripheral equipment. The radiation sources based on the Smith–Purcell radiation (SPR) [6] can avoid the disadvantages of both VEDs and FELs, affording promising ways for developing compact terahertz sources [7–11].

Since its first experimental observation in 1953, Smith–Purcell radiation (SPR), which is generated when a uniformly moving electron beam passes over a periodic surface, has been an attractive research topic for its applications in radiation generation, beam acceleration and nondestructive particle diagnostics etc. [6, 12–15]. It is characterized by the following well-known dispersion relation:

$$\lambda = -\frac{L}{n}\left(\frac{1}{\beta} - \cos\theta\right), \tag{1}$$

where λ is the radiation wavelength, θ the radiation direction, L the structural period, β the ratio of the beam velocity to the speed of light, and n a negative integer indicating the harmonic order.

Unfortunately, the efficiency of conventional SPR in practice is usually not high enough, which restricts the power and efficiency of the terahertz generating sources based on it [16]. Enhancing the efficiency of SPR will substantially improve the performances of the related terahertz sources. Developing new mechanisms or radiation schemes to improve the efficiency and power of SPR terahertz sources is the major goal of the present chapter, which is organized as follows. In Section 2, we will first introduce a unique kind of SPR, so-called Special Smith-Purcell Radiation (S-SPR), which can enhance the efficiency of SPR. And then several variants of S-SPR will be proposed and investigated. In Section 3, a kind of terahertz free-electron laser based on the mechanism of S-SPR will be illustrated, which can generate terahertz radiation with higher power than ordinary SPR devices. Section 4 concludes this chapter.

2. Special Smith-Purcell radiation (S-SPR)

Traditionally, there are two theoretical models dealing with the mechanism of SPR. The first one is the diffraction model [17, 18], by which SPR is considered as the diffraction of the periodic surface to the evanescent self-field of an electron beam. The other one is the surface-current model based on the image-charge approximation [19, 20], according to which SPR is generated from the surface-current, induced by a moving electron beam, on the periodic metallic structure. Both models can deduce the SPR relation.

In the past years, we revealed a class of S-SPR, which cannot be perfectly involved in the previous theoretical models mentioned above. We have found that it is generated due to the coherent interference of the radiation from an one-dimensional or two-dimensional array of resonant modes [21–23]. In this section, a series of variants of S-SPR are proposed and investigated.

2.1 S-SPR from one-dimensional gratings

Figure 1 shows one of the schemes that we are going to study. At first sight, it is a rectangular optical grating that is commonly used in ordinary SPR (O-SPR). For comparison purpose, let us first briefly visit O-SPR. According to the diffraction

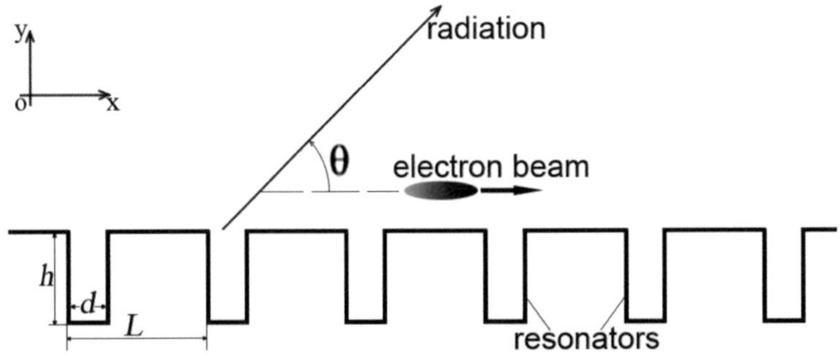

Figure 1.
Schematic diagram of the special Smith–Purcell radiation from the resonator array.

model, the diffraction waves from a grating consist both surface waves and radiation waves. The surface waves, the frequencies of which are below the threshold of SPR, can only propagate along the grating and cannot radiate except for the abrupt change of structure [24, 25], and the radiation waves, which are the negative harmonics of diffraction, can radiate into the upper half-space, and the dependence of radiation frequency on direction satisfies Eq. (1) [26]. In other words, the surface waves and radiation waves are independent from each other: the surface waves are bound to the periodic structure while the radiation waves extend to all directions in the upper half-space, as shown by the simulated results given in **Figure 2(a)**. The simulated radiation spectrum, shown in **Figure 2(b)**, covers a wide frequency band, which is because the radiation direction is continuously changing while the electron beam is moving. **Figure 2(c)** shows that the frequency of the surface wave is below the threshold of SPR.

Now we reduce the gap width d (all other parameters are kept unchanged, this simulated results are given in **Figure 3**, which shows that the radiation spectrum becomes a narrow band one and the spectrum density is 1.5 times enhanced, see **Figure 3(b)**. In addition, the radiation is almost focused at a specific direction defined by Eq. (1), see **Figure 3(a)**. Hence, we obtain a special kind of SPR with monochromatic spectrum and with enhanced intensity at the specific direction determined by SPR relation, which is exactly the S-SPR that will be discussed in the following.

As is known that the surface waves on a periodic structure are formed by the coupling of resonator modes in the periodic resonator array [27, 28]. When we reduce the gap width d, the distance between the adjacent resonators will be increased, which will obviously weaken the coupling of the resonator modes. The surface waves will no longer exist when the coupling of the resonator modes are eliminated. Under this circumstance, the grating changes into an array of independent resonators. Following the above analysis, the mechanism of S-SPR can then be stated as follows. As an electron beam skims over the resonator array, the resonant modes in the array will be excited one by one. These resonant modes then generate radiation through the apertures one after another. And the radiation frequencies are just the eigenfrequencies of the resonator modes. In the direction given by Eq. (1), the phase shift from every adjacent resonator is $2n\pi$ (n is a integer), indicating that the radiation from all resonators is coherent. So as that the radiation in this direction will be enhanced, while in all other directions the radiations from different resonators will counteract each other, and the radiation cannot occur. This is just what have been shown in **Figure 3**.

The resonator modes play an essential role in S-SPR since they determine both radiation frequency and direction. For the case that the gap width is much less than the radiation wavelength ($\lambda \ll d$), the radiations are largely from the transverse electromagnetic (TEM) modes of resonators. To estimate the eigenfrequencies of these modes, we make the approximation thatE_x reaches a maximum at the aperture of the groove, so as that theE_x distributions in the resonator can be illustrated by **Figure 4**, based on which the oscillation wavelength λ of the resonator modes can be expressed by:

$$\lambda = h / \left(\frac{m}{2} + \frac{1}{4} \right) \tag{2}$$

where m is a non-negative integer, indicating the mode number, and h is the depth of the groove (resonator).

To realize S-SPR, the following conditions should be satisfied. 1) The period L and gap width d should be well matched to prevent the coupling of electromagnetic

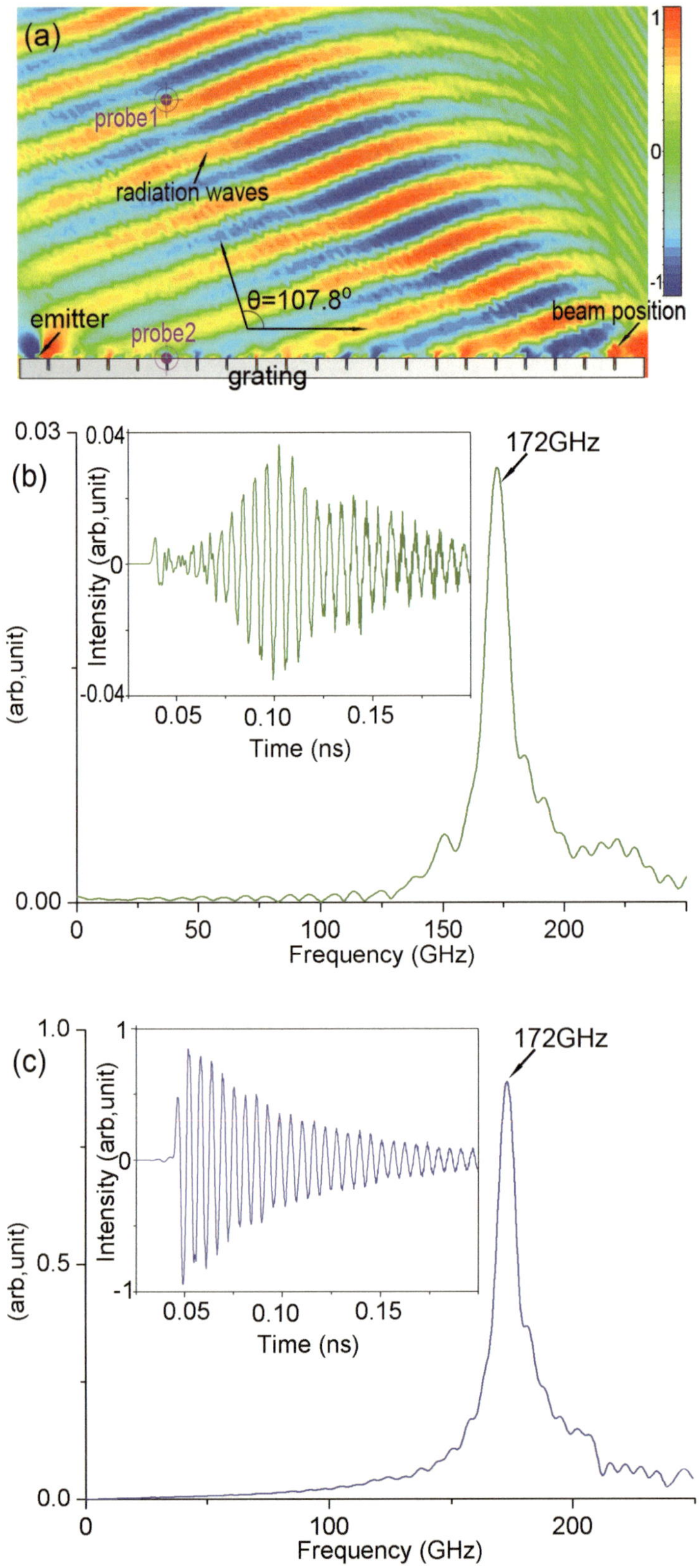

Figure 2.
(a) Simulated contour map of the E_x field of the O-SPR. (b) Simulated radiation spectrum and the waveform in the time-domain. (c) Simulated spectrum of the surface waves and the waveform in the time-domain.

modes in adjacent resonators. Namely, the resonators are deep-narrow-rectangular grooves (DNRGs). 2) The depth of the resonators should be neither too small, otherwise the resonator modes cannot be effectively excited, nor be too large,

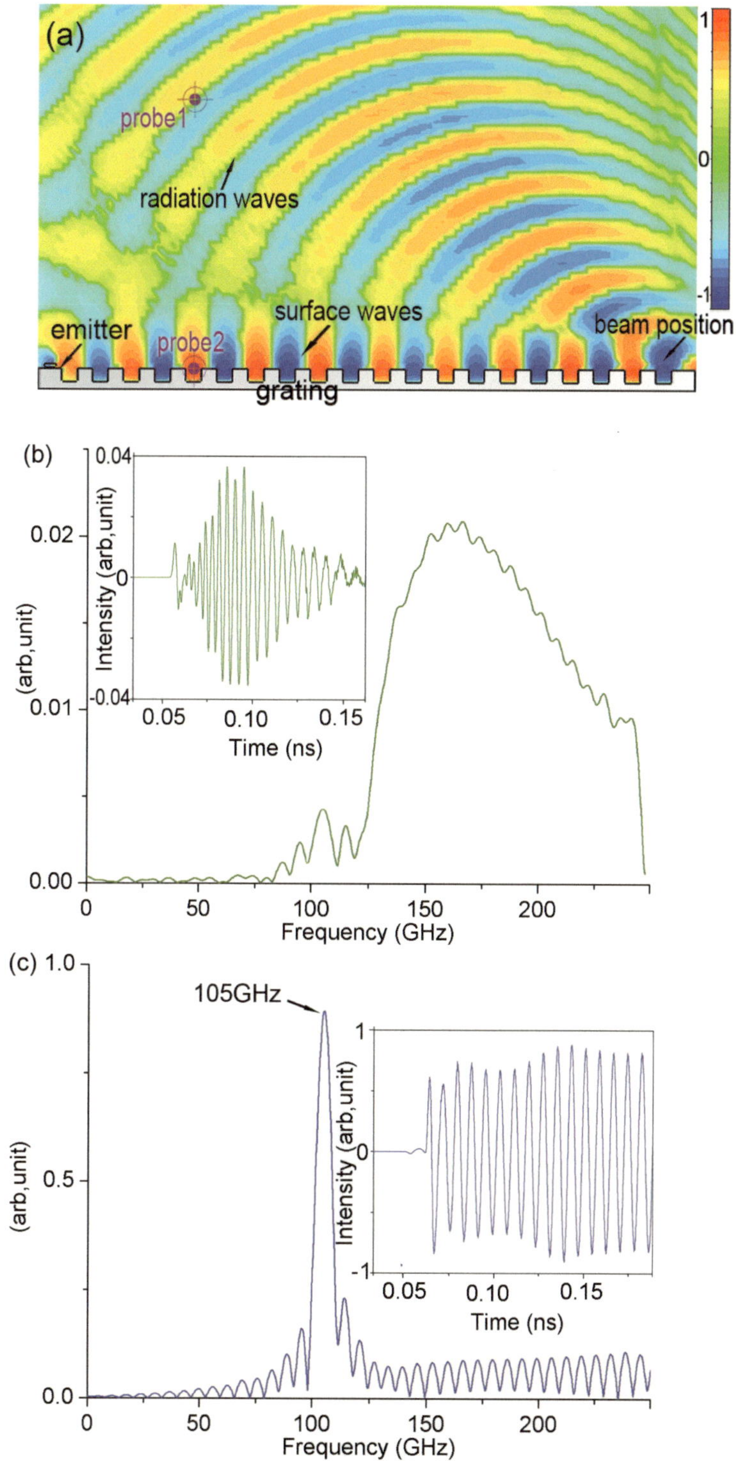

Figure 3.
(a) Simulated contour map of the E_x field of S-SPR. (b) Simulated radiation spectrum and the waveform in the time-domain. (c) Simulated spectrum of the surface waves and the waveform in the time-domain.

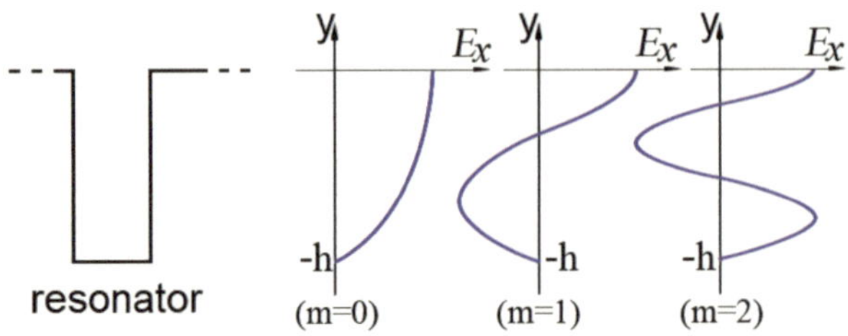

Figure 4.
Diagram of a single resonator and the distribution of the E_x component in the y direction for three resonator modes.

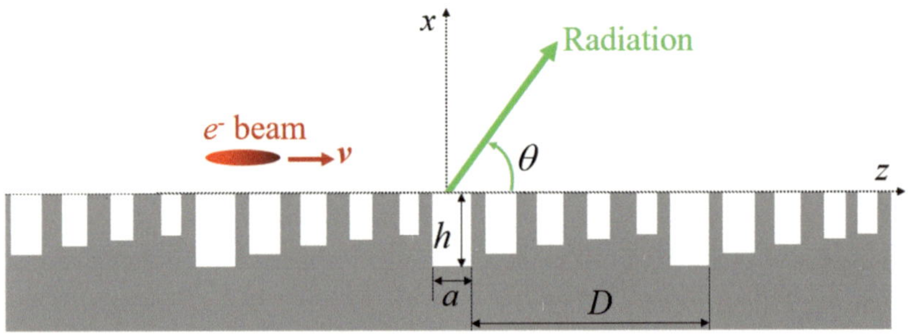

Figure 5.
Diagram of the free-electron beam exciting an array of DNRG-clusters with gradient sizes.

otherwise the SPR cannot be realized since the frequency is below the SPR threshold. 3) The frequency of resonator modes, the grating period, and the beam velocity should be associated to satisfy the SPR relation.

For the S-SPR from an array of DNRGs, only a single radiation frequency can be efficiently obtained. In order to realize several radiation frequencies simultaneously, namely to get multi-color radiation, we proposed to a scheme in which every unit of the array consists several DNRGs with different sizes (gradient DNRGs) as shown in **Figure 5** [29]. The radiation from different DNRGs cannot be coherent since their resonant frequencies are different, however, the radiation from all DNRGs of the same shape in the array will be coherent, which is exactly the S-SPR illustrated above. Thus, we can get an improved S-SPR, which has multiple radiating frequencies. The DNRGs with different shapes will radiate different frequencies at different directions (θ_i). So as that, by using the gradient DNRGs, the spectrum of SPR can be discretized to be a series of peak frequencies with enhanced intensities. Not only spectrum is discretized, but also the radiation directions are also discretized as shown in **Figure 6**. By integrating the field energy in the spectrum band, we obtain that the total radiation intensity is enhanced by more than an order of magnitude.

In the S-SPR illustrated above, all the DNRGs have high aspect-ratios with h/a>20. The manufacturing of the metal structures with high aspect-ratios in the terahertz region is a challenge in practices. In addition, the energy capacities of such grooves/holes are essentially restricted by the structure volume, which reduces the efficiency of terahertz emission and manipulation. We find that, for a single groove with a small aspect-ratio, the field has a broad spectrum. In contrast, for the coupled grooves, the spectrum shows the feature of resonance: having a sharp peak resonating frequency and a narrow bandwidth, see **Figure 7**. As the number of groove (N_g) increases, the peak becomes sharper and the spectrum bandwidth becomes narrower [30]. In other words, a cluster of grooves can operate as an effective electromagnetic resonator, holding a series of Fabry–Pérot (FP)-like modes.

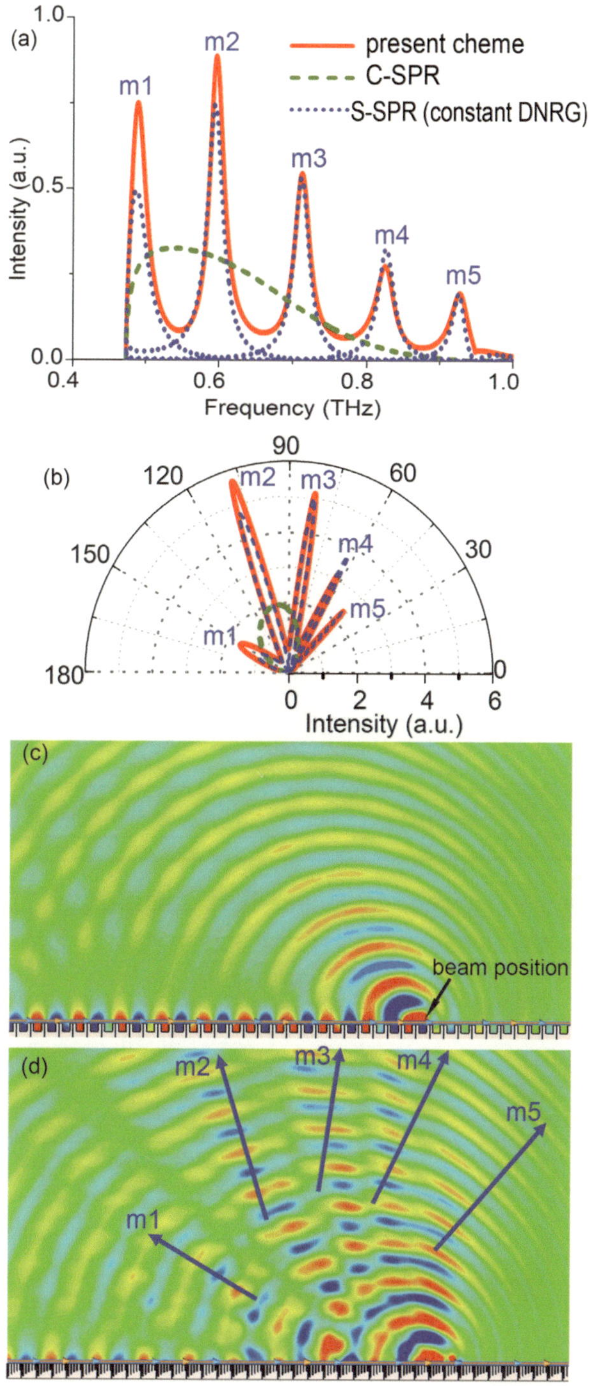

Figure 6.
(a) Calculated radiation spectra of the SPR from an array of gradient DNRGs (present scheme), of O-SPR (green dash line), and of S-SPRs with constant size DNRGs (blue dot lines). (b) Calculated angular distribution of the radiation. (c) Simulated E_z contour map in x-z section for the case of C-SPR. (d) Simulated E_z contour map for the case of gradient DNRGs.

The field intensities within the coupled grooves are remarkably higher than that within a single uncoupled groove, indicating that coupled grooves have much higher energy capacity, and can interact with the electron beam more efficiently. From the insets we can see that the electric fields within adjacent grooves are in the

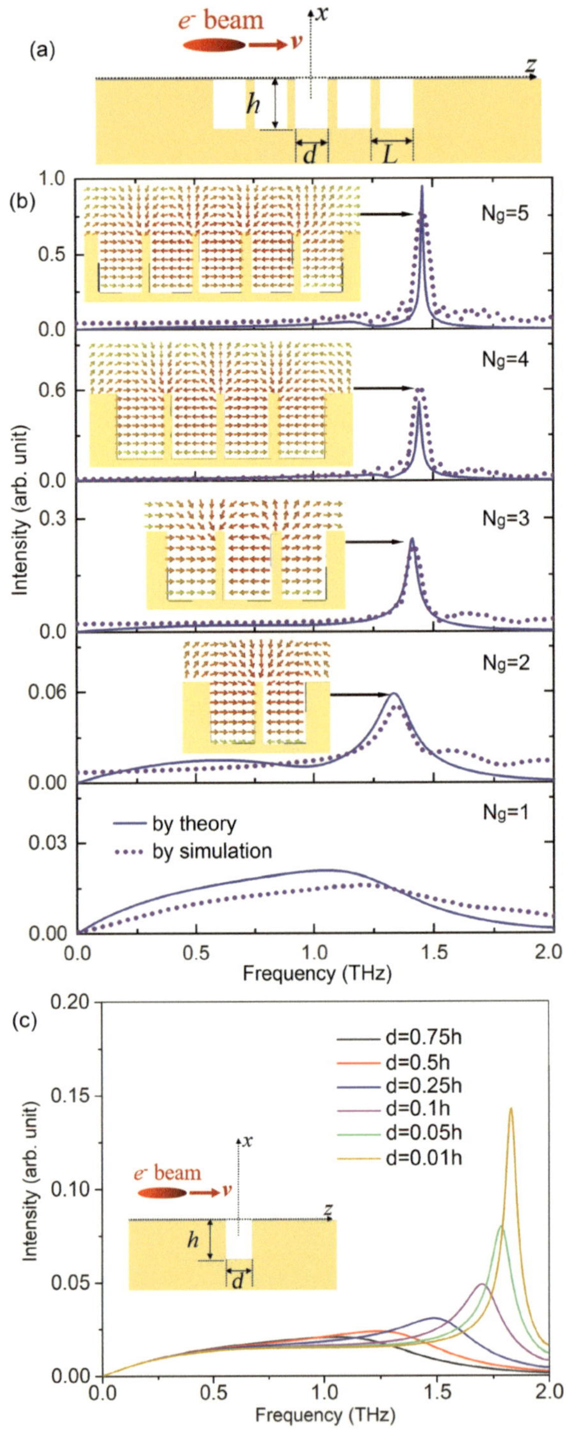

Figure 7.
(a) Schematic diagram of the coupled grooves excited by free-electron beam. (b) Calculated and simulated field spectra at the groove-apertures for different N_g*. The insets are simulated snapshots of the electric field distributions.*

opposite directions, indicating that the phase difference between adjacent grooves is π. We would refer to this kind of FP resonant mode as π mode.

Based on the above knowledge, we propose to use electron beam to excite an array of coupled grooves, the calculated and simulated results are shown in **Figure 8**. For

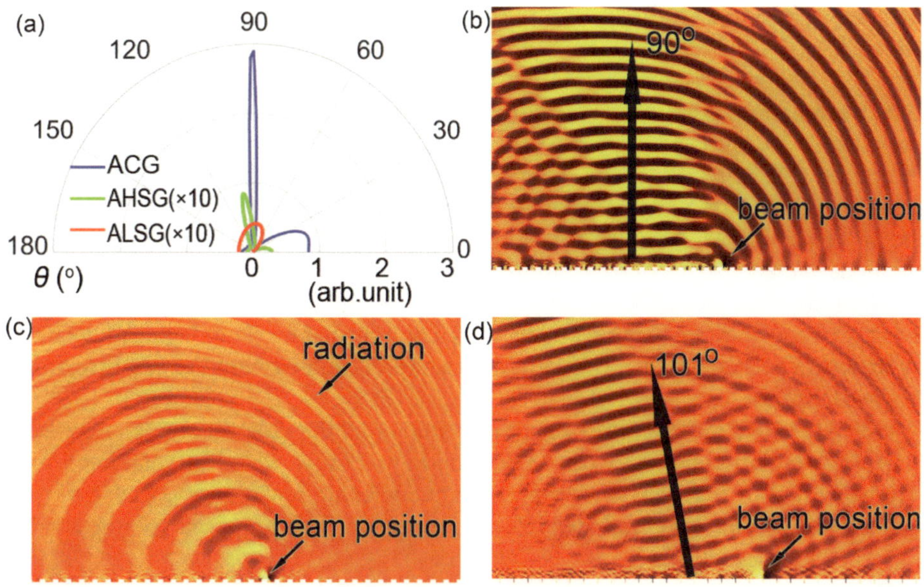

Figure 8.
(a) Calculated angular distribution of radiation from three cases (ACG, ALSG, and AHSG). (b), (c), and (d) are simulated E_z contour-maps for the cases of ACG, ALSG, and AHSG, respectively.

the purpose of comparison, other two cases—an array of low-aspect-ratio single-grooves (ALSG) and an array of high-aspect-ratio single-grooves (AHSG)—are also simulated. For the case of ALSG, the radiation is at all directions, illustrating a typical feature of O-SPR. For AHSG, the radiation is mainly concentrated at a specified direction (θ=101°), indicating the feature of S-SPR [22]. For the proposed array of couple-grooves (ACG), the main radiation wave is at the specified direction ($\theta = 90°$). Here the direction changes due the changing of the period of the array. The figure shows that by using the coupled grooves, the radiation intensity can be enhanced by more than an order of magnitude compared with that from ALSG and AHSG.

2.2 S-SPR from two-dimensional sub-wavelength hole arrays

Sub-wavelength hole array (SHA) is a unique open period structure, which has tremendous interesting properties and applications [31–33]. Here we propose a modified S-SPR by using a sheet FEB to excite two-dimensional (2D) SHAs, see **Figure 9** [34]. The 2D SHA is formed by periodically etching rectangular sub-wavelength holes on a planar conductor plate. All the one-dimensional (1D) SHAs are parallel, and every adjacent two arrays have a certain deviation in the longitudinal (z) direction, forming a parallelogram pattern of holes on the plate. We set the longitudinal width of the hole to be much less than the spatial period of the array ($d \ll \lambda$), such that each hole is an independent resonant unit with specific resonant modes. In light of the shapes and boundary conditions of sub-wavelength holes, the frequencies of the resonant modes can be approximately expressed as:

$$f_{l,m,n} = \frac{c}{2}\sqrt{\left(\frac{l}{d}\right)^2 + \left(\frac{n}{h}\right)^2 + \left(\frac{m}{w}\right)^2} \quad (3)$$

in which h, w and d are the hole parameters shown in **Figure 9**. l, m and p are non-negative integers signifying the mode orders. These resonant modes are successively excited by the FEB and then generate coherent radiation via S-SPR effect, namely, the radiations from all sub-wavelength holes constructively interfere. The frequency is exactly the frequency of resonant modes defined by Eq. (3), and the radiation direction is determined by SPR relation. Since the 1D SHAs are periodically arranged not only in the longitudinal (z) direction but also in the lateral directions (z' and z" directions shown in the **Figure 9**), the coherent radiation can be realized when the constructive interference is realized at specified directions in the 3D space. **Figure 10** shows the simulated results. When the 2D SHAs are arranged rectangularly ($\alpha_1 = 0$), there is only one dominant radiation direction. As the deviation increases, the radiation gradually deflects to the lateral directions, and two radiation lobes are obtained. Sinceθ_0 can be adjusted by changing beam velocity or structure periodicity (L), the coherent radiation can be steered to any directions in the 3D space.

Since the total radiation is the superposition of all the radiating units in the 2D SHA, high radiation intensity can be obtained by setting a wide FEB together with a large 2D SHA. In addition, since several frequencies are obtained and each frequency component could radiate at several directions, the proposed scheme can be

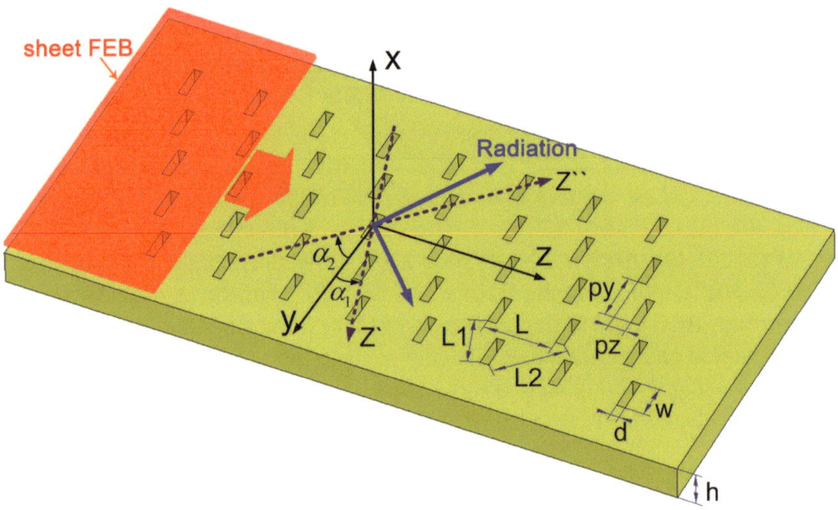

Figure 9.
3D schematic diagram of the sheet free-electron beam driving a 2D SHA.

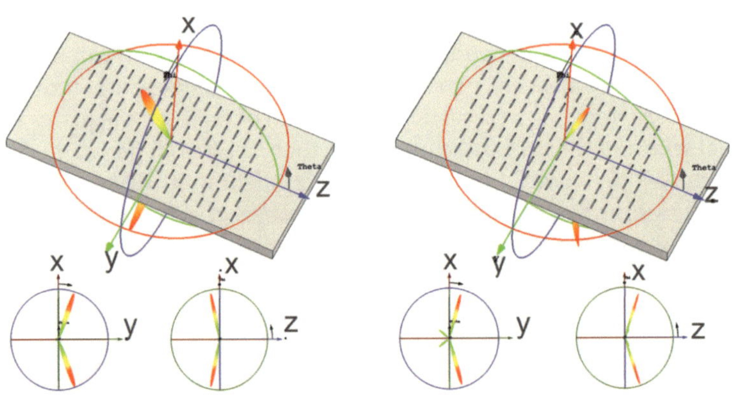

Figure 10.
3D and 2D radiation directivity diagrams of the far-field from a 2D SHA.

developed as radiation sources simultaneously used for several objectives. By rearranging the alignment of SHA, the spatial distribution of CR can also be shaped, which is an increasing hot research topic [35–38].

In addition, we find that, not only S-SPR but also a modified Cherenkov radiation (CR) can be realized from a 2D SHA excited by a sheet FEB, which is illustrated as follows. When a sheet FEB skims over the SHA, the resonant modes within adjacent holes will be successively excited with a time delay of

$$\Delta_t = \frac{L_z}{v_e} = \frac{L}{v_p} \tag{4}$$

in which v_e is the velocity of the FEB in z direction, $v_p = v_e / \sin\alpha_1$ is the projective (effective) velocity of the FEB in z' direction. The CR in the vacuum can be realized if $v_p > c$, which is exactly the requirement of CR, namely the velocity of the radiation source is greater than the speed of light. Different from that of the transition radiation (TR) [35, 38] and that of the Smith-Purcell radiation (SPR) [36, 37, 39], which are frequency dependent, the spectral and spatial shaping of CR is independent from the frequency, indicating that the radiations of all frequencies are at the same direction.

This CR can be more effectively manipulated and even be focused at specific spots in the space by re-arranging the subwavelength holes in the array. When the subwavelength holes are lined hyperbolically on the conductor plate, the CR will be focused at a specific spot in the space as shown in **Figure 11** [40].

2.3 S-SPR from surface plasmonics

In above S-SPR schemes, the resonator arrays are all realized by perfect electric conductor (PEC). When PEC is replaced by semiconductors or metamaterials with terahertz surface plasmonics (SPs) taken into account, the electromagnetic properties of the rectangular sub-wavelength hole (RSH) will change significantly [41–45]. In this sub-section, we propose to use free-electron beam to excite an array of open resonators covered by an array of meta-films, the schematic diagram of which is shown in **Figure 12**.

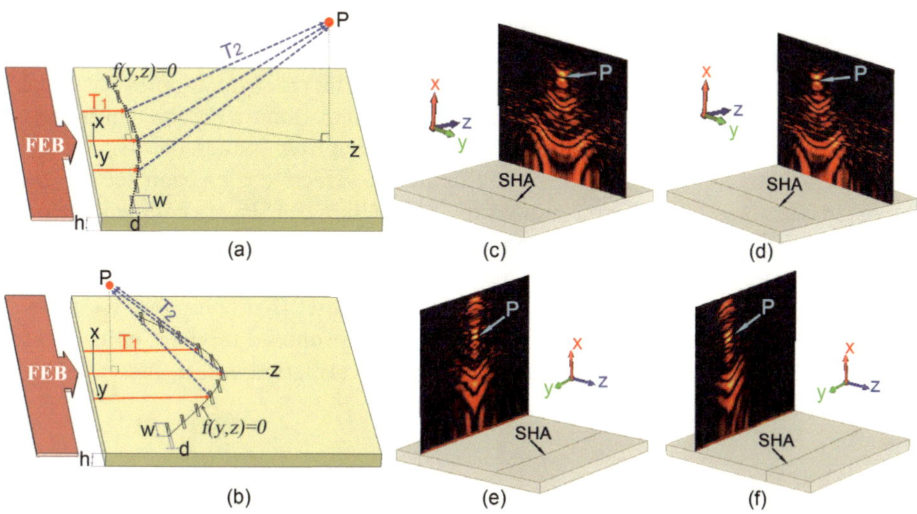

Figure 11.
(a) and (b) The proposed scheme using a sheet FEB to excite a SHA, which is lined along the curve. (c–f) Simulated field intensity distributions in the focusing plane.

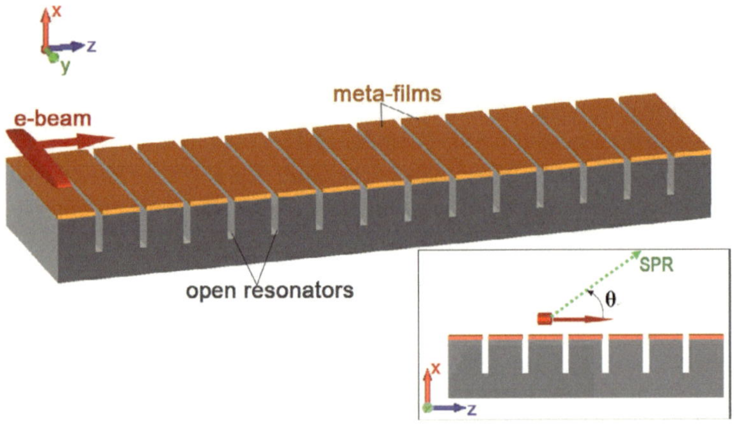

Figure 12.
3D diagram of the proposed scheme and its cross-section (in the inset).

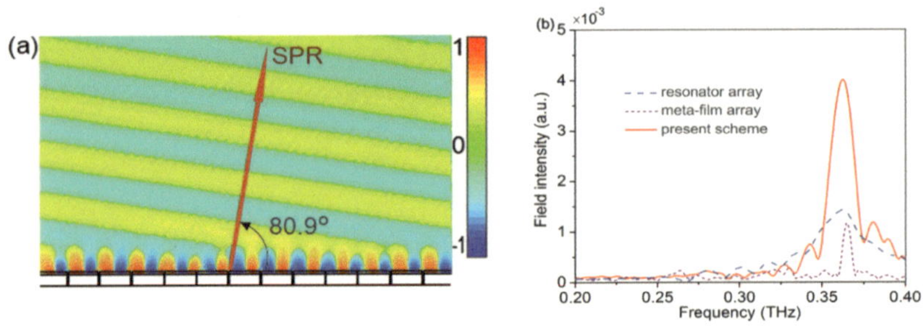

Figure 13.
(a) Simulated field (E_z) contour map. (b) Simulated radiation spectra of three cases.

We find that, when there is only an array of meta-films, the fields are largely confined on the meta-films array, and only a small portion can radiate into the upper space; while there is only open-resonators array, the waves can efficiently radiate into space, however, the surface waves in the open-resonator array is much weaker than that in the meta-films array, so as that the radiation intensity is much weaker. When the open-resonators array and the meta-films array are combined, the FEB first excites the SPs on the meta-films, which then induce the resonator modes within the open-resonators. The SPs are the effective excitation sources for the resonant modes in the open-resonators [46]. When the SPs are coupled with the resonator modes, namely, the resonant frequency of the SPs matches that of the resonator mode, the intensity of the resonator mode, together with the radiation intensity, will be greatly enhanced [47]. **Figure 13** shows that compared with the un-coupled cases, the radiation field intensity of the proposed model is enhanced by more than three times, indicating that the power of radiation will be increased by about an order of magnitude.

3. Terahertz free-electron sources based on S-SPR

In the section, we introduce a class of terahertz free-electron sources based on the mechanism of S-SPR illustrated in Section 2. The schematic diagram of the proposed source is shown in **Figure 14**. It looks similar to a conventional Orotron, which is an electronic oscillator with open resonator and reflecting grating [48, 49]. A dc electron beam passes over a rectangular grating, from which S-SPR is generated in the perpendicular direction. A planar conductor mirror is placed above the grating, which reflects the radiation waves. The electromagnetic waves then propagate back and forth between the radiation grating and the conductor mirror, forming a series of oscillation modes between them, which is in effect an electromagnetic cavity with open boundary. These oscillation modes will interact with the electron beam and will be amplified via gaining energy from the electron beam. Unlike the conventional Orotron based on O-SPR, here the S-SPR is applied. Since the intensity of S-SPR is remarkably higher than that of O-SPR, the oscillation modes with higher intensity will be excited in the grating-mirror cavity, so as that the output power will be remarkably enhanced.

The distance between the grating and the reflection mirror should be chosen as follows. Since the tangential electric field should reach the maximum at the beam position and vanish at the surface of the conductor mirror, the wavelength of the oscillation waves between the grating and mirror can be evaluated by the equation

$$H \approx \frac{\lambda}{2}\left(p + \frac{1}{2}\right) \tag{5}$$

in which p is a non-negative integer. When the wavelength of the oscillation modes defined by Eq. (5) is close to that of the S-SPR defined by Eq. (1), the oscillation modes will be coupled by S-SPR and be enhanced, which also defines the operation requirement of the proposed device.

The simulated results are shown in **Figure 15**, in which **Figure 15(a)** shows the distribution of the modulated electron beam in the 'x-energy' phasespace. We can see that the beam-to-wave efficiency of the model is over 10%. **Figure 15(b)** shows that the oscillation waves are effectively excited in the cavity, and **Figure 15(c)** illustrates that the frequency of the oscillating field matches that of S-SPR. **Figure 15(d)** shows that the output power of the model reaches 4 W/mm at the frequency of 0.3 THz. The simulated results of the conventional Orotron are also shown for the comparison purpose. We can see that the output power of the proposed model is several orders of magnitude higher than that of conventional Orotron based on O-SPR.

In order to further increase the output power, we propose a double-grating scheme as illustrated in **Figure 16**. Two identical gratings are symmetrically set face

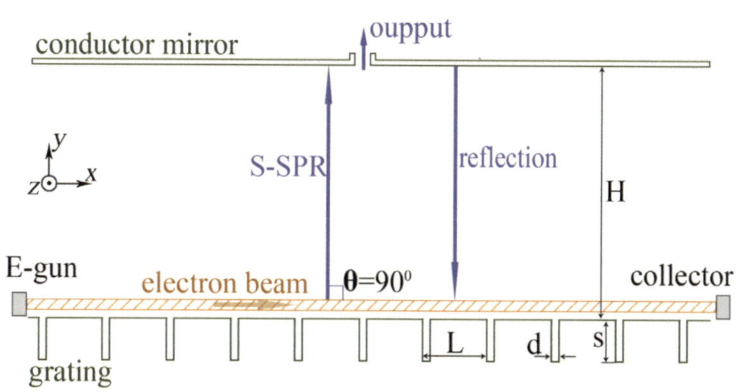

Figure 14.
Diagram of the proposed model of the THz-Orotron based on the S-SPR.

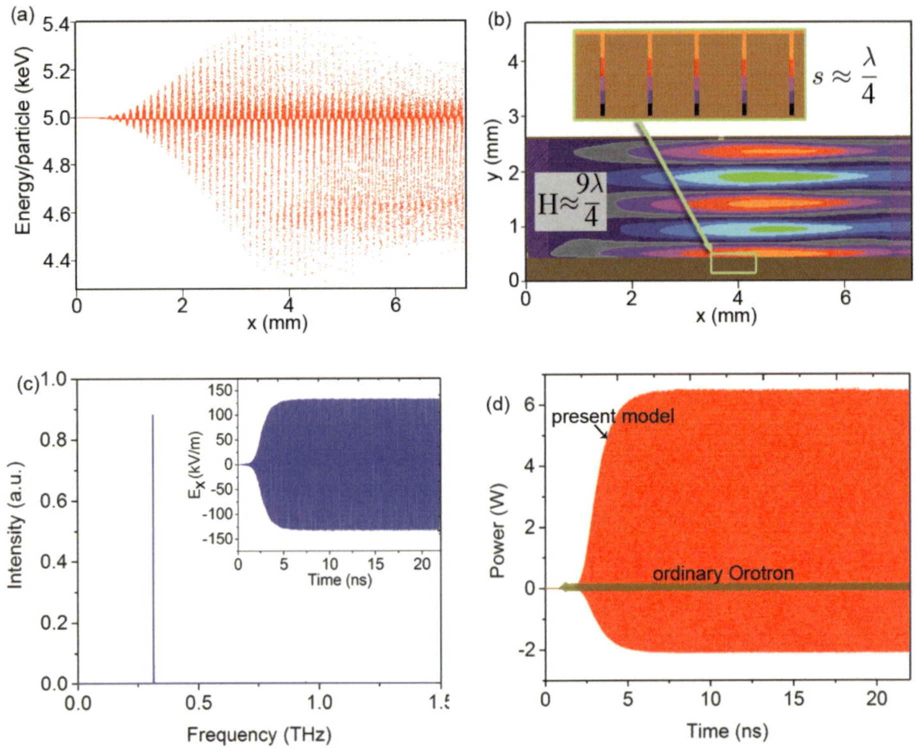

Figure 15.
Simulated snapshot of the phasespace of the modulated electron beam. (b) Simulated contour map of the E_x field. (c) the electric field and its spectrum in the cavity. (d) the output power in the time domain.

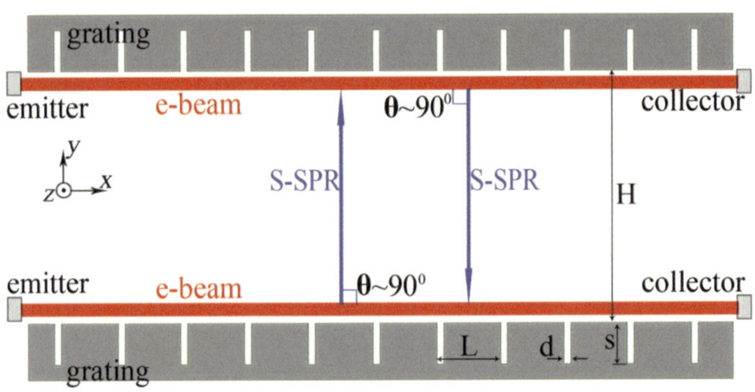

Figure 16.
Schematics of the proposed novel terahertz free-electron laser.

to face, which forms a modified cavity with open boundaries. Two electron beams pass over those gratings, respectively. The S-SPR from each grating are generated in the opposite directions [50], and will interfere in the cavity, forming electromagnetic modes as that in the previous case. Here we appropriately choose grating parameters and the beam velocity, letting the radiation direction of the S-SPR be exactly in the perpendicular direction of the gratings.

Following the analyses of the previous model, the wavelength of the operation modes in the cavity formed by two symmetric gratings can be approximately expressed as:

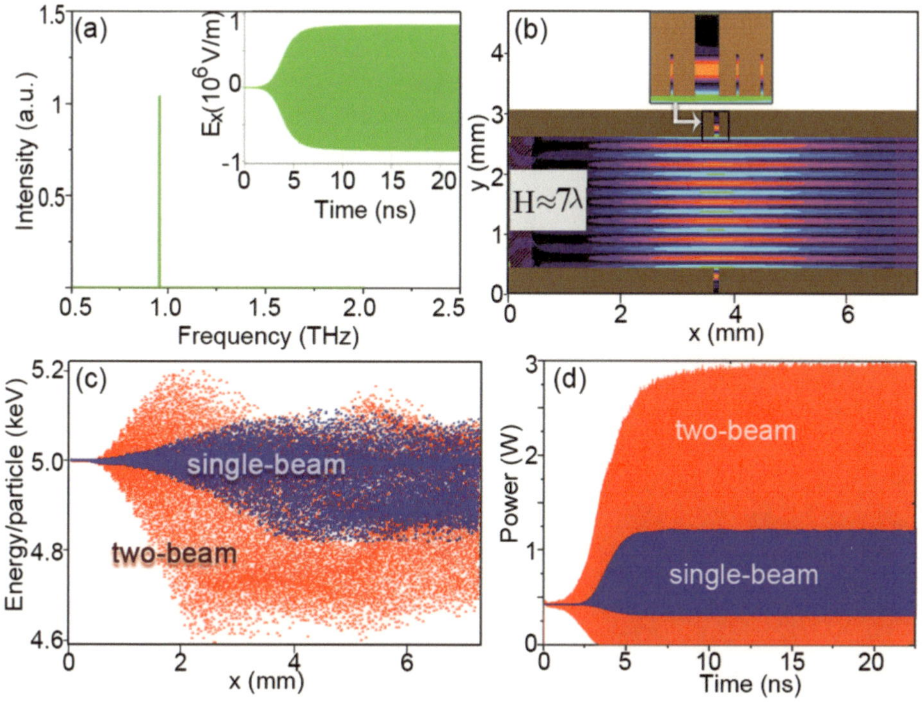

Figure 17.
Simulated results of the proposed terahertz source based on S-SPR (double-beam case). (a) the simulated electric field and its spectrum detected at the port. (b) Snapshot of the contour map of the E_x field. (c) the phase space distributions of the modulated electron beams. (d) Simulated power in the time domain.

$$H \approx p\lambda/2 \quad (6)$$

Figure 17 shows the simulated results of the model, in which the results of the single-beam case is also shown for comparison. One can see that the efficiency and the output power are about 2.5 times than that of the single-beam case. By using an electron beam with moderate current-density (less than 50A/cm^2), the output power reaches 3 watt at the frequency of 0.96 THz. This power level is higher than that of majority terahertz sources in this frequency region, affording a promising option for developing high-power terahertz sources.

4. Conclusions

Using free-electron beams to interact with periodic structures via Smith-Purcell effect is an effective option for developing high-power terahertz sources, which have broad application prospects. In this chapter, we introduced the basic mechanism and several forms of a modified Smith-Purcell radiation–the special Smith-Purcell radiation. Based on the mechanism of the special Smith-Purcell radiation, a set of Smith-Purcell free-electron lasers in the terahertz region were proposed and investigated, promises high-power terahertz wave sources for practical applications.

Acknowledgements

This work was supported by the Natural Science Foundation of China (Grants Nos. 12075239, U1632150, and 61471332).

Conflict of interest

The authors declare no conflicts of interest.

Author details

Weihao Liu[1]*, Zijia Yu[2] and Zhi Tao[1]

1 College of Electronic and Information Engineering, Nanjing University of Aeronautics and Astronautics, Nanjing, Jiangsu, China

2 National Synchrotron Radiation Laboratory, University of Science and Technology of China, Hefei, Anhui, China

*Address all correspondence to: liuwhao@nuaa.edu.cn

References

[1] M. Tonouchi. Cutting-Edge Terahertz Technology. Nature Photonics. 2007; 1, 2: 97-105.

[2] C. G. Wade and N. Šibalić, et al. Real-Time Near-Field Terahertz Imaging with Atomic Optical Fluorescence. Nature Photonics. 2017; 11, 1: 40-43.

[3] A. J. Seeds and H. Shams, et al. Terahertz Photonics for Wireless Communications. Journal of Lightwave Technology. 2015; 33, 3: 579-587.

[4] J. H. Booske and R. J. Dobbs, et al. Vacuum Electronic High Power Terahertz Sources. Ieee Transactions On Terahertz Science and Technology. 2011; 1, 1: 54-75.

[5] S. Bielawski and C. Evain, et al. Tunable Narrowband Terahertz Emission From Mastered Laser–Electron Beam Interaction. Nature Physics. 2008; 4, 5: 390-393.

[6] S. J. Smith and E. M. Purcell. Visible Light From Localized Surface Charges Moving Across a Grating. Physical Review. 1953; 92, 4: 1069-1069.

[7] D. Li and Y. Wang, et al. Coherent Radiation at the Fundamental Frequency by a Smith-Purcell Free-Electron Laser with Dielectric Substrate. Applied Physics Letters. 2017; 110, 15.

[8] J. Gardelle and P. Modin, et al. Radiation at 100 and 200 Ghz From a Compact Planar Smith–Purcell Free-Electron Laser. Ieee Transactions On Terahertz Science and Technology. 2017; 7, 2: 151-163.

[9] K. B. Oganesyan. Smith–Purcell Radiation Amplifier. Laser Physics Letters. 2015; 12, 11.

[10] M. Cao and W. Liu, et al. Three-Dimensional Theory of Smith-Purcell Free-Electron Laser with Dielectric Loaded Grating. Journal of Applied Physics. 2014; 116, 10.

[11] J. Gardelle and P. Modin, et al. Observation of Copious Emission at the Fundamental Frequency by a Smith-Purcell Free-Electron Laser with Sidewalls. Applied Physics Letters. 2012; 100, 13.

[12] P. A. Molenaar and P. V. D. Straten, et al. Diagnostic Technique for Zeeman-Compensated Atomic Beam Slowing: Technique and Results. Physical Review a. 1997; 55, 1: 605-614.

[13] M. Castellano and V. A. Verzilov, et al. Measurements of Coherent Diffraction Radiation and its Application for Bunch Length Diagnostics in Particle Accelerators. Physical Review E. 2001; 63, 5.

[14] A. S. Kesar. Smith-Purcell Radiation From a Charge Moving Above a Grating of Finite Length and Width. Physical Review Special Topics Accelerators and Beams. 2010; 13, 2.

[15] M. J. Moran. X-Ray Generation by the Smith-Purcell Effect. Physical Review Letters. 1992; 69, 17: 2523-2526.

[16] W. Liu and Y. Liu, et al. Enhanced Smith–Purcell Radiation by Coupling Surface Plasmons On Meta-Films Array with Resonator Modes On a Grating. Journal of Physics D. 2019; 52, 7.

[17] O. Haeberlé and P. Rullhusen, et al. Calculations of Smith-Purcell Radiation Generated by Electrons of 1-100 Mev. Physical Review E. 1994; 49, 4: 3340-3352.

[18] P. M. V. D. Berg. Smith–Purcell Radiation From a Point Charge Moving Parallel to a Reflection Grating. Journal of the Optical Society of America. 1973; 63, 12: 1588-1597.

[19] D. V. Karlovets and A. P. Potylitsyn. Comparison of Smith-Purcell Radiation Models and Criteria for their Verification. Physical Review Special Topics Accelerators and Beams. 2006; 9, 8.

[20] J. H. Brownell and J. Walsh, et al. Spontaneous Smith-Purcell Radiation Described through Induced Surface Currents. Physical Review E. 1998; 57, 1: 1075-1080.

[21] W. Liu and Y. Lu, et al. A Multimode Terahertz-Orotron with the Special Smith–Purcell Radiation. Applied Physics Letters. 2016; 108, 18.

[22] W. Liu and Z. Xu. Special Smith–Purcell Radiation From an Open Resonator Array. New Journal of Physics. 2014; 16, 7.

[23] W. Liu and L. Sun, et al. Improved Terahertz Smith–Purcell Radiation Using Side-Opening Cylindrical Gratings. Journal of Applied Physics. 2020; 127, 18.

[24] J. Gardelle and L. Courtois, et al. Observation of Coherent Smith-Purcell Radiation Using an Initially Continuous Flat Beam. Physical Review Special Topics Accelerators and Beams. 2009; 12, 11.

[25] J. T. Donohue and J. Gardelle. Simulation of Smith-Purcell Radiation Using a Particle-in-Cell Code. Physical Review Special Topics Accelerators and Beams. 2005; 8, 6.

[26] S. Liu and M. Hu, et al. Electromagnetic Diffraction Radiation of a Subwavelength-Hole Array Excited by an Electron Beam. Physical Review E. 2009; 80, 3: 36602-36602.

[27] H. Liu and A. Yariv. Designing Coupled-Resonator Optical Waveguides Based On High-Q Tapered Grating-Defect Resonators. Optics Express. 2012; 20, 8: 9249-9263.

[28] I. G. Tigelis and J. L. Vomvoridis, et al. High-Frequency Electromagnetic Modes in a Dielectric-Ring Loaded Beam Tunnel. Ieee Transactions On Plasma Science. 1998; 26, 3: 922-930.

[29] W. Li and Y. Liu, et al. Spectrum Discretization of Smith–Purcell Radiation for Efficient Multicolor Terahertz Emission. Journal of Physics D. 2020; 53, 18.

[30] W. Liu and L. Zhang, et al. Coherent Terahertz Emission From Free-Electron-Driven Fabry-Perot Resonators with Coupled Grooves. Journal of Physics D. 2021; 54, 5.

[31] F. J. García-Vidal and L. Martín-Moreno, et al. Light Passing through Subwavelength Apertures. Reviews of Modern Physics. 2010; 82, 1: 729-787.

[32] W. L. Barnes and A. Dereux, et al. Surface Plasmon Subwavelength Optics. Nature. 2003; 424, 6950: 824-830.

[33] H. F. Ghaemi and T. Thio, et al. Surface Plasmons Enhance Optical Transmission through Subwavelength Holes. Physical Review B. 1998; 58, 11: 6779-6782.

[34] L. Liang and W. Liu, et al. Multi-Color and Multidirectional-Steerable Smith-Purcell Radiation From 2D Sub-Wavelength Hole Arrays. Applied Physics Letters. 2018; 113, 1.

[35] B. P. Clarke and K. F. MacDonald, et al. Direction-Division Multiplexed Holographic Free-Electron-Driven Light Sources. Applied Physics Letters. 2018; 112, 2.

[36] R. Remez and N. Shapira, et al. Spectral and Spatial Shaping of Smith-Purcell Radiation. Physical Review a. 2017; 96, 6.

[37] Z. Wang and K. Yao, et al. Manipulating Smith-Purcell Emission

with Babinet Metasurfaces. Physical Review Letters. 2016; 117, 15.

[38] G. Li and B. P. Clarke, et al. Holographic Free-Electron Light Source. Nature Communications. 2016; 7, 1: 13705-13705.

[39] D. Y. Sergeeva and A. A. Tishchenko, et al. Smith-Purcell Radiation From Concave Dotted Gratings. Journal of Instrumentation. 2018; 13, 2.

[40] Y. Liu and W. Liu, et al. Threshold-Less and Focused Cherenkov Radiations Using Sheet Electron-Beams to Drive Sub-Wavelength Hole Arrays. Optics Express. 2018; 26, 26: 34994-35002.

[41] Y. Liu and W. Liu, et al. Efficient and Tunable Terahertz Emission by Transforming Surface Waves to Coherent Smith–Purcell Radiation. Ieee Transactions On Plasma Science. 2021; 49, 1: 455-459.

[42] W. Liu and L. Liang, et al. Two-Dimensional Reversed Cherenkov Radiation On Plasmonic Thin-Film. Optics Express. 2017; 25, 15: 18216-18229.

[43] W. Liu. Dispersive 2D Cherenkov Radiation On a Dielectric Nano-Film. Scientific Reports. 2017; 7, 1: 5787-5787.

[44] W. Liu. Free-Electron Exciting and Steering of the Radiation From Surface Plasmons On Nanowire Array. Ieee Photonics Technology Letters. 2016; 28, 23: 2657-2660.

[45] W. Liu. Coherent and Tunable Light Radiation From Nanoscale Surface Plasmons Array Via an Exotic Smith-Purcell Effect. Optics Letters. 2015; 40, 20: 4579-4582.

[46] S. Liu and P. Zhang, et al. Surface Polariton Cherenkov Light Radiation Source. Physical Review Letters. 2012; 109, 15: 153902-153902.

[47] Z. Yu and Y. Liu, et al. Transforming Terahertz Plasmonics within Subwavelength Hole Arrays Into Enhanced Terahertz Mission Via Smith-Purcell Effect. Optics Express. 2020; 28, 7: 9501-9509.

[48] V. L. Bratman and B. S. Dumesh, et al. Terahertz Orotrons and Oromultipliers. Ieee Transactions On Plasma Science. 2010; 38, 6: 1466-1471.

[49] Y. A. Grishin and M. R. Fuchs, et al. Pulsed Orotron - A New Microwave Source for Submillimeter Pulse High-Field Electron Paramagnetic Resonance Spectroscopy. Review of Scientific Instruments. 2004; 75, 9: 2926-2936.

[50] W. Liu and Y. Lu, et al. A Compact Terahertz Free-Electron Laser with Two Gratings Driven by Two Electron-Beams. Physics of Plasmas. 2017; 24, 2.

Chapter 9

New Types of Dissipative Streaming Instabilities

Eduard V. Rostomyan

Abstract

Two new, previously unknown types of dissipative streaming instabilities (DSI) are substantiated. They follow from new approach, which allows solving in general form the classical problem of an initial perturbation development for streaming instabilities (SI). SI is caused by relative motion of the streams of plasma components. With an increase in level of dissipation SI transforms into a DSI. The transformation occurs because dissipation serves as a channel for energy removal for the growth of the negative energy wave of the stream. Until recently, only one type of DSI was known. Its maximal growth rate depends on the beam density n_b and the collision frequency ν in the plasma as $\sim \sqrt{n_b/\nu}$. All types of conventional beam-plasma instabilities (Cherenkov, cyclotron, etc.) transform into it. The solution of the problem of the initial perturbation development in systems with weak beam-plasma coupling leads to a new type of DSI. With an increase in the level of dissipation, the instability in these systems transforms to the new DSI. Its maximal growth rate is $\sim \sqrt{n_b}/\nu$. The second new DSI develops in beam-plasma waveguide with over-limiting current of e-beam. Its growth rate $\sim n_b/\nu$. In addition, the solutions of abovementioned problem provide much information about SI and DSI, significant part of which is unavailable by other methods.

Keywords: beam-plasma instability, dissipative instability, development of initial perturbation, growth rate, absolute/convective instability

1. Introduction

Streaming instabilities (SI) occupy a prominent place among other plasma instabilities. They are caused by a motion of some plasma components relative to others. An example is the well-known beam–plasma instability [1]. With this instability, the directed motion of a group of fast electrons passing through the background plasma excites potential oscillations with a large growth rate near the plasma frequency. Particular attention to this instability is mainly due to the idea of creating sources of powerful electromagnetic radiation on its basis. At present, these sources have many advantages over the known vacuum sources [2, 3]. One more example (we mention these two only) is the Buneman instability [4], in which plasma electrons move relative to ions.

In the overwhelming majority of investigations beam–plasma interaction is considered without any noticeable dissipation. It, actually, was assumed that the dissipation is small and cannot have any noticeable effect on the physical processes. In this case, the development of instability leads to an increase in the amplitude of electromagnetic oscillations in the plasma, as well as their energy at the expense of beam kinetic energy. In the absence of dissipation, the level of excited oscillations may be quite high, and their energy can even be comparable to the initial energy of the beam [5].

However, generally speaking, dissipation in the system (collisions between plasma particles, heating of metal surfaces due to their complex impedance, etc.) can play an essential role in plasma–beam interaction. It can become not only a decisive factor in limiting the spatial and temporal growth, determining the field amplitude and the mode structure and limits the growth rates. In addition to these properties, which are common to all systems, it is necessary to pay special attention to the unique role of dissipation in systems with a stream of charge particles: dissipation of high level does not suppress the SI completely. Strong dissipation transforms each SI to instability of other type – to dissipative streaming instability (DSI) [1]. This type of instabilities is due to the presence of the negative energy wave (NEW) in a stream of charge particles [6, 7]. In fact, dissipation serves as a channel for energy removal for excitation of this wave. This leads to instabilities of a new physical nature, to DSI. Dissipation is the cause of this instability.

The physical nature of SI is not as simple as it might seem at first glance. It takes a lot of effort[1] to understand it clearly. This is all the more so, if we are dealing with the transformation of SI into a DSI. The transformation (in general, the transformation of one type of instability into another) makes the behavior of SI in a system with dissipation especially interesting. In addition, there are other reasons that significantly increase interest in the study of problems associated with dissipation and the DSI caused by it. Some of them are as follows.

Modern high-frequency microwave electronics, both plasma and vacuum, have two basic trends of development: an increase in the frequency and power of the output radiation [2]. With increasing frequency, the thickness of the skin layer on the resonators' walls decreases. This, in turn, leads to an increase in active energy losses. Actual dissipation in the system increases.

The second trend – an increase in the power of output radiation – leads to the need to increase the beam current. The role of space charge phenomena increases also, as well as the role of the NEW. In these circumstances it becomes important to take into account all factors that also lead to the buildup of the same wave i.e. to dissipation. In a sense, dissipation becomes associated with the space charge phenomena. In addition with an increase in the beam current, the return current increases also. With account the decrease in the skin layer and the finite conductivity of metallic surfaces, this leads to an increase in the level of dissipation in the system. All this indicates that dissipation, along with the space charge of the beam plays an important role in microwave electronics. A detailed understanding of the role of all these phenomena is vital for many problems aimed at achieving high-intensity beams and their applications.

Until recently, only one DSI was known in beam–plasma interaction theory [1]. Its maximal growth rate depends on collision frequency ν in plasma and on the beam density n_b as $\sim \sqrt{n_b/\nu}$. All types of the beam–plasma instabilities (Cherenkov, cyclotron, etc.), with an increase in the level of dissipation, transform into it. This only known DSI has a number of specific features in comparison with other (no-dissipative) instabilities: relatively low level of excited oscillations, relatively small growth rate, etc. Many investigations have been devoted to its study. It was

[1] The instability of low density e-beam in plasma is a vivid example demonstrating this sense. It is discovered in 1948, experimentally proven in early sixties; however its physical meaning became finally clear in the middle of seventies (see [8]).

assumed that various phenomena in space plasma and in plasma of controlled fusion can be explained on the basis of this instability.

However, recent studies have shown that there are other DSI also [9–11]. The interaction of the stream with the background plasma critically depends on some basic parameters of the system and/or on its geometry. Their changes lead to new physics of the beam-plasma interaction and to previously unknown types of DSI. The parameters are: the level of correlation between the beam and the plasma fields and the value of the beam current.

Available methods of instability investigation do not allow getting complete information on the process of transformation of given instability into another type. Is known the most complete information on instability can be obtained by solving the problem of the evolution of fields in space and time during the development of an initial perturbation. This problem is classical in theory of instabilities [12]. Its results can clear up how the fields of given instability transform to the fields of another one along with many other accompanying details. The character of the space–time evolution of an initial perturbation is an important issue in many branches of physics. However, the results of this problem are hardly achievable. Ultimately its mathematical solution reduces to calculation of the integral with complete dispersion relation (DR) in the denominator of the integrand. For the result the DR should be specified and solved before integration. This sharply reduces generality of results. And even in the special cases, it is not always possible to carry out the integration. In [13] an approach is presented that allowed overcome difficulties and obtain analytical expression for the fields' space–time structure for all types of conventional beam-plasma instabilities. Results show that with increase in level of dissipation all types of beam-plasma instabilities transform to the only known type of DSI.

This review shows that the number of DSI is not limited by the above-mentioned DSI. Two new types of DSI are substantiated. They follow from solution of the same classical problem of initial perturbation development. One of the DSI manifests itself in the results of solving the problem in systems with weak beam-plasma coupling. Weak interaction realizes if the beam and the plasma are spatially separated by a considerable distance. Under weak coupling the beam actually is left to its own and its proper oscillation come into play. Moreover, among them is the NEW. Its interaction with plasma causes instability, the growth rate of which reaches maximum at resonance of the plasma wave with the NEW. This resonance of wave–wave type was called "Collective Cherenkov effect" [14]. An increase in the level of dissipation leads to a new DSI with the growth rate $\sim \sqrt{n_b}/\nu$. Actually the new approach to solution of the classical problem has detected this new DSI.

The second new DSI appears in results of solving of the same problem in uniform cross-section beam-plasma waveguide with over-limiting e-beam. With an increase in the beam current the fields of its space charge affects more and more on the beam-plasma interaction. This manifests itself in two ways. Along with the increasing of the role of space charge oscillations, static fields of the beam space charge set an upper limit on the beam current that can pass through a given vacuum electro-dynamical system. The limit can be overcome by plasma filling. Plasma neutralizes the space charge of the beam. Plasma-filled waveguides can transmit e-beams with a current that is several times higher than the limiting current in vacuum waveguide. The fields of overlimiting e-beam space charge changes the character of its instability. The instability of over-limiting beams is not associated with any radiation mechanism [9, 14]. Its growth rate reaches maximum at the point of exact Cherenkov resonance and depends on the beam density as $\sqrt{n_b}$ [9, 14, 15], With an increase in the level of dissipation, one more new type of DSI develops [9]. Its growth rate depends on the parameters as $\sim n_b/\nu$.

In present review special attention is paid to systems, the geometry of which is similar to geometry of plasma microwave sources. These devices are a cylindrical waveguide with thin annular plasma and spatially separated thin annular e-beam. In this geometry the new types of DSI manifest themselves also [10].

In order to dispel all possible doubts about the correctness of the results, both new DSI are also substantiated by conventional analysis of the corresponding DR. To obtain a geometry-independent result for weak beam-plasma coupling we use perturbation theory based on smallness of the coupling parameter.

2. The only known DSI and transition to it

For the beginning we shortly present rezoning, from which follow: all types of beam-plasma instabilities (Cherenkov, cyclotron, beam instability in spatially periodical structure) transform to the only known DSI with the maximal growth rate $\sim \omega_b/\sqrt{\nu}$ (ω_b is the Langmuir frequency of the beam, ν is the collision frequency in plasma). The transition takes place with an increase in the level of dissipation. This help us to reveal a criterion for identification of DSI type.

In general, the dispersion relation (DR), describing a plasma system penetrating by an electron beam can be written as

$$D(\omega, \mathbf{k}) = D_0(\omega, \mathbf{k}) + D_b(\omega, \mathbf{k}) = 0 \tag{1}$$

where $D_0(\omega, \mathbf{k}) = \operatorname{Re} D_0(\omega, \mathbf{k}) + i\operatorname{Im}D_0(\omega, \mathbf{k})$ describes the plasma (without beam), but $D_b(\omega, \mathbf{k})$ describes the beam contribution in the system dispersion

$$D_b(\omega, \mathbf{k}) = -\frac{\omega_b^2 A(\omega, \mathbf{k})}{\gamma^3(\omega - \mathbf{k}\mathbf{V}_b - f)^2}, \tag{2}$$

ω is the frequency, $\mathbf{k}$ is the wave vector of perturbations, ω_b is Langmuir frequency of the e-beam, $\mathbf{V}_b$ is the velocity of the beam electrons (directed along z axis), $A(\omega, \mathbf{k})$ is a polynomial with respect to ω and $\mathbf{k}$, $\gamma = \left(1 - V_b^2/c^2\right)^{-1/2}$. It is assumed that $|\operatorname{Im}D_0| < < |\operatorname{Re} D_0|$ and $|D_b(\omega, \mathbf{k})| < < |D_0(\omega, \mathbf{k})|, f = 0$ with the Cherenkov interaction, with the cyclotron interaction $f = n\Omega/\gamma$, (Ω is the cyclotron frequency, n is the harmonic number), and $f = k_{\text{cor}} V_b$ when e-beam interacts with the periodical structure, $k_{\text{cor}} = 2\pi/l$, l is the length of spatial period.

The beam electrons interact with the proper oscillations of the system and the interaction leads to instability. Developing instability manifests itself most effectively at frequencies and wavelengths close to the proper frequencies of the system in the absence of the beam, and, at the same time, close to the beam natural frequencies. In fact, along with (1) following condition is met

$$\omega - kV_b - f = 0. \tag{3}$$

All (conventional) beam-plasma instabilities, including DSI, follow from (1)–(3). With an increase in level of dissipation all types of no-dissipative instabilities (Cherenkov, cyclotron etc) transform into the well-known DSI. If one searches the solutions of DR (1) in the form $\omega = \omega_0 + \delta$ (ω_0 satisfies (1) and (3); this case called resonance instability) he arrives to the expression

$$\delta\left(\frac{\partial D_0}{\partial \omega}\right)_{\substack{\omega = \omega_0, \\ k = k_0}} + i\operatorname{Im}D_0(\omega_0, k_0) = \frac{(\omega_b^2/\gamma^3)A(\omega_0, k_0)}{\delta^2}. \tag{4}$$

All types of no-dissipative instabilities follow the first and the right-hand side term. In this case the dissipative (second) term in (4) is small. The DSI follows from the second term (when it is greater than the first term) and the right-hand side term. The relation between the respective growth rates $\delta^{(\nu=0)}$ and $\delta^{(\nu\to\infty)}$ is

$$\delta^{(\nu\to\infty)} = \sqrt{\frac{\left\{\delta^{(\nu=0)}\right\}^3}{2\mathrm{Im}D_0}\frac{\partial D_0}{\partial\omega}} \sim \sqrt{\frac{\left\{\delta^{(\nu=0)}\right\}^3}{\nu}} \sim \sqrt{n_b/\nu} \tag{5}$$

where the frequency of collisions in plasma ν is introduced ($\mathrm{Im}D_0 \sim \nu$). The expression (5) presents relation between the growth rates of no-dissipative and dissipative instabilities. Below we use (5) and its analogs as a criterion for identification of DSI type.

3. Weak beam-plasma coupling. New type of DSI

3.1 Solution of the problem of initial perturbation development under weak beam-plasma coupling

The best way to study an instability in detail and its possible transformation to that of other type is the solving of the problem of initial perturbation development. The information obtained by other ways is insufficient and does not give any details. Here we present general (geometry independent) solution of the problem for weakly coupled beam-plasma systems.

Consider a system consisting of a mono-energetic rectilinear electron beam and cold plasma. To begin with, suppose the following: the plasma and the beam are weakly coupled (e.g. in a consequence of a sufficiently large distance between them). Let an initial perturbation arises at a point $z = 0$ (the electron beam propagates in the direction $z > 0$) at the instant $t = 0$ and the instability begins developing. Our goal is to obtain the fields' space–time distribution at an arbitrary instant $t > 0$ and investigate in detail the instability behavior by analyzing obtained expression. In the process, we interest only the longitudinal structure of the fields, i.e., their dependence on the longitudinal coordinate z and time t. The transverse structure of the fields can be obtained by expanding in terms of the system's eigenfunctions. In accordance with this, only two arguments are highlighted below: frequency and longitudinal component of the wave vector. Other arguments are irrelevant in the consideration below. To avoid overburdening the formulas, they are omitted.

In given case of weak beam-plasma coupling the instability is the result of the interaction of the beam negative energy wave (NEW) and the slowed down wave in the plasma. The interaction is of Collective Cherenkov type. We proceed from the theory of wave interaction in plasma [16]. In terms of this theory the problem of the initial perturbation evolution under instability development in non-equilibrium plasma can be considered based on the set of partial differential equations for the amplitudes of the interacting waves: beam charge density wave $E_b(z,t)$ and the slowed down electromagnetic wave $E_w(z,t)$ in the plasma

$$\left(\frac{\partial}{\partial t} + V_{\mathrm{b}}\frac{\partial}{\partial z}\right)E_b(z,t) - i\delta^2 E_w(z,t) = J(z,t) \tag{6}$$

$$\left(\frac{\partial}{\partial t} + V_p\frac{\partial}{\partial z} + \nu^*\right)E_w(z,t) - iE_b(z,t) = 0$$

where t is the time, z is the coordinate along the beam propagation direction, $J(z,t)$ is a function determined by the initial conditions, V_b is the directed velocity of the beam, V_p is the group velocity of the resonant wave in plasma, $V_b > V_p$, ν^* describes dissipation in plasma and is proportional to the frequency of collisions in it. The meaning of the denotation δ will be cleared up below. Note, the set (6) is meaningful irrespective of the problem of development of any instability. Generally, it describes resonant interactions between two waves in unstable medium. One only condition should be satisfied: the growth rate attains maximum under Collective Cherenkov Effect. If the maximum is attained under conventional Cherenkov Effect, as for conventional beam-plasma instabilities, the interaction should be described by other set of Equations [16].

The solution of the set (6) gives the dependence of the field's amplitude on longitudinal coordinate and time under instability development. Applying the Laplace transformation with respect to time t and the Fourier transformation with respect to the spatial coordinate z, we obtain following expressions for the transform $E_w(\omega,k)$:

$$E_w(\omega,k) = \frac{J(\omega,k)}{D(\omega,\mathrm{k})}$$

$$D(\omega,k) = (\omega - kV_b)(\omega - kV_p + i\nu^*) + \delta^2 \tag{7}$$

The field's amplitude $E_w(z,t)$ can be found by inverse transformation

$$E_w(z,t) = \frac{1}{(2\pi)^2}\int\limits_{C(\omega)} d\omega \int\limits_{-\infty}^{\infty} \frac{dk\ J(\omega,k)\exp(-i\omega\ t + ikz)}{(\omega - kV_b)(\omega - kV_p + i\nu^*) + \delta^2} \tag{8}$$

where $C(\omega)$ is the contour of integration with respect to ω. For given case it is a straight line that lies in the upper half plane of the complex plane $\omega = \mathrm{Re}\,\omega + i\mathrm{Im}\omega$ and passes above all singularities of the integrand.

Thus, the problem has been reduced to the problem of integration in (8). It is somewhat simpler in comparison to the integral, which represents classical solution. Instead of full DR its analog stands. The analog is determined by interaction of the waves, participating in the instability development. This replacement simplifies integration. However, it remains difficult and many authors use roundabout methods carry out an expression for possible estimation of the fields behavior [17, 18]. Presented here method easily leads to the desired result i.e. to expression for space–time distribution of the fields. We merely transform the variables ω and k to another pair ω and $\omega' = \omega - kV_b$. The first integration (over ω) may be carried out by the residue method and the integration contour must be closed in the lower half-plane. The first order pole is

$$\omega = -(1 - V_p/V_b)^{-1}\{\delta^2/\omega' + i\nu + \omega' V_p/V_b\} \tag{9}$$

The second integration (over ω') cannot be carried out exactly, and we are forced to restrict ourselves to the approximate steepest descent method [19]. This method gives result in the limit of relatively large t. According to this method, the contour of integration should be deformed to pass through the saddle point in the direction of the steepest descent. The saddle point is found from the condition

$$\frac{d}{d\omega'}(\omega(\omega')t + i\omega' z/V_b) = 0 \tag{10}$$

and is equal to

$$\omega_s' = i\delta\{(V_b t - z)/(z - V_p t)\}^{1/2} \tag{11}$$

As a result we arrive to the following expression for the field's space time structure under development of the instability in spatially separated beam-plasma system

$$E_w(z,t) = -\frac{J_0}{2\sqrt{\pi}} \frac{\exp\chi_\nu^{(wk)}(z,t)}{(V_b - V_p)^{1/2}\delta^{1/2}(V_b t - z)^{1/2}} \tag{12}$$

$$\chi_\nu^{(wk)} = \chi_0^{(wk)} - \nu^* \frac{V_b t - z}{V_b - V_p}; \chi_0^{(wk)} = \frac{2\delta}{V_b - V_p}\sqrt{(z - V_p t)(V_b t - z)}$$

$$J_0 = J(\omega = \omega(\omega_s'), \omega' = \omega_s')$$

3.2 Analysis of the instability development

The expression (12) looks very complicate. At first glance it is impossible to extract any information on the instability behavior from it. However, it turned out, the expression may be easily analyzing. Moreover, the results are obtained from scratch, i.e. they are not based on prior research. Substantial part of the information is unavailable by other way. In particular, the analysis clearly shows that with increase in level of dissipation the no-dissipative instability turns to a new type of DSI and provides detailed information on both instabilities.

The properties of the instability is determined mainly by the exponential factor

$$\exp\chi_\nu^{(wk)}(z,t) = \exp\left\{\frac{2\delta}{V_b - V_p}\sqrt{(z - V_p t)(V_b t - z)} - \nu^* \frac{V_b t - z}{V_b - V_p}\right\}, \tag{13}$$

which provides many information: the temporal and the spatial growth rates, the spread of the unstable perturbations' velocities, the nature of the instability (absolute or convective), the effect of dissipation on instability, etc.

First consider some general properties of the instability, which follow from (13).

It is easily seen that in the absence of dissipation unstable perturbations have velocities in the range from V_p to V_b. The wave packet moves in the beam propagation direction and, along with exponential growth of the fields, expands. Its length increases over time $l \sim (V_b - V_p)\ t$. The knowledge of the boundary velocities of unstable perturbations allows at once determining the nature of the instability (convective/absolute) based on the definition only, without reference to additional studies (we mean the Sturrock's laws [20]). It is clearly seen that the instability is convective in the laboratory frame and other frames moving at velocities $V > V_b$ and $V < V_p$. However, if the observer's speed is within the range $V_p < V < V_b$, then the same instability is absolute (see **Figure 1**).

Now we turn to determination of the meaning of the denotation δ in (6). For this we consider case $\nu = 0$ and find the point of the field's maximum from expression

$$\frac{\partial\chi_0^{(wk)}(z,t)}{\partial z} = 0 \tag{14}$$

Its root is $z_m = w_{pk}^{(\nu=0)} t$ i.e. the point of the field's maximum moves at velocity

$$w_{pk}^{(\nu=0)} = (1/2)\left(V_b + V_p\right). \tag{15}$$

In the wave theory the velocity (15) is called convective velocity. It characterizes the spatial convection of the fastest growing perturbations. (15) shows that the peak of the wave packet disposes in its middle. The packet is symmetric with respect to its peak. Substitution of z_m into the $\chi_0^{(wk)}(z,t)$ determines the field's behavior in the maximum as $E_0(z_m,t) \sim \exp\,(\delta\ t)$, i.e. δ represents the maximal growth rate of the instability, which develops in absence of dissipation in systems with weak beam-plasma coupling. At the point $z_m = w_{pk}^{(\nu=0)}t$ the peak forms, because here the growth rate of perturbations is maximal.

The meaning of the parameter δ may also be determined from the DR (7) only, bypassing the results of integration (12). The general expression for the group velocity $V_{gr}(\omega,k)$ obtained from DR (7) has the limit (15) under $k = 0$. The same limit (note that $\nu = 0$) leads to DR in form $\omega^2 + \delta^2 = 0$, i.e. the parameter δ is the imaginary part of complex frequency (the growth rate). In this case (absence of dissipation) the instability is due to interaction of the NEW with the plasma. To emphasize the important role of δ we add the respective indexes $\delta \equiv \delta_{NEW}^{(\nu=0)}$. Its dependence on specific parameters is found out below.

At a fixed point z the field first grows up to the value $\sim \exp\left\{\delta_{NEW}^{(\nu=0)}z/(V_b V_0)^{1/2}\right\}$ that is reached at the instant $t = z/w_a$ where

$$w_a = 2V_b V_p/\left(V_b + V_p\right). \tag{16}$$

Then the field decreases, and at the time $t \geq z/V_0$ the wave packet completely passes given point. The exponent $\delta_{NEW}^{(\nu=0)}z/(V_b V_0)^{1/2}$ is, in fact, the maximal spatial growth rate. At a given point, the field reaches its maximum at the moment when the peak has already passed it (see **Figure 1**). The reason is that perturbations moving at lower velocities reach the point for a longer time, and they have time to grow more. w_a is the velocity of the most effectively amplified perturbations.

Thus, the solution of the problem of initial perturbation development along with other detailed information, gave results of conventional initial and boundary problems. This coincidence confirms correctness of developed approach (initial assumptions, mathematics, etc.). An additional advantage of the approach is in its geometry-independence. At first glance, the presented approach seems more complicated than traditional approaches, but this complexity is only apparent.

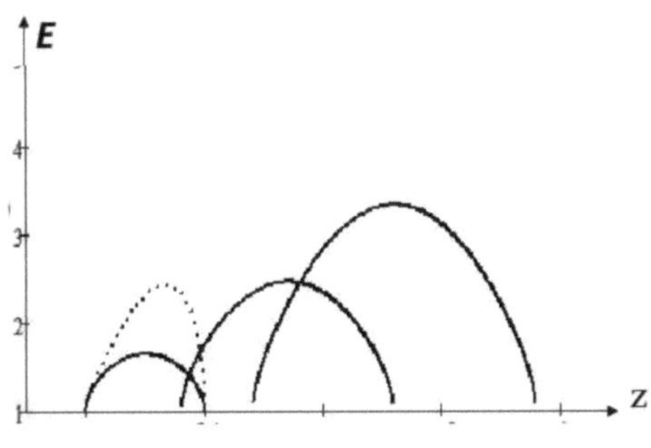

Figure 1.
Asymptotic shapes of the instability development under weak beam-plasma coupling vs. longitudinal coordinate z at instants $t_1 = 0, 5/\delta_{NEW}^{\nu=0}$ $t_2 = 0, 9/\delta_{NEW}^{\nu=0}$ $t_3 = 1, 2/\delta_{NEW}^{\nu=0}$. *The dotted line gives the shape of the wave packet for strong beam-plasma coupling.*

3.3 The influence of dissipation. New type of DSI

Dissipation significantly influences on the presented picture of the instability development and changes it. First of all, it suppresses slow perturbations. The wave packet shortens. The threshold velocity $V_{th}^{(\mathrm{wk})}$ is determined from the condition $\chi_0^{(\mathrm{wk})} = \nu^* (V_b t - z)/(V_b - V_0)$ and is equal

$$V_{th}^{(\mathrm{wk})} = \frac{\lambda'^2 V_\mathrm{b} + V_0}{1 + \lambda'^2} > V_0; \lambda' = \nu^* / \left(2\delta_{NEW}^{(\nu=0)}\right) \tag{17}$$

Only high-velocity perturbations (in the range $V_{th}^{(wk)} < v < V_b$) grow. The change in the velocity of the trailing edge shortens the packet's length and can affects the nature of instability (convective/absolute) if the frame's velocity lies in the range $V_p \le v \le V_{th}^{(wk)}$. Also, dissipation limits the growth rates of perturbations with velocity v. Substituting $z = vt$ we have for the field $E(z = vt, t) \sim \exp G(v)t$, where

$$G(v) = \frac{2\delta_{NEW}^{(\nu=0)}}{V_b - V_p} \sqrt{(V_b - v)(v\text{-}V_p)} - \nu^* \frac{V_b - v}{V_b - V_p} \tag{18}$$

As expected, the growth rates fall down. Dissipation distorts the symmetry of the induced wave packet. In presence of dissipation the dynamics of the fields can be obtained from the same Eq. (14) accounting for dissipation. It has the form

$$(z - w_g t)^2 = \lambda'^2 (V_\mathrm{b} t - z)(z - V_0 t). \tag{19}$$

The solution of (18) gives the point of the field maximum $z_{pk}^{(\nu)} = w_{pk}^{(\nu)} t$, where

$$w_{pk}^{(\nu)} = \frac{1}{2} \left\{ (V_b + V_p) + \sqrt{\frac{\lambda'^2}{1 + \lambda'^2}} (V_b - V_\mathrm{p}) \right\} > w_{pk}^{(\nu=0)} \tag{20}$$

This expression shows that with an increase in the level of dissipation, the peak shifts more and more to the front of the wave packet. This takes place along with the decreasing of the wave packet's length. Substitution of $w_{pk}^{(\nu)}$ into $\chi_\nu^{(\mathrm{wk})}$ gives the field value in the peak and shows the respective growth rate as the function on the level of dissipation

$$E_0 (z = z_{pk} t, t) \sim \exp\left\{\delta_{NEW}^{(\nu)} t\right\}; \delta_{NEW}^{(\nu)} = \delta_{NEW}^{(\nu=0)} f(\lambda'^2); f(x) = \sqrt{1 + x} - \sqrt{x} \tag{21}$$

The function $f(x)$ presents the dependence of the growth rate on the level of dissipation (see **Figures 2** and **3**). In the limit $\nu^* \to \infty$ we have $E_0 \to \sim \exp\left\{\delta_{wk}^{(\nu\to\infty)} t\right\}$, where

$$\delta_{wk}^{(\nu\to\infty)} = \left[\delta_{NEW}^{(\nu=0)}\right]^2 / \nu^* \sim \sqrt{n_b}/\nu^* \tag{22}$$

As a criterion for the type of DSI this relation between the growth rates of DSI $\delta_{NEW}^{(\nu\to\infty)}$ and the growth rate of SI $\delta_{NEW}^{(\nu=0)}$ sharply differs from that for the conventional case (5). Actually the expression (22) shows that with an increase in level of dissipation in weakly coupled beam-plasma systems the instability, caused by the

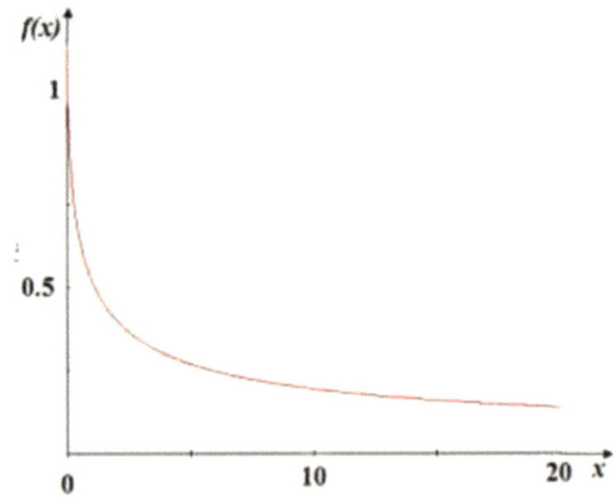

Figure 2.
The function $f(x)$ presents the dependence of the growth rate of the instability, caused by NEW excitation on the level of dissipation. Here $x = \nu/\delta_{NEW}^{\nu=0}$.

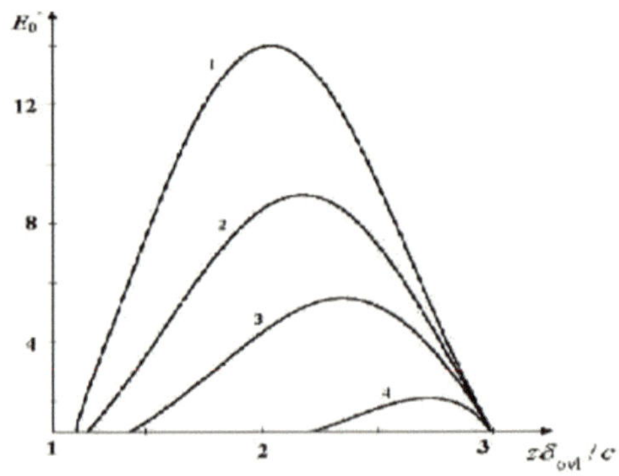

Figure 3.
Shapes of developing waveform versus longitudinal coordinate at fixed instant $3/\delta_{NEW}^{(\nu=0)}$ for various values of dissipation (parameter $k = \nu/\delta_{NEW}^{(\nu=0)}$): $k_1 = 0$, $k_2 = 1$, $k_3 = 2$, $k_4 = 4$.

beam's NEW interaction with the plasma transforms to a new type of DSI. Its characteristic peculiarity is in new, previously unknown, ***inverse proportional*** dependence of the growth rate on dissipation. Below this result is confirmed by conventional electro-dynamical analysis of the DR for weakly coupled beam-plasma system.

3.4 Substantiation of the new DSI by conventional analysis of the DR

From electro-dynamical point of view, a spatially separated beam-plasma system is nothing, but a multilayer structure. The traditional analytical consideration of such systems leads to a very cumbersome DR, which, in addition, is highly dependent on the geometry and greatly complicates with an increase in the number of layers. However, the importance of the problem and the need for its analytical investigation has led to development of specific methods. Here an approach is presented that allows avoiding abovementioned difficulties. Also, the approach has an important advantage: the procedure for obtaining the DR does not depend on specific shape/geometry. In other words, obtained results can be adapted to systems of any geometry. The approach considers the problem of weak beam-plasma interaction by perturbation theory. The small parameter, which underlies the theory, is the parameter of weak beam-plasma coupling. We briefly present here the basics of this approach accounting for dissipation [11].

Consider a system consisting of a mono-energetic rectilinear electron beam and cold plasma. To begin with, suppose the following: the plasma and the beam are weakly coupled (e.g. a consequence of a sufficiently large distance between them). We also assume their homogeneity in the cross section. The geometry of the system is not specified. It also is assumed that the beam current is less than the limiting current in the vacuum waveguide. Dissipation in the system is taken into account by the introduction the collisions in plasma. For simplicity, consideration is limited to the case of a strong external longitudinal (to the beam propagation direction) magnetic field, which prevents the transverse motion of the beam and plasma particles.

The small parameter underlying the perturbation theory is the parameter of weak coupling between the beam and the plasma (that is, the smallness of the integrals describing the overlap of beam and plasma fields). In the zero order approximation, the perturbation theory assumes independence of the beam and plasma. In the first-order approximation, the theory leads to the DR [11, 14].

$$D_p(\omega,k)D_b(\omega,k) = G\left(\kappa^4\delta\varepsilon_p\delta\varepsilon_b\right)_{\omega=\omega_0,k=k_0} \tag{23}$$

$$D_{p,b}(\omega,k) = k^2_{\perp p,b} - \kappa^2\delta\varepsilon_{p,b}; G < <1$$

$$\kappa^2 = k^2 - \frac{\omega^2}{c^2}; \delta\varepsilon_p = \frac{\omega_p^2}{\omega(\omega+i\nu)}; \delta\varepsilon_b = \frac{\omega_b^2}{\gamma^3(\omega-kV_b)^2},$$

ω and k are the frequency and longitudinal component of the wave vector, $\omega_{p,b}$ are Langmuir frequencies for the plasma and the beam respectively, ν is the collision frequency in the plasma, V_b is the velocity of the beam electrons, $\gamma = \left(1-V_b^2/c^2\right)^{-1/2}$, c is speed of light, G is the coupling parameter, the point $\{\omega_0, k_0\}$ is the intersection point of the beam and the plasma dispersion curves, the values $k_{\perp p}$ and $k_{\perp b}$ play role of transverse wave numbers. Analytically, G as well as $k_{\perp p}$ and $k_{\perp b}$ are expressed through the integrals of eigenfunctions of the zero order problem [11, 14]. The integral for G represents overlap of the beam and plasma fields. It shows how far the plasma field penetrates the beam and vice versa. The specific expressions for $k_{\perp p}$, $k_{\perp b}$ and G are not essential for the subsequent presentation and are not presented here (see [11, 14]).

The expressions $D_{p,b}(\omega,k) = 0$ are the zero order DR for the plasma and the beam respectively. Their solutions are assumed to be known. The form of the DR (23) is comparatively simple. It shows the interaction of beam and plasma waves. Using (23) with small G, it is easy to describe instabilities in given system. The main result of a decrease in the beam–plasma coupling is in the increase in role of the beam NEW. Its interaction with plasma leads to instability. The spectra of slow (−) and fast (+) beam waves follow from the roots of $D_b(\omega,k) = 0$. If one searches them in form $\omega_\pm = kV_b(1+x_\pm)$, $|x_\pm| < <1$, the roots become [11, 14].

$$x_\pm = \pm\left(\sqrt{\alpha}/\gamma\right)\left(\sqrt{\beta^4\gamma^2\alpha+1} + \beta^2\gamma\sqrt{\alpha}\right), \tag{24}$$

where $\alpha = \omega_b^2/k^2_{\perp b}V_b^2\gamma^3$, $\beta = V_b/c$. The interaction of the NEW (x_-) with the plasma leads to instability. If one looks for the solutions of (23) in the form $\omega = kV_b(1+x)$, ($|x| < <1$) it becomes [11].

$$(x+q+i\nu/kV_b)(x-x_+)(x-x_-) = G\alpha/2\gamma^4 \tag{25}$$

where $q = (2\gamma^2)^{-1}\left(k_{\perp p}^2 V_b^2 \gamma^2/\omega_p^2 - 1\right)$. Mathematically, the instability is due to corrections to the expression for the slow beam wave $x = x_- + x'$. Under collective Cherenkov resonance $q = -x_-$ [11], the equation for $x'$is

$$\left(x' + i\nu/\left(2\gamma^2 kV_b\right)\right)x' = -G\sqrt{\alpha}/\left(4\gamma^3\right) \tag{26}$$

In absence of dissipation the instability is due to NEW interaction with the plasma. Its growth rate is

$$\delta_{\text{NEW}}^{(\nu=0)} = kV_b \text{Im} x' = (kV_b/2\gamma)\sqrt{\left(G\sqrt{\alpha}\right)/\gamma}. \tag{27}$$

We emphasize unusual dependence on the beam density as $n_b^{1/4}$ (for strong coupling this dependence is $\sim n_b^{1/3}$). With ordinary Cherenkov resonance the system is stable. Under collective Cherenkov resonance dissipation manifested itself as an additional factor that enhances NEW growth and the instability gradually transforms to that of dissipative type. The Eq. (26) gives an expression for the growth rate as a function on level of dissipation

$$\delta(\lambda) = \delta_{\text{NEW}}^{(\nu=0)} f\left(\lambda^2\right); \quad \lambda = \left(1/2\gamma^2\right)\left(\nu/\delta_{NEW}^{(\nu=0)}\right). \tag{28}$$

where $f(x)$ is the function given in (21). The dependence of the growth rate on the level of dissipation in (28) coincides to that in (21). In limit $\lambda \to 0$ (28) coincides to (27). In the opposite limit of strong dissipation $\lambda \to \infty$ (28) represents the growth rate of the new type of DSI (it also follows from (26) by neglecting the first term in brackets)

$$\delta_{\text{NEW}}^{(\nu\to\infty)} = \frac{2\gamma^2\left(\delta_{\text{NEW}}^{(\nu=0)}\right)^2}{\nu} = \frac{G\sqrt{\alpha}}{2\gamma}\frac{(kV_b)^2}{\nu} \sim \frac{\omega_b}{\nu} \tag{29}$$

We arrive to the same new type of DSI presented in (22). The expression (28) shows a gradual transition of the growth rate of no-dissipative instability caused by NEW interaction with plasma into the growth rate of new type of DSI. It develops under weak coupling and differs from the conventional DSI (with an growth rate $\sim \omega_b/\sqrt{\nu}$). In [21] the same new DSI is substantiated in a finite external magnetic field.

4. Uniform cross section beam-plasma waveguide. One more new type of DSI

4.1 Evolution of the initial perturbation in plasma waveguide with over-limiting electron e-beam

One more new DSI arises under consideration of the problem of the initial perturbation development for the instability of over-limiting beam (OEB) in uniform cross-section plasma waveguide.

Consider a cylindrical waveguide, fully filled with cold plasma. A monoenergetic relativistic electron beam penetrates it. The external longitudinal magnetic field is assumed to be strong enough to freeze transversal motion of the beam and the plasma electrons. We also assume that the beam and plasma radii coincide with the waveguide's radius and consider only the symmetrical E-modes with nonzero components E_r, E_z, and B_φ. The development of resonant instability in this system is described by the DR and resonant condition those are [1].

$$D_0(\omega,k) + D_b(\omega,k) = 0; \omega = kV_b \tag{30}$$

$$D_0 = k_\perp^2 + \kappa^2\left(1 - \frac{\omega_p^2}{\omega(\omega+i\nu)}\right); D_b = -\kappa^2\frac{\omega_b^2/\gamma^3}{(\omega - kV_b)^2}; \kappa^2 = k^2 - \frac{\omega^2}{c^2}$$

ω and k are the frequency and the longitudinal (along beam propagation direction that is z axis) wave vector, $k_\perp = \mu_{0s}/R$, R is the waveguide's radius, μ_{0s} are the roots of Bessel function J_0: $J_0(\mu_{0s}) = 0$, s = 1,2,3 ..., $\omega_{p,b}$ are the Langmuir frequencies for the plasma and the beam, V_b is the beam velocity, $\gamma = \left(1 - V_b^2/c^2\right)^{-1/2}$, ν is the frequency of collisions in plasma, c is the speed of light.

The character of the beam-plasma interaction changes depending on the beam current value. If the beam current is less than the limiting current in vacuum waveguide the instability is due to induced radiation of the system eigenwaves by the beam electrons. But, if the beam is over-limiting, its instability has the same nature as the instability in medium with negative dielectric constant [9, 14, 15]. We introduce a parameter $\alpha = \omega_b^2/k_\perp^2 V_b^2\gamma^3$, which represents the beam current value and the character of beam-plasma interaction. It corresponds (correct to the factor γ^{-2}) to the ratio of the beam current to the limiting current in vacuum waveguide [14] $I_0 = mV_b^3\gamma/4e$, i.e. $\alpha = (I_b/I_0)\gamma^{-2}$ (I_b is the beam current). The values $\alpha < <\gamma^{-2}$ correspond to under-limiting beam currents $I_b < <I_0$, but the values $\gamma^{-2} < <\alpha < <1$ correspond to over-limiting beam currents. This is possible under comparatively high values of the relativistic factor γ. Here we consider development of an initial perturbation in the system, when the beam current slightly exceeds the limiting vacuum value. In this case the instability is due to a-periodical modulation of the beam density in medium with negative dielectric constant. Its growth rate attaints maximum under exact Cherenkov resonance and is equal [15].

$$\delta_{\text{ovl}}^{(\nu=0)} = \frac{\omega_b V_b}{c\sqrt{\gamma(1+\mu)}}, \quad \mu = \gamma^2\frac{k_\perp^2 V_b^2}{\omega_p^2 - k_\perp^2 V_b^2\gamma^2} \tag{31}$$

However, the resonant frequency, which is determined by the expressions (30), remains unchanged [15].

In order to show the variety of possible approaches to the solution of the problem of the initial perturbation development, in given case we solve it by other way. We turn to the set of origin equations, which describes e-beam instability in magnetized plasma waveguide

$$\begin{gathered}\frac{\partial E_r}{\partial z} - \frac{\partial E_z}{\partial r} = -\frac{1}{c}\frac{\partial B_\varphi}{\partial t}; \quad \hat{L}\,\mathrm{v}_\mathrm{b}' = \frac{e}{m}E_z; \quad \frac{\partial \mathrm{v}_\mathrm{p}'}{\partial t} = \frac{e}{m}E_z - \nu \mathrm{v}_\mathrm{p}' \\ \frac{\partial B_\varphi}{\partial z} = -\frac{1}{c}\frac{\partial E_r}{\partial t}; \quad \hat{L}n_b' = -n_0\frac{\partial \mathrm{v}_\mathrm{b}'}{\partial t}; \quad \frac{\partial n_p'}{\partial t} = -n_{p0}\frac{\partial \mathrm{v}_\mathrm{p}'}{\partial z} \\ \frac{1}{r}\frac{\partial}{\partial r}rB_\varphi = \frac{1}{c}\frac{\partial E_r}{\partial t} + 4\pi e\left(n_{p0}\mathrm{v}_\mathrm{p}' + n_{b0}\mathrm{v}_\mathrm{b}' + n_b' V_b\right); \quad \hat{L} \equiv \frac{\partial}{\partial t} + V_b\frac{\partial}{\partial z};\end{gathered} \tag{32}$$

where t is time, z and r are the cylindrical coordinates, E_r, E_z and B_φ are the fields' components which are coupled with the beam, $\mathrm{v}'_{b,p}$ and $n'_{b,p}$ are the perturbations of velocity and density for the beam and the plasma respectively, n_0 and n_{p0} are the unperturbed densities for beam and plasma respectively. In the process, we interest only the longitudinal structure of the fields, i.e., their dependence on the longitudinal coordinate and time. The transverse structure of the fields can be obtained by expansion on series of the system's eigenfunctions. For given case those are the Bessel functions. We use the expansions

$$E_z(r,z,t) = \sum_s E_z^{(s)}(z,t) J_0(\mu_{0s} r/R), B_\phi(r,z,t) = \sum_s B_\phi^{(s)}(r,t) J_1(\mu_{1s} r/R) \qquad (33)$$

where J_0 and J_1 are the Bessel functions; μ_{os} and μ_{1s} their roots in ascending order, $J_0(\mu_{0s}) = 0, J_1(\mu_{1s}) = 0, s = 1, 2, 3, \ldots$. The quantities $\mathrm{v}_{p,b}$ and $n_{p,b}$ should be expanded by analogy to E_z, but E_r – by analogy to B_φ. From here on we deal with the expansion coefficients and mention arguments z and t only.

The fields' growth in the linear stage reveals itself most effectively on frequencies, closely approximating to roots of the DR and, simultaneously, to kV_b (resonant instability). The conditions (30) hold. In this connection it is reasonable to assume that originated perturbations form a wave packet of following type (e.g. for $E_z^{(s)}(z,t)$):

$$E_z^{(s)}(z,t) = E_0(z,t) \exp(-i\omega_0 t + ik_0 z), \qquad (34)$$

where the carrier frequency ω_0 and wave vector k_0 satisfy the conditions (30). We also assume that the amplitude of the wave train $E_0(z,t)$ varies slowly in space and time as compared to k_0 and ω_0 that is

$$\left|\frac{\partial E_0}{\partial t}\right| < < |\omega_0 E_0| \ ; \quad \left|\frac{\partial E_0}{\partial z}\right| < < |k_0 E_0|. \qquad (35)$$

Thus, the problem of the initial pulse behavior reduces to determination of the slowly varying amplitude (SVA) $E_0(z,t)$. The equation that $E_0(z,t)$ satisfies can be derived from the set of origin Eqs. (32). The expansions (33) reduce it to a set of the equation for the amplitudes of expansions. In its turn the resulting set can be reduced to one equation for $E_0(z,t)$. We write it in form similar to the DR

$$\left(\hat{\omega} - \hat{k}V_b\right)^2 D_0\left(\hat{\omega}, \hat{k}\right) E_z^{(s)}(z,t) = \omega_b^2 \gamma^{-3} \kappa^2 E_z^{(s)}(z,t) \qquad (36)$$

where $\hat{\omega}$ and $\hat{k}$ are differential operators $\hat{\omega} \equiv i\frac{\partial}{\partial t}; \hat{k} \equiv -i\frac{\partial}{\partial z}$. The DR in form (30) follows from (36). To derive the equation for $E_0(z,t)$ one should expanding (36) in power series near resonant values of frequency ω_0 and wave vector k_0 by using the relations $\hat{\omega} \to \omega_0 + i\frac{\partial}{\partial t}$ and $\hat{k} \to k_0 - i\frac{\partial}{\partial z}$ with account of OEB existence condition. As a result we arrive to the following second-order partial differential equation for $E_0(z,t)$

$$\left(\frac{\partial}{\partial t} + V_b \frac{\partial}{\partial z}\right)\left(\frac{\partial}{\partial t} + V_{\mathrm{P}} \frac{\partial}{\partial z} + \nu'\right) E_0(z,t) = \delta_{\mathrm{ovl}}^2 E_0(z,t) \qquad (37)$$

where $\nu' = \mathrm{Im} D_0 (\partial D_0/\partial \omega)^{-1}$, $V_0 = -\left\{(\partial D_0/\partial k)(\partial D_0/\partial \omega)^{-1}\right\}_{\substack{\omega = \omega_0 \\ k = k_0}}$ and the expression for δ_{ovl} is obtained from the relation $\delta_{\mathrm{ovl}}^3 = \kappa^2 \omega_b^2 \gamma^{-3} (\partial D_0/\partial \omega)^{-1}$ accounting the condition for OEB. It is important to emphasize that this denotation (as well as V_0) is introduced for reasons of simplicity of the resulting Eq. (36) only.

The solution of (37) is, actually, known. If one returns to the set (6) and transforms it (under $J(z,t) = 0$) to one equation for $E_w(z,t)$ then the equation will completely coincide to (37). This means that we already have the solution of (36) and its analysis. It only remains to rewrite the solution (12) in new denotations and, where needed, re-interpret results. This shows that the instability in uniform cross-section beam-plasma waveguide develops in space and time in the same manner as the instability in weakly coupled beam-plasma system, and δ_{ovl} is its growth rate in limit $\nu \to 0$, that is $\delta_{\text{ovl}} \equiv \delta_{\text{ovl}}^{(\nu=0)}$. However there is a very important quantitative difference. In present case the growth rate $\delta_{\text{ovl}}^{(\nu=0)}$ depends on the beam density as $\sim n_b^{1/2}$ (for the case of weak beam-plasma coupling the dependence is $\sim n_b^{1/4}$ (see (27)). The criterion for determining the type of DSI takes the form

$$\delta_{\text{ovl}}^{(\nu\to\infty)} = \left[\delta_{\text{ovl}}^{(\nu=0)}\right]^2 / \nu' \sim \omega_b^2/\nu' \tag{38}$$

Comparison of (38) with (22) indicates one more new type of DSI. It develops in uniform cross section beam-plasma waveguide under over-limiting beam current and high level of dissipation. Its growth rate depends on the beam density and collision frequency as $\sim n_b/\nu'$.

4.2 Substantiation of the second new DSI by conventional method

Now we substantiate the second new DSI by solving the DR (30). We look for its roots in the form $\omega = kV_b + \delta$, $\delta << kV_b$. The DR (30) reduces to [1, 14].

$$x^3 + i\frac{\nu}{\omega_0}\frac{\omega_p^2 \text{v}_0}{V_b\gamma^2\omega_\perp^2}x^2 + \frac{\alpha \text{v}_0 V_b}{\gamma^2 c^2}x = \frac{\alpha}{2\gamma^4}\frac{\text{v}_0}{V_b} \tag{39}$$

where $x = \delta/kV_b$, $\alpha = \omega_b^2/k_\perp^2 V_b^2 \gamma^3$, $\beta = V_b/c$, $\omega_\perp^2 = k_\perp^2 V_b^2 \gamma^2$, $\text{v}_0 = \mu V_b/(1+\mu)$, is the group velocity of the resonant wave in the system without beam, $\mu = \gamma^2\omega_\perp^2/\omega_0^2$; $\omega_0 = \left(\omega_p^2 - \omega_\perp^2\right)^{1/2}$ is the resonant frequency of the plasma waveguide.

The solutions of (39) depend on the beam current value that is on the value of parameter α. If $\alpha << \gamma^{-2}$ (under-limiting e-beams) one can obtain the growth rates of conventional instability under $\nu = 0$ (first and right-hand side terms) and in limit $\nu >> \delta_{\text{und}}$ i.e. DSI

$$\delta_{\text{und}} = \frac{\sqrt{3}}{2}\frac{\omega_0}{\gamma}\left(\frac{\omega_b^2}{2\omega_0^2(1+\mu)}\right)^{1/3}; \quad \delta_{\text{und}}^{(\nu)} = \frac{\omega_0^{3/2}}{2\gamma^{3/2}\omega_p}\sqrt{\frac{\omega_b}{\nu}} \tag{40}$$

If the beam current increases and become comparable or higher than the limiting vacuum current i.e. $\gamma^{-2} \le \alpha << 1$, the physical nature of the instability changes. It becomes due to a-periodical modulation of the beam density in medium with negative dielectric constant. The distinctive peculiarity of this instability is in following: its growth rate attains maximum under exact Cherenkov resonance and is equal to (31) [9, 11, 14, 15]. If, along with the beam current, dissipation also increases the instability turns to DSI of over-limiting beam with growth rate [9].

$$\delta_{\text{ovl}}^{(\nu)} = \frac{\beta^2}{\gamma}\frac{\omega_b^2}{\omega_p^2}\frac{\omega_0^2}{\nu} \sim \frac{\omega_b^2}{\nu}. \tag{41}$$

We emphasize new dependences on ν and on the beam density. This, actually, substantiates one more new type of DSI. It develops in uniform cross section beam-plasma waveguide if the beam current is higher than the limiting vacuum current.

5. The role of the new DSI in no-uniform-cross-section beam-plasma waveguide

5.1 Statement of the problem. Dispersion relation

In this section we pay special attention to systems, the geometry of which is similar to the geometry of plasma microwave sources and possible development of the new types of DSI in such systems. The simplest theoretical model of plasma microwave generators assumes relativistic e-beam propagating along axis of a plasma filled waveguide of radius R. The beam and plasma are assumed to be completely charge and current neutralized. In the waveguide cross-section the plasma and beam are annular, with mean radii r_p and r_b. Their thicknesses Δ_p and Δ_b are much smaller, than the mean radii. Strong external longitudinal magnetic field is assumed to freeze transversal motion of beam and plasma electrons.

For theoretical study of the problem we use an approach [10], which gives result for arbitrary level of beam-plasma coupling. This condition is obligatory for obtaining comprehensive results. The DR, which follows from the approach, has a form, which clearly shows interaction of the beam and plasma waves. The approach proceeds from equation for polarization potential ψ

$$\frac{\partial}{\partial t}\left(\Delta_\perp + \hat{L}\right)\psi = -4\pi\left(J_{bz} + J_{pz}\right), \hat{L} = \frac{\partial^2}{\partial z^2} - \frac{1}{c^2}\frac{\partial^2}{\partial t^2} \tag{42}$$

Here $J_{bz}(\mathbf{r}_\perp, z, t) = p_b(\mathbf{r}_\perp) j_{bz}(z,t)$ and $J_{pz}(\mathbf{r}_\perp, z, t) = p_p(\mathbf{r}_\perp) j_{pz}(z,t)$ are perturbations of the longitudinal current densities in the beam and plasma. Functions $p_{b,p}(\mathbf{r}_\perp)$ describe transverse density profiles of the perturbations of the longitudinal currents in the beam and the plasma. For homogeneous beam/plasma $p_{b,p} \equiv 1$ for infinitesimal thin $p_{b,p} \sim \delta(r - r_{b,p})$ (δ is Dirac function), $\Delta_\perp$ is the Laplace operator over transverse coordinates, z is the longitudinal coordinate, t is the time, c is the speed of light. The longitudinal electric field expresses as $E_z = \hat{L}\psi$. The equations for j_{bz} and j_{pz} are

$$\left(\frac{\partial}{\partial t} + V_b\frac{\partial}{\partial z}\right)^2 j_{bz} = \frac{\omega_b^2\gamma^{-3}}{4\pi}\frac{\partial}{\partial t}E_z; \left(\frac{\partial}{\partial t} + \nu\right) j_{pz} = \frac{\omega_p^2}{4\pi}E_z, \tag{43}$$

where $\omega_{p,b}$ are the Langmuir frequencies for plasma and beam respectively, ν is the effective collision frequency in plasma, $\gamma = \left(1 - V_b^2/c^2\right)^{-1/2}$, V_b is the velocity of beam electrons.

The DR, which follows from the statement, is still very cumbersome (of integral type). To reduce the DR to a simple algebraic form one should make following expedient for theoretical model assumption: the plasma and the beam are not just thin but infinitesimal thin. In this case the DR becomes

$$D_p(\omega,k)D_b(\omega,k) = G\kappa^4\delta\varepsilon_p\delta\varepsilon_b, \tag{44}$$

where $D_{p,b}(\omega,k) = k_{\perp p,b}^2 - \kappa^2\delta\varepsilon_{p,b}$, $\delta\varepsilon_p = \frac{\omega_p^2}{\omega(\omega+i\nu)}$, $\delta\varepsilon_b = \frac{\omega_b^2}{\gamma^3\left(\omega - kV_b\right)^2}$, $\kappa^2 = k^2 - \omega^2/c^2$, k is the wave vector along axis, ω is the frequency, $k_{\perp p}$ and $k_{\perp b}$ play role of the zero order transversal wave numbers for plasma and beam [11, 14].

$$k_{\perp p,b}^{2}=\left\{r_{p,b}\Delta_{p,b}I_l\left(\kappa r_{p,b}\right)\left[\frac{K_l\left(\kappa r_{p,b}\right)}{I_l\left(\kappa r_{p,b}\right)}-\frac{I_l(\kappa R)}{K_l(\kappa R)}\right]\right\}^{-1} \tag{45}$$

(I_l and K_l are modified Bessel and Mac-Donald functions, $l=0,\ 1,\ 2\dots$ is the azimuthal wave numbers). G is the coupling parameter. It depends on the overlap of the plasma and the beam fields and shows efficiency of their interaction

$$G=\begin{cases}\dfrac{I_l(\kappa r_b)K_l\left(\kappa r_p\right)I_l(\kappa R)-K_l(\kappa R)I_l\left(\kappa r_p\right)}{I_l\left(\kappa r_p\right)K_l(\kappa r_b)I_l(\kappa R)-K_l(\kappa R)I_l(\kappa r_b)} & r_b\le r_p\\[2ex] \dfrac{I_l\left(\kappa r_p\right)K_l(\kappa r_b)I_l(\kappa R)-K_l(\kappa R)I_l(\kappa r_b)}{I_l(\kappa r_b)K_l\left(\kappa r_p\right)I_l(\kappa R)-K_l(\kappa R)I_l\left(\kappa r_p\right)} & r_p\le r_b\end{cases} \tag{46}$$

An important property of G is: G =1 for $r_p=r_b$ and $G<1$ in other cases. In long wavelength limit (for definiteness $l=0$ and $r_b\le r_p$) we have $G\approx\ln\left(R/r_p\right)/\ln\left(R/r_b\right)$, but in opposite limit $G\approx\exp\left(-2\kappa\left|r_p-r_b\right|\right)$ (for arbitrary l).

5.2 Growth rates

The DR (44) determines proper oscillations of transversally no uniform beam-plasma waveguide. The changes of the physical character of beam-plasma interaction must reveal themselves on its solutions. $D_{p,b}(\omega,k)=0$ are the DR for wave-guide with thin annular plasma and e-beam respectively. The spectra of fast (+) and slow ($-$) waves are

$$\omega_{\pm}=kV_b(1+x_{\pm});x_{\pm}=\left(\sqrt{\alpha}/\gamma\right)\left(\pm\sqrt{\beta^4\gamma^2\alpha+1}-\beta^2\gamma\sqrt{\alpha}\right); \tag{47}$$

where $\beta=V_b/c$. The parameter $\alpha=\omega_b^2/k_{\perp b}^2V_b^2\gamma^3$ is familiar (see above). It determines the beam current value: $\alpha=I_b/\left(\gamma^2I_0\right)$ (I_b is the beam current, I_0 is the limiting current in vacuum waveguide). In the limit of under-limiting beams $x_{\pm}\to\pm\sqrt{\alpha}/\gamma$. In opposite limit of over-limiting beam $x_+=1/2\beta^2\gamma^2$ and $x_-=-2\beta^2\alpha$. If one looks for solutions of (44) in form $\omega=kV_b(1+x)$, $x<<1$ it becomes

$$\left(x+q+i\frac{\nu}{ku}\frac{1-2\beta^2\gamma^2x}{2\gamma^2}\right)(x-x_+)(x-x_-)=G\frac{\alpha}{2\gamma^4}\left(1-2\beta^2\gamma^2x\right)^2, \tag{48}$$

where $q=\left(k_{\perp p}^2u^2\gamma^2/\omega_p^2-1\right)/2\gamma^2$. The Eq. (48) presents sound way to study instabilities in given system. First of all, it is easily seen that in conditions of growing negative energy wave $x\approx x_-$ and collective Cherenkov resonance $q\approx-x_-$ the role of dissipation increases. For under-limiting e-beams $\alpha\le1/\gamma^2$ and in case of strong coupling $G\sim1$ the DR (44) leads to the well-known conventional beam instabilities of no-dissipative and dissipative type. The growth rates of these instabilities have well-known dependencies on beam density $\sim n_b^{1/3}$ and on dissipation ($\sim1/\sqrt{\nu}$). Both for these instabilities proper oscillations of the beam are neglected. Only for explanation of the physical meaning of the DSI the conception of NEW should be invoked. However, if $G<<1$ (weak coupling) the growing of the NEW plays dominant role. In this case the growth rate of no-dissipative instability reaches its maximum under Collective Cherenkov resonance $q=\sqrt{\alpha}/\gamma$ and is equal

$$(\mathrm{Im}\omega)_{\text{und}}^{(\nu=0)} = (kV_b/2\gamma)\left(G\sqrt{\alpha}/\gamma\right)^{1/2}. \tag{49}$$

This expression coincides to (27). Dissipation coming into interplay transforms this instability to DSI of new type with growth rate (coincides to (29))

$$(\mathrm{Im}\omega)_{\text{und}}^{(\nu\to\infty)} = G\sqrt{\alpha}(kV_b)^2/2\gamma\nu \tag{50}$$

As it should be, this is the instability discovered under consideration of the classical problem of the initial perturbation development in weakly coupled beam-plasma systems.

Of particular interest are limit of high, over-limiting currents of e-beam $\gamma^{-2} << \alpha << 1$. In this case the DR (44) takes the form

$$\left(x + q + i\frac{\nu}{ku}\frac{1-2\gamma^2 x}{2\gamma^2}\right)(x + 2\alpha) = -G\frac{\alpha}{\gamma^2}\left(1 - 2\gamma^2 x\right) \tag{51}$$

For $\nu = 0$ the analysis of (51) leads to following. Under single particle resonance we have either instability of negative mass type (under $G \sim 1$) with the growth rate $\mathrm{Im}\omega = ku\sqrt{\alpha}/\gamma$, or stability (under $G << 1$). But under collective Cherenkov effect $q = 2\alpha$ the growth rate of developing instabilities is

$$(\mathrm{Im}\omega)_{\text{ovl}}^{(\nu=0)} = \begin{cases} \sqrt{3}kV_b\alpha & \text{for} \quad G \sim 1 \\ 2kV_b\alpha\sqrt{G} & \text{for} \quad G << 1 \end{cases} \tag{52}$$

The instability (52) under $G \sim 1$ has mixed mechanism: it is caused simultaneously (i) by a-periodical modulation of the beam density in media with negative dielectric constant and (ii) by excitation of the NEW. But the lower expression is the growth rate of instability caused only by excitation of the NEW of overlimiting e-beam. The presence of dissipation intensifies the growing of the slow beam wave. Instability turns to be of dissipative type with growth rate that again is inverse proportional to dissipation.

$$(\mathrm{Im}\omega)_{\text{ovl}}^{(\nu)} = 2(ku)^2 G\alpha/\nu \sim \omega_b^2/\nu \tag{53}$$

However, the dependence on the beam density is completely different. This is the same DSI, which develops in uniform cross-section beam-plasma waveguide under over-limiting currents. Instabilities of the same type may be substantiated for finite thicknesses of the beam and plasma layers in waveguide. In this case one must use perturbation theory based on smallness of coupling coefficient.

As follows from this section, in the geometry of microwave plasma sources, the development of both new DSI is possible. Basic parameters of the both new DSI, and the conditions of their development should be taken into account upon design of the high power, high frequency plasma microwave devices.

6. Conclusion

Thus, based on very general initial assumptions, we have found out that the number of DSI in the beam-plasma interaction theory is not limited by the only previously known type. Two new, previously unknown types of DSI are presented. The new DSI reveal themselves in the analysis of solution of the problem of initial perturbation development. This problem is classical in the theory of instabilities.

The first new DSI is the dissipative instability under weak beam-plasma coupling. In absence of dissipation the instability in these systems is caused by the interaction of the beam NEW with the plasma. With an increase in the level of dissipation this instability gradually transforms to the new type of DSI. Its maximal growth rate depends on the beam Langmuir frequency ω_b and the frequency of collisions in plasma ν as ω_b/ν. This, more critical (as compared to conventional), inverse proportional dependence on ν is a result of superposition of two factors those lead to growth of the beam NEW: weak coupling and dissipation.

The second new type of DSI is dissipative instability of over-limiting e-beam in uniform cross section waveguide. With increase in the beam current, its space charge and inner degrees of freedom reveal themselves more efficiently. If the beam current becomes higher than the limiting current in vacuum waveguide then the instability mechanism changes. In uniform cross section beam-plasma waveguide the instability becomes due to a-periodical modulation of the beam density in medium with negative dielectric constant. In this case the increase in the level of dissipation leads to one more new type of DSI with the maximal growth rate $\sim \omega_b^2/\nu$.

The same types of DSI develop in systems having geometry, similar to microwave sources: cylindrical waveguide with thin annular beam and thin annular plasma. If the coupling between the beam and the plasma hollow cylinders is weak and the beam current is under-limiting the first type of DSI develops, but under over-limiting currents – the second. However, if the coupling of the beam and the plasma cylinders is strong, conventional type of DSI develops with well-known growth rate $\sim \omega_b/\sqrt{\nu}$.

Both new DSI are confirmed by conventional analysis of the respective DR.

Some words about the approach used. It has many advantages. First of all, it is based on very general initial assumptions and gives results regardless on geometry and specific parameters. The same approach is used for solving the same problem for conventional beam-plasma instabilities of all types (Cherenkov type, cyclotron type etc) [13], for the Buneman instability [22] etc. Obtained expressions for the spatial–temporal distribution of growing fields clearly show that with increase in the level of dissipation in background plasma, all these SI transform into DSI of conventional type. In addition, the analysis of obtained expressions gives much more detailed information on SI than other methods give. Part of the information on SI is not available in any other ways. The coincidence of other information to the results of conventional analysis confirms the validity of the approach (initial assumptions, mathematics etc).

Also, the presented approach shows that the DR describing the SI of given type can serve not only for solving of the initial/boundary problems and obtaining the dispersion curves. This point of view is very simplified. The approach shows that much more additional information is available from the DR. It, in fact, provides results on the initial perturbation development.

Summarizing, one can state that the presented approach can serve as an independent and very effective method for studying of any SI. There is no need to solve the problem again. One should only substitute the parameters of given instability in general expression for the field's space–time distribution. The usage of this approach instead of traditional initial/boundary problems gives complete picture of the instability development. At first glance, it might seem that this method of analyzing instabilities is more complicated. However, this complexity is only apparent. In addition, this complexity, if any, is overlapped by the completeness of the information received.

Author details

Eduard V. Rostomyan
Institute of Radiophysics and Electronics Armenian National Academy of Sciences, Ashtarak, Armenia

*Address all correspondence to: eduard_rostomyan@mail.ru

References

[1] Aleksandrov A.F., Bogdankevich L. S., Rukhadze A.A. Principles of Plasma Electrodynamics Springer-Verlag: Berlin, Germany, 1984

[2] Kuzelev M.V., Rukhadze A.A. Plasma Phys. Rep. 2000, v. 26, p 231

[3] Kuzelev M.V., Loza O.T., Rukhadze A.A.; Strelkov, P.S., Shkvarunets A.G. Plasma Phys. Rep. 2001, v. 27, p. 669.

[4] Buneman O. Phys. Rev. 1959, 115, 503.

[5] Matsiborko, N.G.; Onischenko, I.N.; Shapiro, V.D.; Shevchenko, V.I. Plasma Phys. 1972, 14, 591.

[6] Briggs, R.J. Advances in Plasma Phys; Symon, A.; Tomson, W., Eds.; Wiley: New York, NY, USA, 1971; v. 3, p. 132.

[7] Nezlin, M.V. Physics of Intense Beams in Plasmas; IOP Publishing Bristol, 1999.

[8] Kurilko V.I. Doklady Sov Akad, 1974, v208, p. 1059.

[9] Rostomyan E.V. IEEE Trans. Plasma Sci. 2003, 31, 1278.

[10] Rostomyan E.V. EPL 2007, 77, 45001.

[11] Rostomyan E.V. J. Plasma Phys. 2012 **2012, 78, 531.**

[12] Bers A. In "Basics of Plasma Physics" Eds Rosenbluth, M.N., Sagdeev, R.Z. North Holland: Amsterdam, The Netherlands, 1983; v. 1, p. 451

[13] Rostomyan E.V. Phys Plasmas, 2000, v. 7,p 1595.

[14] Kuzelev, M.V., Rukhadze, A.A. Plasma Free Electron Lasers; Frontieres: Paris, France, 1995.

[15] Aizatski N.I. Sov. J. Plasma Phys. 1980, 6, 597.

[16] Weiland J, Wilhelmsson H. Coherent Nonlinear Interaction of Wave in Plasma. Pergamon Press. Oxford. 1977.

[17] Kondratenko A.N., Kuklin V.M. Basics of Plasma Electronics. Atomizdat. Moscow, 1988.

[18] Kuzelev M.V. Plasma Phys Reports, 2006, v. 32, p. 572.

[19] F. W. J. Olver, Asymptotics and Special Functions (Academic, New York, 1974;

[20] Sturrock P.A. Phys. Rev 1958 v. 112 p. 1488

[21] Rostomyan E.V. Phys Lett A. 2009. v. 373, p. 2581

[22] Rostomyan E.V. Phys Plasmas, 2017, v. 24, 102102.

Section 4

Hall Thrusters and Tokamak

Chapter 10

Numerical Investigations of Electromagnetic Oscillations and Turbulences in Hall Thrusters Using Two Fluid Approach

Sukhmander Singh, Bhavna Vidhani and Ashish Tyagi

Abstract

The first part of the contributed chapter discuss the overview of electric propulsion technology and its requirement in different space missions. The technical terms specific impulse and thrust are explained with their relation to exhaust velocity. The shortcoming of the Hall thrusters and its erosion problems of the channel walls are also conveyed. The second part of the chapter discuss the various waves and electromagnetic instabilities propagating in a Hall thruster magnetized plasma. The dispersion relation for the azimuthal growing waves is derived analytically with the help of magnetohydrodynamics theory. It is depicted that the growth rate of the instability increases with magnetic field, electron drift velocity and collisional frequency, whereas it is decreases with the initial drift of the ions.

Keywords: electric propulsion, Hall thruster, dispersion, impulse, exhaust velocity, growth rate

1. Introduction

In the past few years, electric propulsion has received widespread attention as an alternative to chemical propulsion for spacecraft. A chemical thruster is capable of producing high thrust, but it provides a relatively low specific impulse. This limitation of chemical thrusters opens up discussion of the use of electric thrusters, which provide much higher specific impulses. A high specific impulse allows the spacecraft to reach the same speed with lower propellant consumption than chemical ones or travel faster with the same propellant mass [1–10]. In addition, the use of a high specific impulse, electric thruster permits a significant reduction in propellant mass on spacecraft, which helps reduce launching cost, extending mission lifetime or increase the payload mass on the spacecraft [2]. Therefore, the benefits that electric thrusters provide make them the best choice for in-space propulsion of spacecraft.

To date, several different types of electric propulsion thrusters have been developed, of which Hall thrusters are the most reliable and widely used. Hall thrusters are simple in design, consisting of an annular discharge channel, an anode, a cathode and a radial magnetic field across the channel [1]. A propellant, usually Xenon, is injected to channel through the hollow anode and a high potential difference is applied between the anode and cathode. The electrons emitted from the cathode and start moving towards the anode due to the potential difference. As they enter the discharge channel, they get trapped in the radial magnetic field generated by electromagnetic coils (or permanent magnets) and start drifting in the azimuthal direction (E x B). Due to this trapping, the residence time of electrons in the discharge channel increases and they move very slowly towards the anode. Electrons, then collide with neutral propellant atoms entering the discharge channel and ionize them. Then, these generated ions are accelerated to high velocity towards the thruster exhaust by the electric field to generate thrust. On the thruster exterior, the ion beam is neutralized by electrons from the virtual cathode so that no charge builds up on the spacecraft's surface [2]. More details on Hall thruster operation and fundamentals can be found in [1–13].

Hall thrusters offer several benefits like simple design, high thrust-to-power ratio, high efficiency, improved performance, etc., which give them a clear advantage over other electric thrusters. For an input power range of 0.1 kW–20 kW, they can produce a few mN to 1 N of thrust and offers a specific impulse in the range of 1000s–3000s with more than 50% efficiency [14]. Hall thrusters can also adjust their thrust level and specific impulse by varying the discharge voltage and propellant mass flow rate, which makes them suitable for applications such as precision maneuvering, attitude control, station keeping and orbital raising [15]. In addition, plasma in a Hall thruster remains quasi-neutral, which eliminates the issue of space charge, allowing the Hall thruster to achieve higher thrust densities. Because of all these characteristics, the space community has shown great interest in Hall thrusters and they have been used successfully on many spacecraft for space missions and maneuvers.

Thrust is caused by a change in a substance's momentum as a result of a chemical reaction or an electrical principle. The thrust indicates how much force, in newtons (N), the propulsion system exert on the vehicle. Let us denote $\dot{m}_p$ as the mass flow rate, the exhaust velocity $\vec{U}_{ex}$ and g is the acceleration due to gravity, then spacecraft's thrust denoted by

$$T = \dot{m}_p \vec{U}_{ex} \tag{1}$$

The performance of thrusters is usually determined by thrust T, which is the total force undergone by the rocket. Thrust also has same unit as a force in newton, which shows the movement of the propulsion system. Thrust, is generated by the burning of fuel or by the electrostatic forces. The thrust $T = \dot{m}_p \vec{U}_{ex}$, if the mass flow rate is constant. The specific impulse I_{sp} is used to compare the efficiencies of different type of propulsion systems [2]. The specific impulse is expressed as $I_{sp} = \frac{T}{\dot{m}_p g}$. In general, the higher the specific impulse the less fuel is required. Therefore the specific impulse simplifies to $I_{sp} = \frac{\vec{U}_{ex}}{g}$. The specific impulse has the dimension of time and is a measure for the effective mission time of the thruster. The high value of the specific impulse reduces the mission time. If we denote the thrust efficiency η and the input power P_t, then these are related by

$$T = \frac{2\eta P_t}{I_{sp} g} \tag{2}$$

Tsiolkovsky rocket Eq. (1) can be read as

$$\Delta\vec{v} = \vec{v}_f - \vec{v}_i = \vec{U}_{ex} \ln\left(\frac{m_f}{m_f + m_p}\right) \tag{3}$$

The above Rocket equation defines the change in velocity of a spacecraft. It is clear that a higher $d\vec{v}$ demands more propellant. In terms of specific impulse, the above equation simplifies to

$$\Delta\vec{v} = \vec{v}_f - \vec{v}_i = I_{sp} g \ln\left(\frac{m_f}{m_f + m_p}\right) \tag{4}$$

We can now say that I_{sp} plays a key role in the design of a space mission propulsion system.

2. Description of the Hall thruster

This description of the Hall thruster is shown in **Figure 1**. The anode and cathode set up an axial electric field and magnetic coils create a radial magnetic field. First cathode starts producing a stream of electrons and these electrons are attracted by positive anode. When electrons are moving to anode, then at the channel exit they face of perpendicular electric and magnetic fields, these fields are more strong at channel exit. Due to these perpendicular fields, electrons trapped in an ExB drift and formed Hall current [1]. At this time Xenon gas is released from the anode, when neutral atom of gas reached at the channel exit, then electron collides with neutral atoms and ionized them. These ions are accelerated out of channel by electric field at the channel exit and this motion of ions imparts a reaction force on the thruster in reverse direction. The cathode is also used to neutral the ions charge so it produce a stream of electron to neutral the ions [1–10].

2.1 Stationary plasma thruster (SPT)

The wall material of this type of thruster is dielectric like borosil (BN-SiO2), boron nitride (BN) and alumina. This type of material has low secondary electron emission coefficient with Xenon ion interaction. Most of SPT use Xenon as a propellant because of its higher mass (131.3amu), lower first ionization potential, less toxicity, ionization cross section of ($\sim$2.310^{-6}cm^2) and of its desirable thermodynamics properties [1, 10].

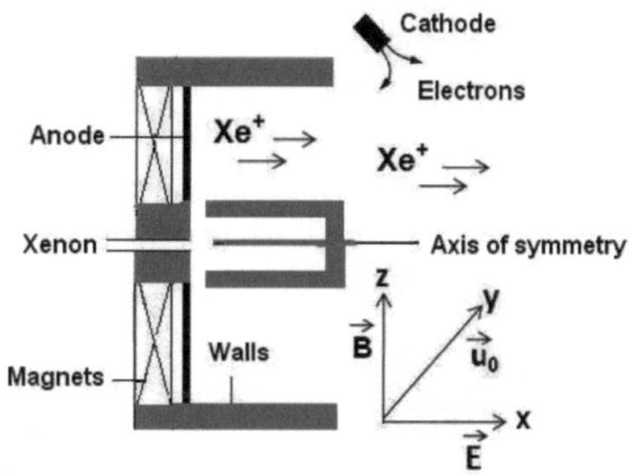

Figure 1.
Schematic diagram of a typical Hall plasma thruster.

2.2 Thruster with anode layer (TAL)

These types of thruster has metallic conducting walls and has narrow acceleration zone associated with the narrow electric field region near the anode. The ratio of channel width and channel depth in these thruster is 2:1. The conducting channel wall is negatively biased and it is a part of magnetic circuit so it prevent electron to move in direction of wall and repel them to ionization region and reduce electron power losses [1–10].

3. Components of a Hall thruster

The main components of a Hall thruster are body, magnetic coils, discharge channel, anode and cathode. The body of the Hall thruster provides better shape of external magnetic field, so it is normally made up of iron (steel). In this thruster a single thick disk shape annular discharge channel exist on the front face. To create better shape of magnetic field two type of magnetic coils are situated in the thruster channel. The inner coil is made of single reel and located radially inward in the channel and the outer coil is made of multiple reels and located radially outward in the channel. The magnetic coils and iron body made a magnetic circuit. The magnetic iron body of the thruster is separated from thruster discharge channel by walls and the material of these walls is metal or ceramic. The material of the anode is normally stainless steel, which avoids magnetic interaction and supply hardiness [1–10]. To set up an axial electric field in discharge channel of thruster a cathode is required and this cathode is normally hollow cathode with lanthanum hexaboride or barium oxide. The cathode completes the discharge circuit and work like an electron source. This description of the Hall thruster provides knowledge about the structure of thruster.

4. Hall current and thrust

As discussed earlier, there is an azimuthal electron drift in a Hall thruster channel, which is normal to both the applied electric field and generates a Hall current [1, 10]. Because of the collisions of the electrons with neutrals, electrons, ions and potential fluctuations in the plasma, generate the axial electron current density. The azimuthal E X B drift is given by

$$v_d = \frac{\vec{E} \times \vec{B}}{B^2} \approx \frac{E_r}{B_z} \tag{5}$$

The Hall current may be approximated by integrating the magnitude of E_r/B_z along the acceleration region and multiplying the result by the electron charge density and width. In mathematically, it can be read as

$$I_H \approx e n_e \int_0^L \frac{E_r}{B_z} dx \approx \frac{e w n_e V_d}{B} \tag{6}$$

Here, w is the channel width and V_d is the voltage at cathode.

5. Channel length and scaling laws for Larmor radius

The ratio of electric and magnetic field is such that the ion's gyro radius is much larger than electron's gyro radius, so that ions are not magnetized inside the channel. The channel length is calculated by determining the Larmor radius of both the plasma species [10]. The motion of the moving charged particle under the electromagnetic fields defined as follows

$$\frac{mv_{\perp}^{2}}{r} = q(v_{\perp} \times B) \tag{7}$$

This gives the radius of gyration

$$r = \frac{mv_{\perp}}{qB} = \frac{v_{\perp}}{\Omega_c} \tag{8}$$

Here, $\Omega_c = \frac{qB}{m}$, is called the cyclotron frequency. If ions are accelerated through electrical potential $V_{\perp}$ (perpendicular to the magnetic field), then energy balance leads to $\frac{mv_{\perp}^{2}}{2} = qV_{\perp}$.

This implies to

$$v_{\perp} = \sqrt{\frac{2qV_{\perp}}{m}} \tag{9}$$

These above equations gives the larmor radius in terms of the applied potential

$$r = \frac{1}{B}\sqrt{\frac{2mV_{\perp}}{q}} \tag{10}$$

The magnitudes of the fields are such that $r_i >> L >> r_e$, where L ($\sim$6 cm) is the length of the acceleration channel. For a typical Hall thruster, Larmor radius for electrons ($\sim$0.13 cm) and ions ($\sim$180 cm) corresponds to the radial magnetic field strength of order 150 G and energy 300 eV.

Despite many successful applications of Hall thrusters, some aspects of their operation are still poorly understood. One notable problem is the anomalous electron mobility [16, 17] and plasma sheath, which is far above the classical collisional values. It has been established that the inhomogeneous plasma under the electromagnetic fields is not in the thermodynamically equilibrium state [17–20]. The equilibrium $E \times B$ electron drift, ion flow rate, plasma and magnetic field gradients are all sources of plasma instabilities in Hall plasmas. Hall plasmas devices with ExB electron drift demonstrate wide range of turbulent fluctuations. These fluctuations are probably the reason of the observed anomaly in the electron transport across the magnetic field [21–23] and other nonlinear phenomena (coherent rotating spoke) [24–26]. Understanding of the mechanisms of the coherent structures and anomalous transport requires the detailed study of linear instabilities in Hall plasma devices. These instabilities are also considered to be a principal source of anomalous transport in toroidal magnetic confinement devices [2].

6. Oscillations and instabilities in Hall thrusters

If free energy is available in the system and even though system is in equilibrium in the sense that all the forces are in balance, then these oscillations can grow at the cost of free energy and hence instabilities can take place. There are different types of instabilities that depend on different conditions. For proper description of a particular instability, one should be able to define the mode of the growing wave, the nature of the growth and the source of the free energy. Instabilities are mainly classified into four groups, namely streaming instabilities, Rayleigh–Taylor instabilities, universal instabilities and kinetic instabilities.

When there is any kind of perturbation, the free energy available excites the waves and the plasma waves no longer remains in thermal equilibrium. Even though there exist an equilibrium because all the forces are balanced and there is no net force and it is possible to find a time independent solution to the wave. Perturbation makes the plasma waves unstable, which is always a motion that brings the plasma closer to true thermodynamic equilibrium by decreasing the free energy. Instabilities may be classified according to the type of free energy available to drive them. Few of them has been explained below.

6.1 Streaming instability

This type of instability occurs, when either a current or a beam of energetic particles is driven through plasma so that the different species of particles have drifts relative to one another. This drift energy attempt to excite waves and oscillation. This energy is acquired at the loss of the drift energy, which is in the unperturbed state.

6.2 Rayleigh–Taylor instabilities

In this kind of instability, the plasma has a sharp boundary or a density gradient, so that it does not remain uniform. In addition to this, an external, non-electromagnetic force is applied into the plasma. This force is responsible for driving the instability in plasma. This analogy can be realized in an inverted glass of water. Though the plane interface between the air and the water is in the state of equilibrium, where the air pressure support the weight of the water. Any ripple arising in the surface of water tend to grow at the loss of potential energy in the influence of gravitational field. Whenever a light fluid supports a heavy fluid this things happen, which is quite known in the field of hydrodynamics [1].

6.3 Universal instabilities

A plasma is hardly present in perfect thermodynamic equilibrium, when it is confined by gravitational field or an electric field (driving forces). The plasma is expanded by plasma pressure and an instability is driven by the expansion energy. Any finite plasma always contains this type of free energy and the waves, which comes out as a result are called universal instabilities.

6.4 Kinetic instabilities

Generally, the velocity distributions are assumed to be Maxwell-Boltzmann in fluid theory. If there is a case, where distributions are not Maxwell-Boltzmann, then there may be a departure from thermodynamic equilibrium and this anisotropy of the velocity distribution drive the kinetic instabilities.

6.5 Streaming instabilities

These kind of instabilities occur, when the plasma particles streams with their relative drift, which may be driven due to the travel of charge particle beam or internal driven current. For example, the two stream instability is due to the relative drift of ion and electron streams.

6.6 Electromagnetic resistive instability

These instabilities have been investigated in various positions of the acceleration channel of the thruster. The effect of radial magnetic field, thermal motions of plasma particles and electron collisions can induce time varying fields in the discharge oscillations. This can prompt an electromagnetic resistive instability in the Hall thruster. In this chapter, we derive the dispersion relation for the electromagnetic instabilities in a Hall thruster and conditions for the instability.

6.7 Oscillations and instabilities in Hall thrusters

There are numerous types of oscillation found in thrusters which propagates in azimuthally and axially direction ranging from a few kHz to tens of MHz [1, 10]. These oscillations may reduce the specific impulse and the efficiency of a Hall thruster. These disturbances can also limit the operating life of a Hall thruster and therefore, suppression of this oscillation has become an essential task for Hall thrusters. These fluctuations are responsible for electron transport across magnetic field lines, performance and ionization of propellant [27–30]. The magnitude of these oscillations strongly depend on the magnetic field, location of the cathodes, the discharge voltage and the mass flow rate.

Low frequency oscillation, frequently referred to as the breathing mode, is intimately tied to the details of propellant ionization and eventual ion acceleration and manifests itself as a strong 10–30 kHz oscillation in the thruster discharge current. Breathing mode models suggest the presence of a propagating ionization front traversing the channel of the thrusters [31]. Benítez and Ahedo investigated the axial-azimuthal instabilities in the global discharge region of a Hall-thruster to identify dominant mode (develops in the near plume at 1–5 MHz) and subdominant mode (develops near the anode100–300 kHz) [32]. Lafleur *et al.* studied transport effects of the electron drift instability in Hall-effect Thrusters [33]. Fan *et al.* studied the effects of the peak magnetic field position on Hall thruster discharge characteristics [34]. Sekine *et al.* investigated the spatially and temporally resolved ion flow measurements inside the plasma source of an inductive radio frequency plasma thruster [35]. Puerta *et al.* generalized non-ideal treatment and growth rates analysis of drift waves instabilities in a collisions-free magnetized dusty plasma [36]. The electrostatic dispersion relation for an unbounded homogeneous plasma in the presence of unmagnetized ions, magnetized electrons and an applied magnetic field has been solved numerically by Mikellides and Ortega in the near-plume of a magnetically shielded Hall thruster [37]. Tomilin and Ivan Khmelevskoi studied the influence of kinetic effects on dispersion properties of high-frequency perturbations in Hall thruster plasmas [38]. Marcovati *et al.* reported the dynamic behavior of gradient-driven drift waves in a strongly magnetron discharge in multi ion plasma [39]. Litvak and Fisch [40] investigated resistive instabilities in a Hall plasma and found that plasma perturbations in the acceleration channel are unstable in the presence of collisions. Fernandez *et al.* [41] did simulations for resistive instabilities. Marusov et al. [42] modeled the stability of gradient-drift waves in a Hall-type plasma using two-fluid ideal magnetohydrodynamics. Ducrocq et al. [43] studied high-frequency electron drift instability, where they derived three-dimensional dispersion relation for a model of a crossed electric and magnetic field configuration existing in the Hall thruster. Litvak and Fisch [44] have analyzed gradient driven Rayleigh type instabilities in a Hall thruster using two fluid hydrodynamic equations. Kapulkin and Guelman [45] investigated low frequency instability in near anode region of a Hall thruster, where they obtained that the instability can be responsible for the enhanced transfer of the electrons between the ionization region and the anode. Choueiri [46] has quantitatively discussed the nature of oscillations in the 1 kHz–60 MHz frequency range observed during operation of Hall thrusters. Various plasma parameters measured inside the accelerating channel of a typical Hall thruster were used to evaluate the various stability criteria and dispersion relations of oscillations [46].

At small amplitudes, the oscillations can be considered a part of normal operation with no significant effect on operation. At large amplitudes, the oscillations can severely and adversely affect operation. The details of these oscillation are given in **Table 1**.

Lakhin *et al.* have developed the effects of finite electron temperature on gradient drift instabilities in partially magnetized plasmas in the frequency range $\omega_{Bi} \ll \omega \ll \omega_{Be}$ driven by the equilibrium current perpendicular to the magnetic field [56]. Romadanov *et al.* [57] studied the structure of nonlocal gradient-drift instabilities in Hall E × B discharge plasma. Koshkarov *et al.* calculated the linear and nonlinear nonlocal instability of axial lower-hybrid modes in plasma under the influence of ion flow rate [58]. Smolyakov *et al.* have used fluid theory and performed the simulations of instabilities, turbulent transport and coherent structures in magnetized plasmas [59]. Singh and Malik [60, 61] investigated that temperature of the ion and drift velocity profiles of the electron modifies the conditions

Type	Description	Ranges of frequencies
Spoke type	Spoke type oscillations are propagating in the azimuthal direction at low discharge voltages [47, 48].	15–35 kHz
Breathing mode	Discharge-current low-frequency oscillations [49]	Range of 10–100 kHz
Contour oscillations	These are longitudinal oscillations connected with an instability in the location of the ionization region and can have very large amplitude [50–53]. Contour oscillations depend on the parameters of the discharge power supply circuit.	1–30 kHz (correspond to the transit time for a neutral propellant atom)
Ionization oscillations	These oscillations are caused from the ionization front propagating irregularly around the circumference of the discharge channel [47, 52].	10–100 kHz.
Flight oscillations	These oscillations are also called ionization flight oscillations, because they are determined by the change of ionization rate due to the delay of particle appearance, from one region to another. The analysis is based on the concept that plasma particles are delayed in being transferred from one region to other [54].	100 kHz up to 10 MHz (correspond with the transit time for an ion)
High frequency oscillations	The amplitude of these oscillations is smaller and these are azimuthal waves generated near the exit part of the thruster [55].	1–100 MHz
Super high frequency oscillations	These oscillations are connected with the development of electron layers in the plasma and the formation of flows of fast overheated electrons and have the smallest amplitude [55].	Few GHz

Table 1.
Types and frequency range of oscillations in Hall thrusters.

for Rayleigh type instability under the effects of thermal motions of ions and plasma resistivity induces resistive instabilities (electrostatic and electromagnetic) [62–65] associated with azimuthal and axial directions. Yadav et al. presented the model for relativistic electron-beam assisted growth of oscillating two-stream instability [66]. Aria and Malik [67] have investigated the propagating modes, instabilities and plasma sheath formed on the outer surface of a spacecraft. The detailed physical picture of the processes in the thruster is very complex. It includes a whole series of phenomena, such as stabilization of the flow in cross-fields. Becatti et al. investigated the properties of plasma oscillations in the exterior region of a high-current hollow cathode [68].

7. Magnetic field profiles and erosion problems

Magnetic field is an important parameter, which affect the performance and operation of Hall thruster. The erosion on the channel walls and on the electrodes can be minimized by accelerated beam of ions to a narrower beam width and by use of appropriate profile of magnetic field. The radial magnetic field causes the erosions of the walls, so new topology of magnetic field 'lens' is proposed by Morozov [69]. By using magnetic shielding technique, the discharge chamber erosion rate can be reduced by orders of magnitude [70]. Morozov and Lebedev proposed a magnetic field lens to focus the plasma beam [71]. Morozov et al. have designed a lens-type magnetic field with a zero magnetic point to make the plume divergence half angle only 10° in SPT-ATON Hall thruster [72]. Hofer et al. showed that the performance of the Hall thruster can be enhanced by reducing the divergence of the plume that could damage the solar panels and other parts of the satellite [73]. Hofer et al. used magnetic field to control the unmagnetized ion beam that formed in Hall thruster. By controlling ion beam the wall erosion was reduced to 2–3 order by ion bombardment [74]. Huang et al. studied the effect of background pressure on performance and plume of NASA's High Voltage Hall Accelerator Hall thruster [75]. Their result shows that discharge current and thrust are increases with pressure and ion beam current, ion energy per charge, plum divergence decreases with increasing background pressure.

Liu et al. studied the effect of azimuthally electron drift on sheath profile and anomalous erosion in thrusters. Their results show that azimuthally electron drift induce sheath oscillations and it can produce an asymmetric sheath structure. To simulate the azimuthally erosion evolution, an erosion modal is used and it has been concluded that the azimuthally asymmetric ion sputtering is responsible for an asymmetric erosion profile [76]. Kim studied the ionization processes and ion dynamics in the accelerating channel to determine the stationary plasma thrusters performance levels [54]. Choueiri theoretically studied the difference between stationary plasma thruster and thruster with anode layer by measuring the temperature of secondary electron emission from the walls of thruster [77]. Hong et al. studied the effect of wall grooves on Hall thruster discharge to improve the performance of Hall thruster. If wall grooves are present in ionization region it increases near wall conductivity and decrease electron transient time, thrust and efficiency [78]. Ding et al. studied the discharge current and stability in graphite walls and BN-SiO_2 walls for magnetic shielded and unshielded thruster. Their result shows that magnetic shielded thruster does not show change in stability and discharge current but 25% discharge current increases in unshielded thruster [79]. Experimental results and PIC simulation show that oblique channel's specific impulse, anode efficiency, thrust, propellant utilization and ionization in plume region is improved 20% rather than the straight channel thruster [80]. Olano et al. studied the effect of magnetic field configuration on thruster performance and plasma discharge by proposing three types of configurations: (A) nominal, (B) orthogonal magnetic field and (C) high magnetic field [80, 81]. Garrigues et al. did PIC simulations and shows that in a Hall thruster electric discharge produce due to wall-plasma interaction and wall of thruster channel is reduced due to interaction with ions [82]. The electrostatic and magnetic probes are used to investigate the electromagnetic fluctuations and coherent magnetohydrodynamic azimuthal modes in thruster [68].

8. Introduction to electromagnetic instabilities and their dispersion relations

The resistive instability is driven by coupling to a dissipative process. Due to the collisions between the electrons and neutral particles, resistive instability can occur in the acceleration channel of the Hall thruster. This instability occurs due to the interaction of the wave with the electrons' flow in the presence of electron collisions. The two major problems in plasma confinement are equilibrium and stability of plasma. The system in an equilibrium state if all the forces, which act on a system are balanced but stability and instability of the equilibrium can be assured by giving some perturbations to the equilibrium state. The stable and unstable state of equilibrium depend on small perturbation whether the perturbations are damped or amplified. For small perturbations, equilibrium is non- linear but stability can be linearized.

9. Plasma model and basic equations for purely azimuthal waves

For the case of small amplitude perturbations and wavelengths much smaller than the scale lengths of inhomogeneities, the analysis of linearization can be applied. In the simplest approach, we consider the x- axis is taken along the axis of the thruster and the z-axis is taken along the radius of the thruster. The direction the magnetic field $\vec{B}$ is along the z- axis. The y- axis is taken correspond to the azimuthal direction. Consistent to the fluid approach, we write below the basic fluid equations

$$\frac{\partial n_j}{\partial t} + \vec{\nabla} \cdot \left(\vec{v}_j n_j \right) = 0 \tag{11}$$

where n_j is the mass density and $\vec{v}_j$ is the velocity of species j (j = i, e). The momentum equations for electrons and ions are

$$\frac{\partial \vec{v}_e}{\partial t} + \left(\vec{v}_e \cdot \vec{\nabla} \right) \vec{v}_e + \nu \vec{v}_e = -\frac{e\vec{E}}{m} - \frac{e}{m} \left(\vec{v}_e \times \vec{B} \right) \tag{12}$$

$$\frac{\partial \vec{v}_i}{\partial t} + \left(\vec{v}_i \cdot \vec{\nabla} \right) \vec{v}_i = -\frac{e\vec{E}}{M} \tag{13}$$

9.1 Linearization of fluid equations

To linearize all the equations, let us write $n_i = n_0 + n_{i1}$, $\vec{v}_i = \vec{v}_{i1} + \vec{v}_0$, $\vec{B} = \vec{B}_1 + \vec{B}_0$ and $\vec{E} = \vec{E}_1 + \vec{E}_0$. The unperturbed density is taken as n_0, electric field (magnetic field) as $\vec{E}_0(\vec{B}_0)$ and the perturbed value of the electric field (magnetic field) is taken as $\vec{E}_1(\vec{B}_1)$. Here, we consider the perturbed densities for ions and electrons as n_{i1} and n_{e1} velocities as $\vec{v}_{i1}$ and $\vec{v}_{e1}$ indicated by subscript 1 along with their unperturbed values as v_0 and u_0 in the x- and y-directions respectively. In view of small variations of both the density and magnetic field along the channel, the plasma inhomogeneities are neglected. The oscillations of the perturbed on and electron densities are taken small enough $(n_{i1}, n_{e1} < < n_0)$ so that the collisional effect due to the velocity perturbations dominate over the one due to the density perturbation. Since $\vec{v}_0$ and u_0 are constant, the terms $\left(\vec{v}_0 \cdot \vec{\nabla}\right)n_0$, $n_0\left(\vec{\nabla} \cdot \vec{v}_0\right)$ and $n_1\left(\vec{\nabla} \cdot \vec{v}_0\right)$ are equal to be zero. Further the terms $\left(\vec{v}_1 \cdot \vec{\nabla}\right)n_1$, and $n_1\left(\vec{\nabla} \cdot \vec{v}_1\right)$ are neglected as they are quadratic in perturbation.

The linearizations of the above eqation leads to

$$\left(\frac{\partial}{\partial t} + u_0\frac{\partial}{\partial y} - v\right)\vec{v}_{e1} + \frac{e}{m}\left(\vec{E}_1 + \vec{v}_{e1} \times \vec{B} + \vec{u}_0 \times \vec{B}_1\right) = 0 \tag{14}$$

$$\left(\frac{\partial}{\partial t} + v_0\frac{\partial}{\partial x}\right)\vec{v}_{i1} - \frac{e\vec{E}_1}{M} = 0 \tag{15}$$

$$\frac{\partial n_{e1}}{\partial t} + u_0\frac{\partial n_{e1}}{\partial y} + n_0\left(\vec{\nabla} \cdot \vec{v}_{e1}\right) = 0 \tag{16}$$

$$\frac{\partial n_{i1}}{\partial t} + v_0\frac{\partial n_{i1}}{\partial x} + n_0\left(\vec{\nabla} \cdot \vec{v}_{i1}\right) = 0 \tag{17}$$

9.2 Dispersion equation and growth rate of azimuthal waves

We take the variation of oscillating quantities in azimuthal direction as $\sim \exp(i\omega t - iky)$. The linearized Eqs. (14)–(17) are used and the density and velocity perturbations are expressed in terms of the electric field as

$$v_{ex1} = \frac{eE_y}{m\Omega} + -i\frac{eE_x\hat{\omega}}{m\Omega^2} \tag{18}$$

$$v_{ey1} = \frac{u_0 B_1}{B} - \frac{eE_x}{m\Omega} - \frac{ieE_y\hat{\omega}}{m\Omega^2} \tag{19}$$

From Eq. (15)

$$v_{ix1} = -\frac{ieE_x}{M\omega} \tag{20}$$

$$v_{iy1} = -\frac{ieE_y}{M\omega} \tag{21}$$

From Eqs. (13) and (14)

$$n_{e1} = -\frac{ekn_0(i\hat{\omega}E_y + \Omega E_x)}{m\Omega^2(\omega - ku_0)} \tag{22}$$

$$n_{i1} = -\frac{ien_0 kE_y}{M\omega^2} \tag{23}$$

The x-component of the perturbed current density

$$J_x = en_0(v_{ix1} - v_{ex1}) + en_{i1}v_0e \tag{24}$$

and the y-component of the perturbed current density

$$J_y = en_0(v_{iy1} - v_{ey1}) - en_{e1}u_0 \tag{25}$$

9.3 Conductivity tensor

From the Maxwell's equation, we have

$$\nabla \times \vec{B} = \mu_0\vec{J} + \mu_0\epsilon_0\frac{\partial\vec{E}}{\partial t} \tag{26}$$

or

$$\vec{J} = \frac{\nabla \times \vec{B}}{\mu_0} - \frac{\partial\vec{E}}{\partial t} \tag{27}$$

Faraday's law gives the relationship between changing electric and magnetic field as

$$\nabla \times \vec{E} = -\frac{\partial\vec{B}}{\partial t} \tag{28}$$

For a plane electromagnetic wave

$$\vec{E} = \overrightarrow{E_0}\, e^{j(\vec{k}.\vec{r}-\omega t)} \tag{29}$$

Where $\vec{E}$ is the electric field, $\vec{B}$ is the magnetic field, j represents the complex number, $\vec{k}$ is the wave vector, t is the time, $\overrightarrow{E_0}$ and $\overrightarrow{B_0}$ is complex magnitude of electric and magnetic field. Then ∇ operation gives ik and $\frac{\partial}{\partial t}$ gives $-i\omega$. Substituting these operators Faraday's law become

$$\vec{k} \times \vec{E} = \omega\vec{B}$$

or

$$\vec{B} = \frac{\vec{k} \times \vec{E}}{\omega} \tag{30}$$

9.4 Permittivity tensor

From Maxwell's equations, we get

$$\vec{D} = -\frac{\vec{k} \times \vec{k} \times \vec{E}}{\omega^2\mu_0} \tag{31}$$

which on expansion will give

$$\begin{bmatrix} D_x \\ D_y \\ D_z \end{bmatrix} = \frac{1}{\omega^2 \varepsilon_0 \mu_0} \begin{bmatrix} k^2 - k_x^2 & -k_x k_y & -k_x k_z \\ -k_y k_x & k^2 - k_y^2 & -k_y k_z \\ -k_z k_x & -k_z k_y & k^2 - k_z^2 \end{bmatrix} \begin{bmatrix} E_x \\ E_y \\ E_z \end{bmatrix} \tag{32}$$

Then permittivity tensor will be

$$\varepsilon_r = \frac{1}{\omega^2 \varepsilon_0 \mu_0} \begin{bmatrix} k^2 - k_x^2 & -k_x k_y & -k_x k_z \\ -k_y k_x & k^2 - k_y^2 & -k_y k_z \\ -k_z k_x & -k_z k_y & k^2 - k_z^2 \end{bmatrix}$$

In general

$$\varepsilon_{rij} = \frac{1}{\omega^2 \varepsilon_0 \mu_0} \left[k^2 \delta_{ij} - k_i k_j \right] \tag{33}$$

But for a pure dielectric medium $k = \frac{\omega}{c} \sqrt{\varepsilon_r}$ and we will get

$$\varepsilon_r = \begin{bmatrix} \varepsilon_{rr} - \varepsilon_{rx} & -\varepsilon_{rx}\varepsilon_{ry} & -\varepsilon_{rx}\varepsilon_{rz} \\ -\varepsilon_{ry}\varepsilon_{rx} & \varepsilon_{rr} - \varepsilon_{ry} & -\varepsilon_{ry}\varepsilon_{rz} \\ -\varepsilon_{\mathrm{rz}}\varepsilon_{rx} & -\varepsilon_{rz}\varepsilon_{ry} & \varepsilon_{rr} - \varepsilon_{rz} \end{bmatrix}$$

Then, the Maxwell's equations are used in view of the perturbed electric and magnetic fields of the electromagnetic wave, and the plasma dielectric tensor ε_{ij} is obtained as

$$\varepsilon_{ij} E_j = E_j \delta_{ij} + \frac{j_i(E_j)}{i\omega\varepsilon_0} \tag{34}$$

Finally, the wave equation $\left(k^2 \delta_{ij} - k_i k_j - \frac{\omega^2}{c^2} \varepsilon_{ij}\right) E_j = 0$ is written in the following form

$$k^2 \varepsilon_{yy} - \frac{\omega^2}{c^2} \left(\varepsilon_{xx}\varepsilon_{yy} - \varepsilon_{xy}\varepsilon_{yx} \right) = 0 \tag{35}$$

Where, the components of the dielectric tensor are obtained from Eq. (34) with the help of the Eqs. (24) and (25)

$$\varepsilon_{xx} = \frac{(\omega - k u_0){\omega_e}^2}{\omega \Omega^2} + 1 - \frac{{\omega_i}^2}{\omega^2} \tag{36}$$

$$\varepsilon_{xy} = \frac{i{\omega_e}^2}{\omega \Omega} - \frac{{\omega_i}^2 k v_0}{\omega^3} \tag{37}$$

$$\varepsilon_{yx} = \frac{{\omega_e}^2}{i\omega\Omega} + \frac{{\omega_e}^2 k u_0 \Omega}{i\Omega^2 \omega(\omega - k u_0)} \tag{38}$$

$$\varepsilon_{yy} = \frac{{\omega_e}^2 k u_0 \hat{\omega} k}{\Omega^2 \omega(\omega - k u_0)} - \frac{{\omega_i}^2}{\omega(\omega - k u_0)} + \frac{{\omega_e}^2 \hat{\omega}}{\omega \Omega^2} + 1 \tag{39}$$

By substituting these components into Eq. (35) we get the following cumbersome analytical expression of the dispersion relation of electromagnetic waves propagating in magnetized plasma [40].

$$\frac{k^2c^2}{\omega^2} = \frac{(\omega - ku_0)\omega_e^{\ 2}}{\omega\Omega^2} + 1 - \frac{\omega_i^{\ 2}}{\omega^2} + \frac{\left\{\frac{\omega_e^{\ 2}}{i\omega\Omega} + \frac{\omega_e^{\ 2}ku_0\Omega}{i\Omega^2\omega(\omega - ku_0)}\right\}\left\{\frac{\omega_e^{\ 2}}{i\omega\Omega} - \frac{\omega_i^{\ 2}kv_0}{\omega^3}\right\}}{\left\{\frac{\omega_e^{\ 2}ku_0\hat{\omega}k}{\Omega^2\omega(\omega - ku_0)} - \frac{\omega_i^{\ 2}}{\omega(\omega - k_y u_0)} + \frac{\omega_e^{\ 2}\hat{\omega}}{\omega\Omega^2} + 1\right\}} \tag{40}$$

9.5 Typical parameters of Hall thrusters

The values and ranges of some typical parameters are given in **Table 2**.

9.6 Numerical results and discussion

We solve the dispersion Eq. (40) to find out complex root by using typical values of the magnetic field, azimuthal wave number, collision frequency, electron drift velocity, ion drift velocity and electron temperature. Then the effect of these parameters on the growth γ of the electromagnetic wave is studied in **Figures 1–4**.

Figure 2 shows that the growth rate of the wave get enhanced for the larger values of the drift velocity of the electrons [64]. Since the drift velocity can be correlated with the discharge voltage, so it can be said that the growth rate is increased with the discharge voltage. Esipchuk and Tilinin [83] also reported the proportionality of the frequency of drift instability to the discharge voltage. The increase in the growth rate may be attributed to the strong coupling between the electric field and electron current [84].

Figure 3 shows the variation of growth rate under the effect of collisional frequency. It is seen that the wave grows at a faster rate in the presence of more electron collisions. Since the stronger resistive coupling of the oscillations to the electrons' closed drift requires the electron collisions, it is obvious that this instability grow faster in the presence of higher collision frequency. Similar results were also reported in the simulation studies of resistive instability by Fernandez *et al.* [41] that the growth rate is directly proportional to the square root of the collision frequency.

The dependence of growth rate on the magnetic field is shown in **Figure 4**, where it is observed that the wave grows faster in the presence of strong magnetic

Property	Typical value	Property	Typical value
Inner diameter	60 mm	Neutral velocity	300 m/s
Outer diameter	100 mm	Electron temperature	5-10 eV
Plasma density	$10^{17}/m^3$	Ion temperature	1-5 eV
Neutral density	$10^{18}/m^3$	Neutral temperature	0.9 eV
Ion velocity	10^4 m/s	Debye length	10^{-5} m
Collision mean free path	1 m		

Table 2.
Typical values of parameters used in Hall thruster.

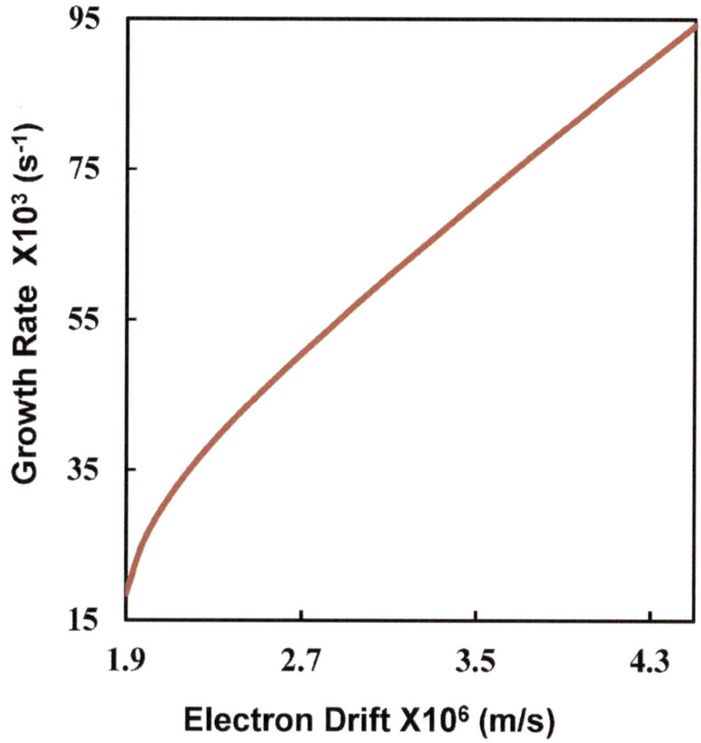

Figure 2.
The dependence of growth rate with electron drift velocity.

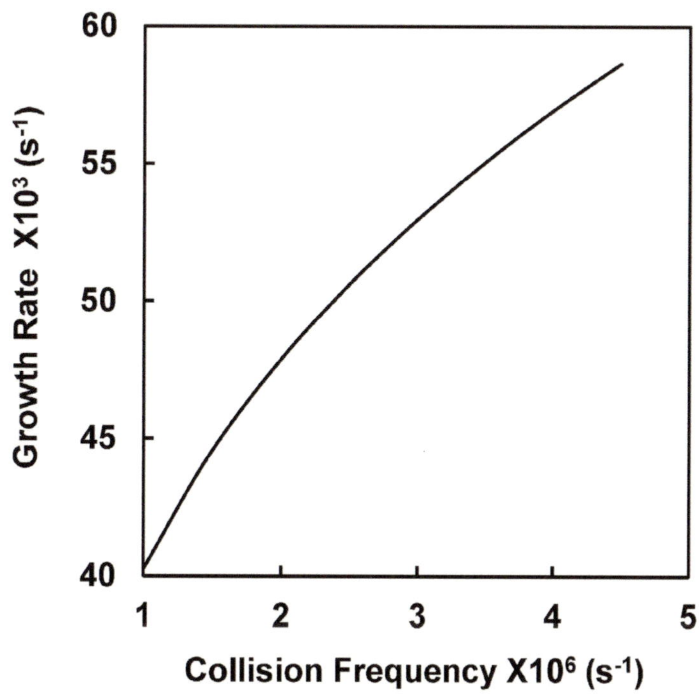

Figure 3.
Variation of growth rate with collision frequency.

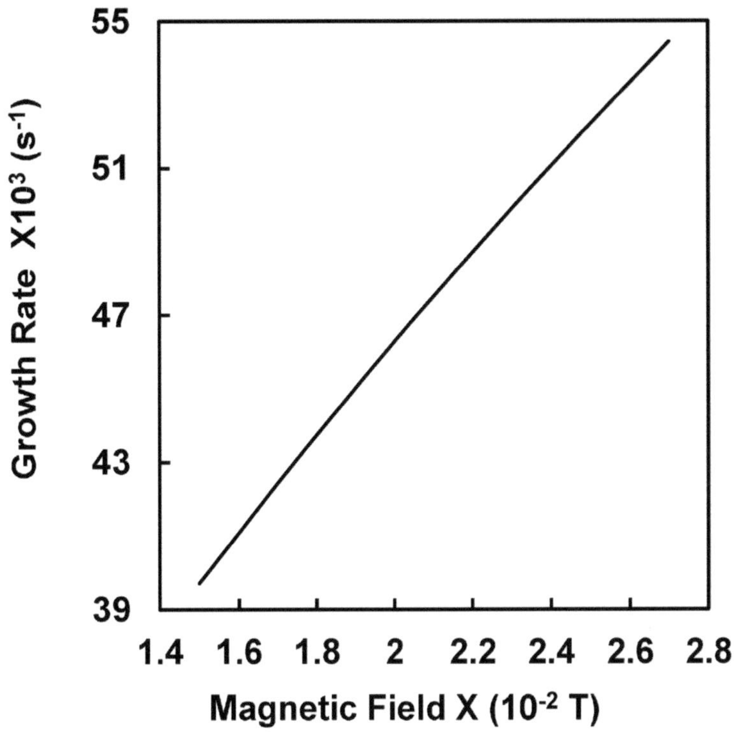

Figure 4.
Variation of growth rate with magnetic field.

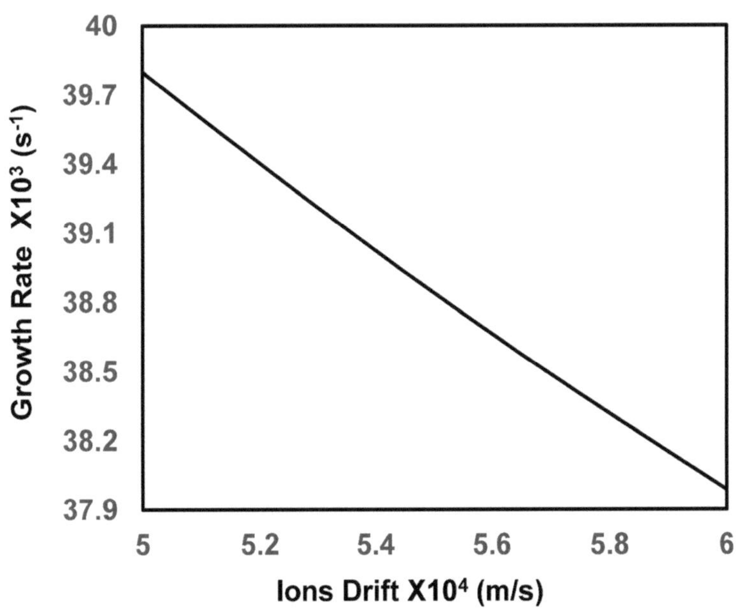

Figure 5.
Dependence of growth on ion drift velocity.

field. The growth rate of the wave gets suppressed under the larger drift of the ions as shown in the **Figure** 5. Since in the presence of their large drift, the ions try to diminish the transverse oscillations of the electrons in the x-direction.

10. Results and the discussion

The different aspects of the Hall thrusters have been reviewed in this chapter along with future challenges related to the performance and efficiency. The different types of turbulence and oscillation with frequency ranges are also summarized. Hall thruster Plasma support electrostatic and electromagnetic waves under the influence of resistive effects. Therefore the waves (electrostatic and electromagnetic) propagating in azimuthal direction in a Hall thruster channel are found to be unstable due to the resistive coupling to the electrons' closed drift in the presence of collisions. The theoretical model for the propagation of electromagnetic instability is derived in a Hall thrusters. The dispersion relation for the electromagnetic instability is derived analytically. The dispersion relation is solved numerically with the help of MATLAB to study the growing and propagating modes. The dominated modes are plotted to observe the behaviors of the instability.

Acknowledgements

The University Grants Commission (UGC), New Delhi, India is thankfully acknowledged for providing the startup Grant (No. F. 30-356/2017/BSR).

Author details

Sukhmander Singh[1]*, Bhavna Vidhani[2] and Ashish Tyagi[3]

1 Plasma Waves and Electric Propulsion Laboratory, Department of Physics, Central University of Rajasthan, Kishangarh, India

2 Department of Physics and Electronics, Hansraj College, University of Delhi, Delhi, India

3 Department of Physics, Swami Shraddhanand College, University of Delhi, Delhi, India

*Address all correspondence to: sukhmandersingh@curaj.ac.in

References

[1] Goebel DM, Katz I. Fundamentals of Electric Propulsion: Ion and Hall Thrusters. New York: Wiley; 2008

[2] Charles C. Plasmas for spacecraft propulsion. J. Phys. D Appl. Phys. 2009; **42**:163001

[3] Mozouffre, S. Electric propulsion for satellites and spacecraft: Established technologies and novel approaches. Plasma Sources Sci.Technol. 25, 033002 (2016).

[4] Levchenko I et al. Recent progress and perspectives of space electric propulsion systems based on smart nanomaterials. Nat. Commun. 2018;**9**:879

[5] Takahashi K. Helicon-type radiofrequency plasma thrusters and magnetic plasma nozzles. Rev. Mod. Plasma Phys. 2019;**3**:3

[6] Rovey JL, Lyne CT, Mundahl AJ, Rasmont N, Glascock MS, Wainwright MJ, Berg SP. Review of multimode space propulsion. Progress in Aerospace Sciences. 2020;118:100627(1-27).

[7] Brown NP, Walker ML. Review of plasma-induced Hall thruster erosion. Applied Sciences. 2020 Jan;10(11):3775 (1-18).

[8] O'Reilly D, Herdrich G, Kavanagh DF. Electric Propulsion Methods for Small Satellites: A Review. Aerospace. 2021; 8:1-30. Doi.org/10.3390/aerospace8010022

[9] Levchenko I, Xu S, Mazouffre S, Lev D, Pedrini D, Goebel D, Garrigues L, Taccogna F, Bazaka K. Perspectives, frontiers, and new horizons for plasma-based space electric propulsion. Physics of Plasmas. 2020; Feb 3;27(2):020601.

[10] Zhurin VV, Kaufman HR, Robinson RS. Physics of closed drift thrusters. Plasma Sources Science and Technology. 1999;**8**(1):R1

[11] Singh S. Evolutions of Growing Waves in Complex Plasma Medium. IntechOpen, London, United Kingdom, Nov: In edited book Engineering Fluid Mechanics; 2020

[12] Singh S. Dynamics of Rayleigh-Taylor Instability in Plasma Fluids. In the edited book Engineering Fluid Mechanics. IntechOpen, London, United Kingdom, April 15th 2020

[13] Singh S. Hall Thruster: An Electric Propulsion through Plasmas. In the edited book Plasma Science IntechOpen, London, United Kingdom, March 2nd 2020

[14] Boeuf, Jean-Pierre. Tutorial: Physics and modeling of Hall thrusters. Journal of Applied Physics 121.1 (2017): 011101.

[15] Warner, Noah Zachary. Theoretical and experimental investigation of Hall thruster miniaturization. Diss. Massachusetts Institute of Technology, 2007.

[16] A. B. Mikhailovskii, Theory of Plasma Instabilities, Vol. 1, Instabilities of a Homogeneous Plasma (Springer, New York, 1974).

[17] O. Buneman, Instability, turbulence, and conductivity in current-carrying plasma, Phys. Rev. Lett. 1(1), 8 (1958).

[18] O. Buneman, Excitation of field aligned sound waves by electron streams, Phys. Rev. Lett. 10(7), 285 (1963).

[19] D. T. Farley, A plasma instability resulting in field-aligned irregularities in ionosphere, J. Geophys. Res. 68(22), 6083 (1963).

[20] A. Simon, Instability of a partially ionized plasma in crossed electric and

magnetic fields, Phys. Fluids 6(3), 382 (1963).

[21] Keidar M, Beilis II. Electron transport phenomena in plasma devices with ExB drift. IEEE Transactions on Plasma Science. 2006 Jun 19;**34**(3): 804-814

[22] Hagelaar GJ, Bareilles J, Garrigues L, Boeuf JP. Role of anomalous electron transport in a stationary plasma thruster simulation. Journal of Applied Physics. 2003 Jan 1;**93**(1):67-75

[23] Adam JC, Boeuf JP, Dubuit N, Dudeck M, Garrigues L, Gresillon D, et al. Physics, simulation and diagnostics of Hall effect thrusters. Plasma Physics and Controlled Fusion. 2008 Nov 5; **50**(12):124041

[24] Ellison CL, Raitses Y, Fisch NJ. Cross-field electron transport induced by a rotating spoke in a cylindrical Hall thruster. Physics of Plasmas. 2012 Jan 12;19(1):013503.

[25] Griswold ME, Ellison CL, Raitses Y, Fisch NJ. Feedback control of an azimuthal oscillation in the E× B discharge of Hall thrusters. Physics of Plasmas. 2012 May 25;19(5):053506.

[26] Alvarez-Laguna A, Magin T, Massot M, Bourdon A, Chabert P. Plasma-sheath transition in multi-fluid models with inertial terms under low pressure conditions: Comparison with the classical and kinetic theory. Plasma Sources Science and Technology. 2020; 29:025003.

[27] Mandal D, Elskens Y, Lemoine N, Doveil F. Cross-field chaotic transport of electrons by E×B electron drift instability in Hall thruster. Physics of Plasmas. 2020; 27:032301.

[28] Smolyakov A, Zintel T, Couedel L, Sydorenko D, Umnov A, Sorokina E, et al. Anomalous electron transport in one-dimensional electron cyclotron drift turbulence. Plasma Physics Reports. 2020;**46**:496-505

[29] Oland E, Kristiansen R, Nicklasson PJ. Combined chemical and electric thruster solution for attitude control. In 4th International Conference on Recent Advances in Space Technologies 2009 Jun 11 (pp. 627-631). IEEE.

[30] Mandal D, Sharma D, Schamel H. Electron hole instability as a primordial step towards sustained intermittent turbulence in linearly subcritical plasmas. New Journal of Physics. 2018 Jul 4;20(7):073004.

[31] Barral S, Makowski K, Peradzyński Z, Dudeck M. Transit-time instability in Hall thrusters. Physics of Plasmas. 2005; 12:073504. Doi.org/10.1063/1.1947796

[32] Bello-Benítez E, Ahedo E. Axial-azimuthal, high-frequency modes from global linear-stability model of a Hall thruster. Plasma Sources Science and Technology. 2021 Jan;**20**

[33] T. Lafleur, S. D. Baalrud and P. Chabert, Theory for the anomalous electron transport in Hall Effect thrusters. II. Kinetic model, Phys. Plasmas 23,053503 (2016).

[34] Fan H, Li H, Ding Y, Wei L, Yu D. Effects of the peak magnetic field position on Hall thruster discharge characteristics. Advances in Space Research. 2020 Oct 15;**66**(8):2024-2034

[35] Sekine H, Koizumi H, Komurasaki K. Electrostatic ion acceleration in an inductive radio-frequency plasma thruster. Physics of Plasmas. 2020 Oct 22;**27**(10):103513

[36] Puerta J, Martin P, Maass F, Puerta JB, Brito J. Generalized non-ideal treatment and growth rates analysis of drift waves instabilities in a collisions-free magnetized dusty plasma. Physics of Plasmas. 2021 Feb 24;28(2):023701.

[37] Mikellides IG, Ortega AL. Growth of the modified two-stream instability in the plume of a magnetically shielded Hall thruster. Physics of Plasmas. 2020 Oct 5;**27**(10):100701

[38] Tomilin DA, Khmelevskoi IA. Influence of kinetic effects on long wavelength gradient-drift instability in high-frequency range in Hall thruster. Physics of Plasmas. 2020 Oct 1;**27**(10): 102103

[39] Marcovati A, Ito T, Cappelli MA. The dynamics of coherent modes of gradient drift instabilities in a small magnetron discharge plasma. Journal of Applied Physics. 2020 Jun 14;**127**(22): 223301

[40] Litvak AA, Fisch NJ. Resistive instabilities in Hall current plasma discharge. Physics of Plasmas. 2001 Feb; **8**(2):648-651

[41] Fernandez E, Scharfe MK, Thomas CA, Gascon N, Cappelli MA. Growth of resistive instabilities in E× B plasma discharge simulations. Physics of Plasmas. 2008 Jan 8;15(1):012102.

[42] Marusov NA, Sorokina EA, Lakhin VP, Ilgisonis VI, Smolyakov AI. Gradient-drift instability applied to Hall thrusters. Plasma Sources Science and Technology. 2019 Jan 11;28(1):015002.

[43] Ducrocq A, Adam JC, Héron A, Laval G. High-frequency electron drift instability in the cross-field configuration of Hall thrusters. Physics of Plasmas. 2006 Oct 20;**13**(10): 102111

[44] Litvak AA, Fisch NJ. Rayleigh instability in Hall thrusters. Physics of Plasmas. 2004 Apr 26;**11**(4):1379-1383

[45] Kapulkin A, Guelman MM. Low-frequency instability in near-anode region of Hall thruster. IEEE transactions on plasma science. 2008 Nov 17;**36**(5):2082-2087

[46] Choueiri EY. Plasma oscillations in Hall thrusters. Physics of Plasmas. 2001; **8**:1411-1426

[47] Esipchuk YB, Morozov AI, Tilinin GN, Trofimov AV. Plasma oscillations in closed-drift accelerators with an extended acceleration zone Sov. Phys.–Tech. Phys. 1974;**18**:928-932

[48] Parker JB, Raitses Y, Fisch NJ. Transition in electron transport in a cylindrical Hall thruster. Applied Physics Letters. 2010 Aug 30;97(9): 091501.

[49] Liqiu W, Ke H, Chunsheng W, Hong L, ChaoHai Z, Daren Y. Study on breathing mode oscillation suppression of self-excited Hall thrusters. Journal of Vacuum Science & Technology A: Vacuum, Surfaces, and Films. 2012 Nov 17;30(6):061304.

[50] Abramkov V V, Izmailov A A and Shishkin G G 1989 Influence of external circuit on characteristics of accelerator with closed electron drift and extended acceleration zone Proc. 7th All-Union Conference on Plasma Accelerators and Ion Injectors (Kharkov, 26–28 September, 1989) pp 72–3 (in Russian)

[51] Vakhnjuk S P, Kapulkin A M and Prisnjakov V F 1990 Stabilization of plasma instabilities in accelerators with closed electron drift by boundary system of feedback pp 78–86 in [24]

[52] Morozov AI, Nevrovskii VA, Smirnov VA. Effect of a feedback system on the plasma flux in an accelerator with closed electron drift. Sov. Phys.-Tech. Phys.(Engl. Transl.), v. 18, no. 3, pp. 344-347. 1973 Sep 1.

[53] Morozov A I 1973 The study of plasma systems with closed electron drift and distributed electric field pp 75–84 in [16]

[54] Kim V 1982 Main physical features of operational processes in modern

accelerators with closed electron drift and extended acceleration zone Proc. 5th All-Union Conf. on Plasma

[55] Bugrova A I and Kim V 1984 Modern state of physical studies in accelerators with closed electron drift and extended acceleration zone pp 107–28 in

[56] V. P. Lakhin, V. I. Ilgisonis, A. I. Smolyakov, E. A. Sorokina and N. A. Marusov, Effects of finite electron temperature on gradient drift instabilities in partially magnetized plasmas, Physics of Plasmas, 25(1) (2018), 012106.

[57] Romadanov I, Smolyakov A, Raitses Y, Kaganovich I, Tian T, Ryzhkov S. Structure of nonlocal gradient-drift instabilities in Hall E×B discharges. Physics of Plasmas. 2016; **23**(12):122111

[58] O. Koshkarov, A. I. Smolyakov, A. Kapulkin, Y. Raitses and I. Kaganovich, Nonlinear structures of lower-hybrid waves driven by the ion beam, Physics of Plasmas, 25(6) (2018), 061209.

[59] A.I. Smolyakov, O. Chapurin, W. Frias, O. Koshkarov, I. Romadanov, T. Tang, M. Umansky, Y. Raitses, I.D. Kaganovich and V.P. Lakhin, Fluid theory and simulations of instabilities, turbulent transport and coherent structures in partially-magnetized plasmas of discharges, Plasma Physics and Controlled Fusion, 59(1) (2016), 014041.

[60] Malik HK, Singh S. Conditions and growth rate of Rayleigh instability in a Hall thruster under the effect of ion temperature. Physical Review E. 2011 Mar 15;83(3):036406.

[61] Singh S, Malik HK. Role of ionization and electron drift velocity profile to Rayleigh instability in a Hall thruster plasma. Journal of Applied Physics. 2012 Jul 1;112(1):013307.

[62] Singh S, Malik H K, Nishida Y. High frequency electromagnetic resistive instability in a Hall thruster under the effect of ionization. Physics of Plasmas.2013; 20: 102109 (1-7).

[63] Singh S, Malik HK. Growth of low frequency electrostatic and electromagnetic instabilities in a Hall thruster. IEEE Transactions on Plasma Science. 2011;**39**:1910-1918

[64] Singh S, Malik HK. Resistive instabilities in a Hall thruster under the presence of collisions and thermal motion of electrons. The Open Plasma Physics Journal. 2011;**4**:16-23

[65] Malik H K and Singh S. Resistive instability in a Hall plasma discharge under ionization effect. Physics of Plasmas.2013; 20: 052115 (1-8).

[66] Yadav P, Gupta DN, Avinash K. Relativistic electron-beam assisted growth of oscillating two-stream instability of a plasma wave. Physics of Plasmas. 2017 Jun 14;24(6): 062107.

[67] Aria AK, Malik HK. Studies on waves and instabilities in a plasma sheath formed on the outer surface of a space craft. Physics of Plasmas. 2008 Apr 11;15(4):043501.

[68] Becatti G, Goebel DM and Zuin M. Observation of rotating magnetohydrodynamic modes in the plume of a high-current hollow cathode. Journal of Applied Physics. 2021 Jan 21; 129(3):033304.

[69] Morozov AI. Focusing of Cold Quasineutral Beams in Electromagnetic Fields. Soviet Physics Doklady. 1966;**10**: 775

[70] Grimaud L, Mazouffre S. Conducting wall Hall thrusters in magnetic shielding and standard configurations. Journal of Applied Physics. 2017 Jul 21;122(3):033305.

[71] Morozov AI, Lebedev SV. Plasma optics. InReviews of Plasma Physics/ Voprosy Teorii Plazmy/Вопросы Теории Плазмы 1980 (pp. 301-460). Springer, Boston. MA.

[72] Morozov A I, Bugrova A I, Desyatskov A V, Ermakov Y A, Kozintseva M V, Lipatov A S, Pushkin A A, Khartchevnikov V K and Banov D V. ATON-Thruster Plasma Accelerator, Plasma Physics Reports, Vol. 23, No. 7, 1997, pp. 587–597 (translated from Fizika Plazmy, Vol. 23, No. 7, 1997, pp. 635–645).

[73] Hofer RR, Peterson PY, Gallimore AD, Jankovsky RS. A high specific impulse two-stage Hall thruster with plasma lens focusing. IEPC-01-036, 27th International Electric Propulsion Conference, Pasadena. CA.

[74] Hofer RR, Jorns B A, Polk J E, Mikellides I G and Snyder JS, Wear test of a magnetically shielded Hall thruster at 3000 seconds specific impuls,IEPC-2013-033,(2013).

[75] W. Huang, Ph.D. thesis, Study of Hall Thruster Discharge Channel Wall Erosion via Optical Diagnostics, The University of Michigan (2011).

[76] Liu H, Niu X and Yu DR. Numerical study of azimuthal sheath structure and asymmetric anomalous erosion in a stationary plasma thruster. Journal of Plasma Physics. 2019 Apr;85(2).

[77] Choueiri EY. Fundamental difference between the two Hall thruster variants. Physics of Plasmas. 2001 Nov;**8**(11):5025-5033

[78] Hong L, Zhongxi N, Daren Y. Hall thruster with grooved walls. Journal of Applied Physics. 2013 Feb;**28**:113(8)

[79] Ding Y, Peng W, Sun H, Xu Y, Wei L, Li H, Zeng M, Wang F and Yu D. Effect of oblique channel on discharge characteristics of 200-W Hall thruster. Physics of Plasmas. 2017 Feb 14;24(2): 023507.

[80] Olano A, Tang H, Ren J, Zhang G, Li J. Modular Design of A Radial Scalled Hall Thruster for Different Magnetic Configurations. In 2019 IEEE Pulsed Power & Plasma Science (PPPS) 2019 Jun 23 (pp. 1-4). IEEE.

[81] Andreussi T, Giannetti V, Leporini A, Saravia MM and Andrenucci M. Influence of the magnetic field configuration on the plasma flow in Hall thrusters. Plasma Physics and Controlled Fusion. 2017 Oct 27;60(1):014015.

[82] Garrigues L, Fubiani G, Boeuf JP. Appropriate use of the particle-in-cell method in low temperature plasmas: Application to the simulation of negative ion extraction. Journal of Applied Physics. 2016 Dec 7;**120**(21): 213303

[83] Esipchuk YV and Tilinin GN. Drift instability in a Hall-current plasma accelerator. Sov. Phys.-Tech. Phys. (Engl. Transl.);(United States). 1976 Apr 1;21(4).

[84] Chable S and Rogier F. Numerical investigation and modeling of stationary plasma thruster low frequency oscillations. Physics of plasmas. 2005 Mar 10;12(3):033504.

Chapter 11

The Emerging Field Trends Erosion-Free Electric Hall Thrusters Systems

Iryna Litovko, Alexey Goncharov, Andrew Dobrovolskyi and Iryna Naiko

Abstract

The Hall-type accelerator with closed Hall current and open (that is unbounded by metal or dielectric) walls was proposed and considered both theoretically and experimentally. The novelty of this accelerator is the use of a virtual parallel surface of the anode and the cathode due to the principle of equipotentialization of magnetic field lines, which allows to avoid sputtering of the cathode surface and preserve the dynamics of accelerated ions. The formation of the actual traction beam should be due to the acceleration of ions with the accumulated positive bulk charge. A two-dimensional hybrid model in cylindrical coordinates is created in the framework of which the possibility of creation a positive space charge at the system axes is shown. It is shown that the ions flow from the hump of electrical potential can lead to the creation of a powerful ion flow, which moves along the symmetry axis in both sides from the center.

Keywords: Hall thruster, accelerator with closed Hall current, plasmaoptical system, plasma lens, low temperature plasma, space charge, ion flow, crossed electric and magnetic fields

1. Introduction

The realization that outer space is becoming an integral part of our earthly existence becomes commonplace. Hundreds of all kinds of spacecraft (maneuverable and marching), plow the vastness of the vast space. The moving force of such spacecraft has become the electric propulsions different kinds. Traditionally, Hall thrusters (plasma accelerators with closed Hall current and metal or dielectric walls [1–4] attract a special attention of the electric propulsion community. Remarkably high efficiency, simplicity, and potential durability make the Hall thruster one of the primary candidates for miniaturization and application in small communications satellites and using for primary propulsion in deep-space scientific missions. Hall thrusters inherit coaxial geometry for which the material of the annular chamber walls significantly affects the plasma discharge properties and erosion of the dielectric walls. In order to avoid erosion developers, try to create modified wall less accelerator constructions [5, 6]. But these attempts while are not completed, quite complicated and sound many skepticisms.

One of the promising ways to avoid erosion issues is the separation of the magnetic and electrical circuits of the accelerator, which is easier to do in the cylindrical geometry. This principle was carried out and tested in the original Hall type erosion-less accelerator on a laboratory stand at the Institute of Physics NAS of Ukraine [7–10]. The novelty of the proposed idea is the use of a virtual parallel surface of the anode to the cathode due to the principle of equipotentialization of magnetic field lines. Which allows to avoid sputtering of the cathode surface. It was shown the potential drop forms at the axis of system that can use for ion beam formation and accelerating.

We describe the Hall-type accelerator with closed Hall current and open (that is unbounded by metal or dielectric) walls that was proposed and considered both theoretically and experimentally. A two-dimensional hybrid model in cylindrical coordinates is constructed in the framework of which the possibility of creation a positive space charge at the system axes is shown. It is shown that the ions flow from the hump of electrical potential can lead to the creation of a powerful ion flow, which moves along the symmetry axis in both sides from the center. The formation of the actual traction beam should occur due to the acceleration of ions by the accumulated positive bulk charge. Thus, such type of accelerator could sound interest in manipulating high-current flow of charge particle as well as can be attractive for many different high-tech applications and for potential elaborations of low cost, compact and durability electric thrusters.

This article is the brief review of the çurrent status an ongoing experimental and theoretical research and simulations results of such kind accelerator based on the axial-symmetric cylindrical electrostatic plasma lens configuration and the fundamental plasma-optical principles of magnetic electron isolation and equipotentialization magnetic field lines.

2. Experimental setup and results

Experiments were carried out on a laboratory stand, the schematic diagram of which is shown in **Figure 1** on the left. Vacuum chamber 1 contained a test mockup of the accelerator with open walls. The accelerator consisted of a magnetic core with permanent magnets 2 and an electrode system with cylindrical anode 3 and cathode 4 formed by a system of pins. The experimental installation allowed a controlled gas input, by usinCHA-2 system. The gas pumping was performed with the use of a vacuum unit with an oil-vapor pump. The photo of the accelerator with anode layer and open wall is shown in **Figure 1** on the right.

As can be seen from the **Figure 1** on the left, the cathode is composed of two parts that are separated in space. The ends of the cathode pins coincide with the magnetic surfaces that in area between them are parallel to the anode surface. The application of the plasmaoptics principles to the design of a cylindrical accelerator with anode layer and open walls allowed to create an accelerator with virtual cathode that is parallel to the anode plane for its entire width and with cathode that is several centimeters wide. That allows to form a wide flow of accelerated ions to the system axis of symmetry. Due to that each part of the cathode is made in the form of pins, the collecting surface area is significantly reduced, that reduces the contribution of the cathode material in the flow.

The discharge in the system burns due to the working gas (argon) ionization by the electrons. Electrons are magnetized and formed stable negative space charge. The created ions are accelerated from the ionization zone towards the cathode.

The formation of the magnetic field is provided by a system of permanent magnets located in the magnetic circuit. By selecting the number and polarity of the

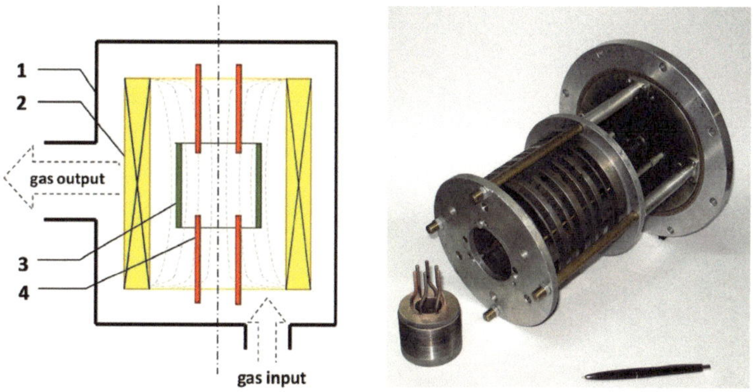

Figure 1.
Left: Experimental setup; vacuum chamber (1), magnetic system (H= 650–750 Oe) (2), anode (3), cathode (4). Right: Photo of the accelerator with anode layer and open walls.

magnets in each system layer, it was possible to change the magnetic field geometry in a wide range. In experiment the system of permanent magnets was arranged in such a way that the magnetic field in the gap between the cathode and the anode was parallel to system's axis as much as possible. It was owing to this configuration of the magnetic field that a system with open walls was created.

2.1 Experimental results

As mentioned above, the cathode is composed of two parts separated in space and their edge coincide with magnetic field lines that are parallel to anode surface. The magnitude of the magnetic field is such that Larmor radius of the electron is much smaller than the system radius. Due to this, the principle of the equipotentialization of magnetic field lines with accuracy up to the electron temperature works [11]. Accordingly, a cylindrical virtual surface is formed between these parts, which potential is close to cathode potential. Through the electrons drift in the crossed electric and magnetic field ExH there is a closed Hall current and corresponding space charge, that creates a layer near the anode surface where the main potential drop occurs [12]. With the appearance of the Hall current in the perpendicular direction, the electron current along E becomes small. In turn, the ions are almost unaffected by the magnetic field, because the Larmor radius of the ion is more than characteristic system size [2, 13]. Ions under the electric field influence, moving in the cathode direction, converge to the center and are pushed out of the system along the axis. The electrons exit speed from the system along the axis is quite small and is compensated by the working gas ionization by these electrons. At low pressure, ionization occurs between the anode and virtual cylindrical cathode. With increasing pressure, the anode layer size can reach the radius of the system. The operation modes of the accelerator also change [9].

The dependence of the current strength on the applied potential to the anode for different working gas pressure were obtained experimentally. The results of the effect of the change in the working gas pressure in the source volume on the discharge current density for the low-current mode are shown in the **Figure 2**. For the presented curves for different voltages, it can be seen that the current density little depends on the gas pressure in the system for potentials up to 1,5 kV in the discharge gap.

This result can be explained by the assumption that the ionized particles concentration under these conditions does not depend on the working gas pressure. This corresponds to the obtained theoretical results also.

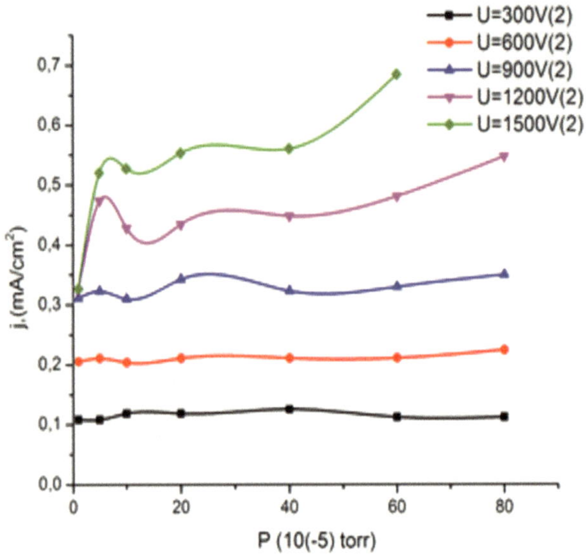

Figure 2.
Dependence of current density on the pressure the chamber at different values of the voltage applied to the discharge gap.

Note that the operation modes of this kind accelerator are very close to those of the classical accelerator with an anode layer. There are two operation modes of this kind accelerator. The first is low-current, with a clearly visible narrow radiating layer between the anode and cathode, in the range of 10^{-4} Torr. The current increases monotonically with increasing applied voltage (see **Figure 3a**). With increasing pressure at a constant applied voltage, this layer gradually occupies the entire volume of the accelerator. During this mode, the ions are mostly formed in a narrow anode layer and move towards the system axis, hardly experiencing the influence of a magnetic field. Accumulating on the system axis, they are pushed out of the volume along it. The study of the distance changing influence between the virtual cathode and anode showed the existence of the optimum. The maximum current on the system axis is fixed at a distance d=10 mm between electrodes, as one can see from **Figure 3a**.

In the second mode – high current, the discharge extends to the entire internal system volume. When the voltage reached a certain value (U > 1,8 kV), the

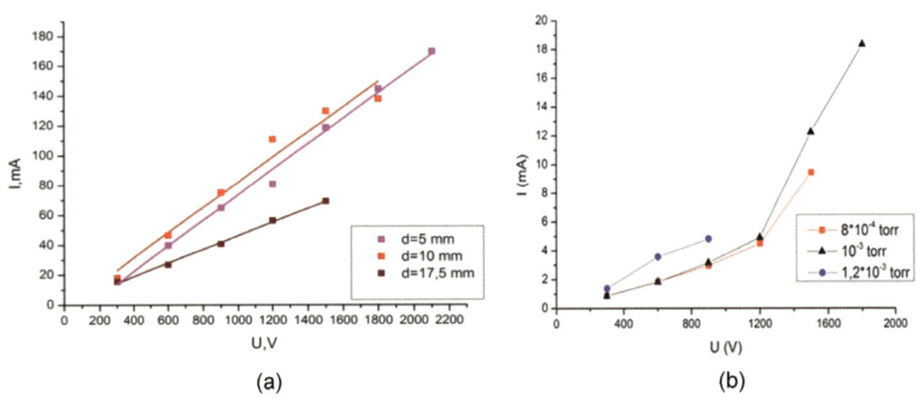

Figure 3.
Volt-ampere characteristic of the accelerator in low-current mode at a) different distance between the anode and cathode at pressure 10^{-4} Torr and b) at different values of pressure in chamber.

discharge current increased in a jump-like manner (see **Figure 4a**), and the discharge transited into the high-current mode, in which the distinct anode layer was absent. In this mode, a typical discharge current was several orders of magnitude higher (up to 2 A, as can be seen from **Figure 4**) than in the low-current mode. Thus, It is evident that the transition into the high-current mode occurs under the influence of two factors: the working gas pressure (see **Figure 3b**) and the voltage applied across the discharge gap (**Figure 4**).

Another characteristic feature of the high-current mode is the formation of a plasma torch. At the discharge voltage U > 1,8 kV bright radiation is observed from the system volume from the ends of the cylindrical channel along the rotation symmetry axis (see **Figure 5a**). In the discharge concerned, ions are accelerated along system's radius towards system's axis. The torches at the ends, on the contrary, are observed along the axis, perpendicularly to the radius and the direction of initial ion acceleration.

Thus, owing to the discharge geometry, in space limited by electrodes of the accelerator there is an accumulation of ion space charge like a lens with a positive space charge, that was proposed earlier for negatively charged particles beam focusing [14–17]. The generated ions reach the system axis and accumulate in the region around it. Ions are stored in the cylinder volume until their own space charge creates a critical electric field. This field forces ions to leave the volume. The main part of the generated ions escapes from the system perpendicularly to its radius. Due to this plasma torches are formed at the edges of the device, which are clearly

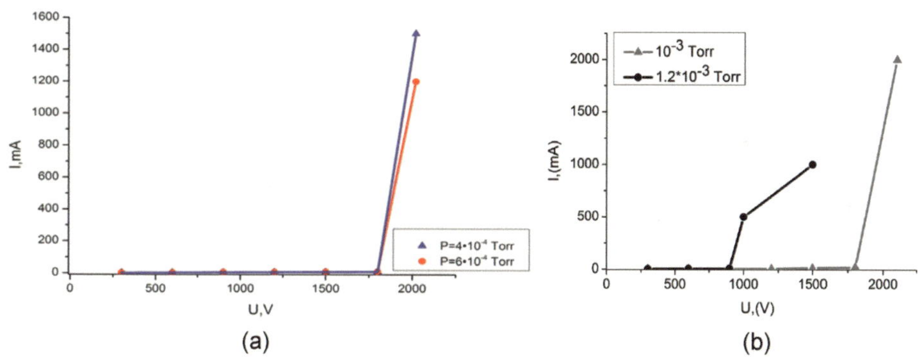

Figure 4.
Volt-ampere characteristic of the accelerator at different pressure in a) low-current mode b) high-current mode.

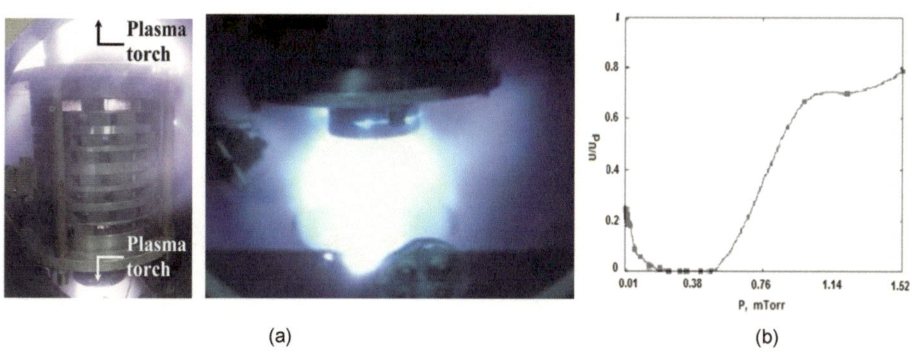

Figure 5.
a) Plasma jet from volume of thruster in high current operation mode. b) Floating potential dependence on the pressure at the system edge.

visible in high-current mode. The results of measuring the floating potential along the axis of the system show that under certain conditions along the plasma torch axis there may be a potential drop (see **Figure 5b**), which can be used to accelerate the generated ions and form a charged particles beam.

Radial studies of the plasma flow coming out along the system axis in this device at different pressures revealed a significant increase in current density on the axis (see **Figure 6**). That fact may indicate the plasma acceleration in this direction. The study of the dependence of the uncompensated current density on the accelerator volume shows the existence of a maximum for a pressure of $6{\cdot}10^{-4}$ Torr. This operation mode of the accelerator is mostly interesting for use as a prototype of the ion-plasma small rocket engine.

Determination of the ion energy distribution function (see **Figure 7a**) in this accelerator was performed by retarding potential method using a 3-line analyzer (discharge current 1.5–2 A; voltage 1.5–2.1 kV; pressure in the range 10^{-4}–10^{-3} Torr). The research results showed the formation of a sufficiently monoenergetic beam with of FWHM (as can be seen from **Figure 7b**), at the level of 10% of the average value which reached two thirds of the anode discharge potential (in particular, at U_{anode} = 1.8 kV, E = 1.2 keV).

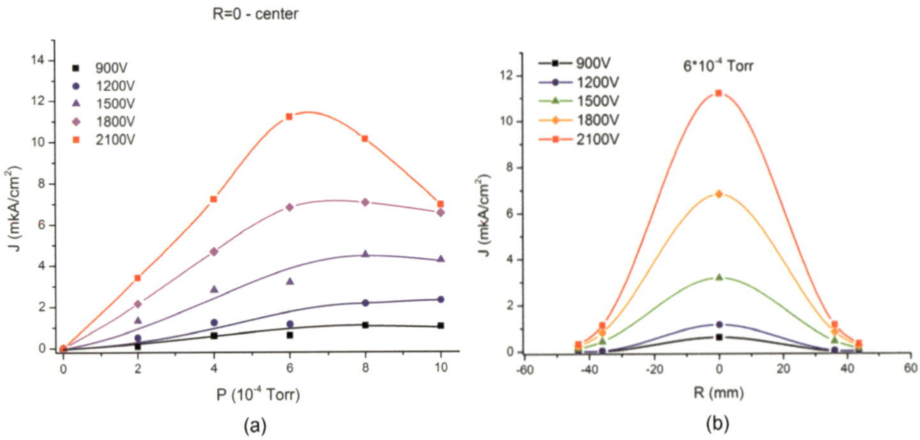

Figure 6.
a) Distribution of the current density along the system radius at the accelerator output. b) the dependence of the current density on the system axis on the pressure in chamber at different values of potential at the anode.

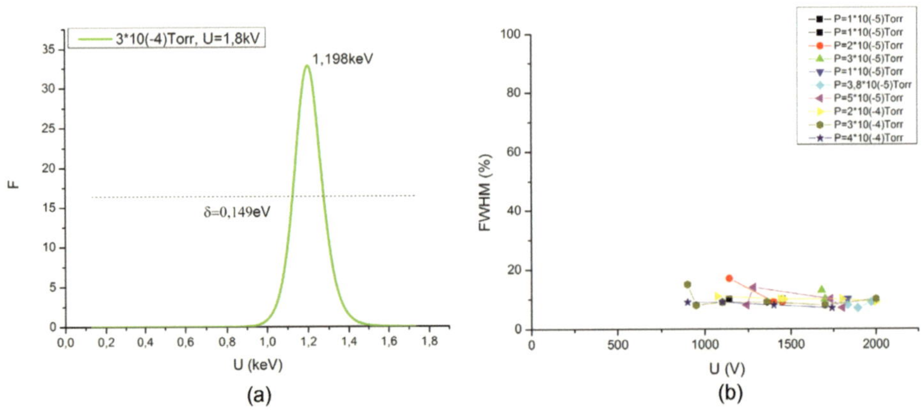

Figure 7.
a) Ion energy distribution function ion energy distribution function (voltage 1.8 kV; pressure $3{*}10^{-4}$ Torr). b) the dependence of FWHM (full width at half maximum) on the voltage applied to the anode.

3. Theoretical model

3.1 One-dimensional hydrodynamic and hybrid model

3.1.1 Hydrodynamic model

To clarify the obtained experimental data, primarily a one-dimensional hydrodynamical theoretical model was developed (see **Figure 8a**). Here x-axis is directed along the radius of the system. We have considered the discharge gap, where ions production appear due to working gas ionization by electrons. Electrons are magnetized, move along magnetic strength lines and drift slowly to anode due to collisions. Ions are free and accelerated by electric field move to the system axis.

The closed system of equations that describes such system in hydrodynamic approximation in stationary state has the form:

$$j_e + j_i = j_d \tag{1}$$

$$j_i(x) = e\nu_i \int_0^x n_e(x)dx \tag{2}$$

$$j_e(x) = e\mu_\perp \left(n_e E(x) - \frac{d}{dx}(n_e T_e) \right) \tag{3}$$

$$T_e(x) = \frac{\beta}{j_e(x)} \int_0^x j_e \frac{d\phi}{ds} ds \tag{4}$$

$$n_i(x) = \sqrt{\frac{M}{2e}} \int_0^x \frac{n_e(s)\nu_i ds}{\sqrt{\phi(x) - \phi(s)}} \tag{5}$$

$$n_e - n_i = \frac{1}{4\pi e} \phi'' \tag{6}$$

here j_i, j_e, n_i, n_e - are ion and electron current density and density consequently, ν_i - is the ionization frequency, $\mu_\perp = \frac{e\nu_e}{m\omega_{eH}^2}$ – electron transverse mobility, E - electric field: $E(x) = -\frac{d\phi}{dx}$, ϕ - potential, ν_e is the frequency of elastic collisions with neutrals and ions, ω_{eH} – the electron cyclotron frequency, T_e – electron temperature.

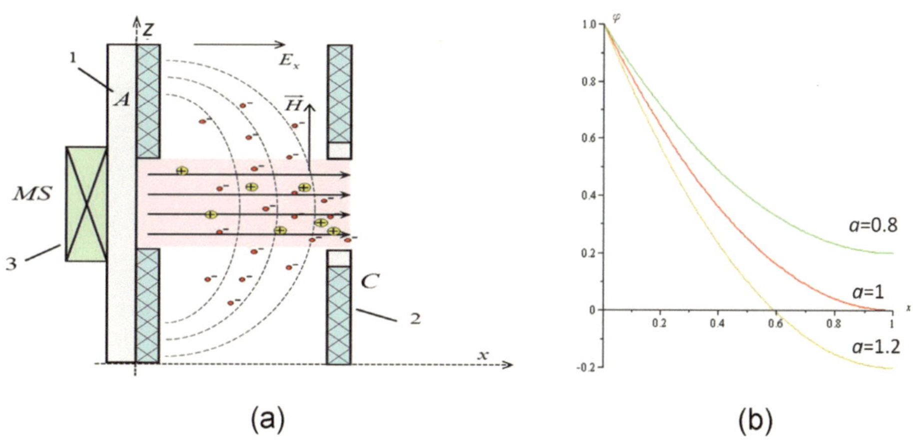

Figure 8.
a) Model of discharge gap: 1- anode, 2- cathode, 3- permanent magnets system; b) potential distribution in the gap for different parameters a *value.*

In the framework of this model, both exact analytical and numerical solutions were obtained [7, 10]. Based on the idea of continuity of current transferring in the system and assumption that the discharge current density in gap volume is the sum of the ion and electron components are found exact analytical solutions describing electric potential distribution along acceleration gap, if T_e = const. It has the next form in low-current mode:

$$\phi = a\left((x-1)^2 - 1\right) + 1, \tag{7}$$

where a is parameter equals $a = \frac{\nu_i d^2}{2\mu_\perp \phi_a}$, d – gap length. The potential distribution in the gap for different value of the parameter a is present in **Figure 8b**. It is found under conditions when all electrons originated within the gap by impact ionization only and go out at the anode due to mobility in transverse magnetic field, the condition of full potential drop in the accelerating gap corresponds to equality gap length to the anode layer thickness. In case when the gap length is less than anode layer thickness, potential drop is not completed, and positive space charge dominated. For case when the gap length is more than anode layer, potential drop exceeds applied potential. This can be due to electron space charge dominated at the accelerator exit. It was shown that potential distribution is parabolic for different operation modes as in low-current mode well as in high current quasi neutral plasma mode and weakly depends on electron temperature. For high-current mode solution has form like expression (7):

$$\phi(x) = 1 + \frac{a^2}{2f^2}\left((x-1)^2 - 1\right) \tag{8}$$

where $f = \nu_i d\sqrt{\frac{M}{2e\phi_a}}$ – describes impact of ion density. Note, that if $a = 2f^2$ we obtain (7). This condition can be rewritten in form:

$$\frac{\tau_{ed}}{\tau_{id}^2} = 2\nu_i \text{ or } \tau_{ed}\nu_i = 2\tau_{id}^2\nu_i^2 = 1 \tag{9}$$

here $\tau_{ed} = \frac{d}{\mu_\perp E}$ – electron lifetime, $\tau_{id} = \frac{d}{v_{id}}$ – ion lifetime. Indeed (9), is some generalization condition of self-sustained discharge in crossed E × H fields taking into consideration both electron and ion dynamic peculiarities.

To clarify the effect of electron temperature on the characteristics of accelerating layer, it was assumed that the electrons received energy from the electric field, thus $T_e = \beta \cdot \varphi$, where $0<\beta\leq 1$. In this assumption, the solution has the form:

$$\phi(x) = \frac{a}{1+\beta}x(x-2) + 1 \tag{10}$$

Consider a more complete description, assuming that heat loss occurs due to different types of collision. Entering the characteristic time τ_0 – energy loss by collision, expression for electron temperature (4) can be represented as:

$$T_e = \frac{j_d E\tau_0}{en_e}\left(1 - e^{-t/\tau_0}\right) \tag{11}$$

If we assume that τ_0 is equal to electron lifetime τ_{ed}, we obtain:

$$T_e = \frac{j_d d}{\mu_\perp e n_e} \tag{12}$$

Thus, the second term in expression (3) disappears and we come back to the solution (7). The numerical solution of system Eq. (1)–(6) showed that the electron density changes extraordinarily little along the gap with typical operating parameters, so our assumption about n_e = const is justified.

So, even in such a simplified model, we obtain a result explaining the appearance a space charge and finding the optimal length of accelerating gap. Nevertheless, although the hydrodynamic model can well describe the dynamics of the electron and ionic components, it does not make allowance for ionization state processes, as well as the influence of neutral atoms in the working gas. A purely kinetic description cannot also be used because of a significant difference between the velocities of electrons and ions. Therefore, a description using the hybrid model may be an optimal decision.

3.1.2 Hybrid model

In the framework of this model [18, 19], the hydrodynamic description is used for the electron component, and the kinetic description for the ionic and neutral ones. This approach also allows the limited stay time of ions in the system to be considered. A one-dimensional model was considered with regard for only the single ionization. In this case, we can write the following equations for neutrals and ions, respectively.

$$\frac{\partial f_o}{\partial t} + v_0 \frac{\partial f_0}{\partial x} = -\langle \sigma_{ie} v_e \rangle n_e f_0 \tag{13}$$

$$\frac{\partial f_i}{\partial t} + v_i \frac{\partial f_i}{\partial x} + \frac{e}{M} E \frac{\partial f_i}{\partial v} = \langle \sigma_{ie} v_e \rangle n_e f_0 \tag{14}$$

here f0, fi – distribution function of neutrals and ions consequently, that satisfy boundary conditions:

$$f_0(0, v, t) = \frac{1}{(2\pi M T)^{\frac{3}{2}}} exp\left(-\frac{Mv^2}{2T}\right), v > 0$$

$$f_0(0, v, t) = 0, v < 0; \tag{15}$$

$$f_i(0, v, t) = 0$$

In (13) and (14) right part expression is:

$$\langle \sigma_{ie} v_e \rangle = \sigma_{max} v_e(T_e)\, exp\left(\frac{-U_i}{T_e}\right) \tag{16}$$

here σ_{max} – maximal ionization cross-section, $v_e(T_e)$ – average electron thermal velocity, U_i – ionization potential.

For electrons we can use hydrodynamic model and to solve system (1)–(6), where instead Eq. (2) and (5), the equations are used:

$$j_i = \int v_i f_i dv \text{ and } n_i = \int f_i dv \tag{17}$$

The numerical solution of the system of Eqs. (13)–(17) showed that in the stationary case the difference between two models is insignificant (see **Figure 9a**). A comparison between the results of both models obtained in model experiments testifies to an insignificant influence of the neutral component of a working gas on the formation of the potential drop across the discharge gap for the examined initial conditions.

To solve the Eqs. (13) and (14), we chose a time step that satisfies the condition $\Delta t < {}^{\Delta x}/_{v_{imax}}$, where Δx – spatial step, v_{imax} – maximal ion velocity. The **Figure 10** shows ion density changing in the gap during the time in high-current mode (under applied anode potential equal 1200 V). At first ions number increases in near anode region and remains almost constant in gap, then it increases almost linearly throughout the gap and finally increases sharply at the cathode region.

Note that for correct description (especially high-current mode), it is necessary to model ionization, collisions, and plasma creation, as well as motion of neutrals and formed ions in the whole volume of accelerator, thus need consider two-dimensional hybrid model.

3.1.3 2D-hybrid model and results

In the framework of this model the kinetic description is used in cylindrical geometry for the ionic and neutral components and the hydrodynamic

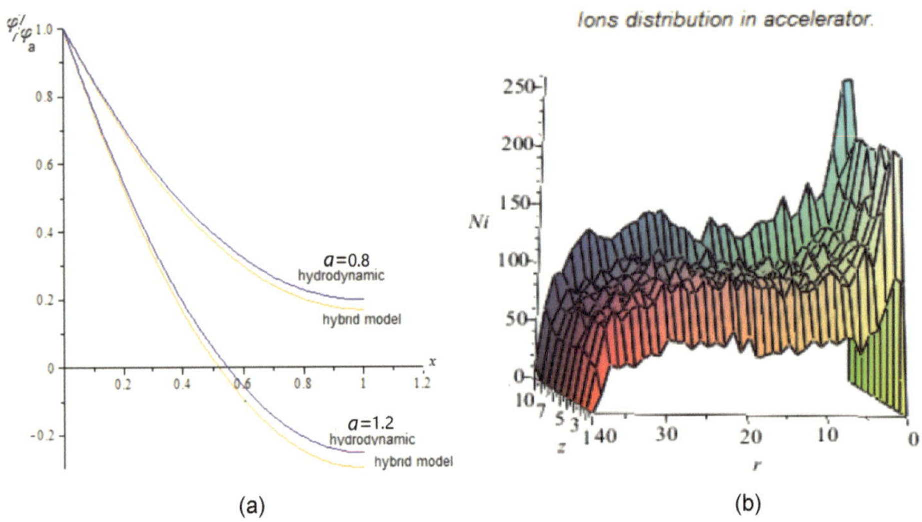

Figure 9.
a) Comparison between the results of numerical calculations for the potential distribution in the discharge gap in the hydrodynamic and hybrid models. b) Ions number dependence on r and z (r = 0, z = 0 – center of the system).

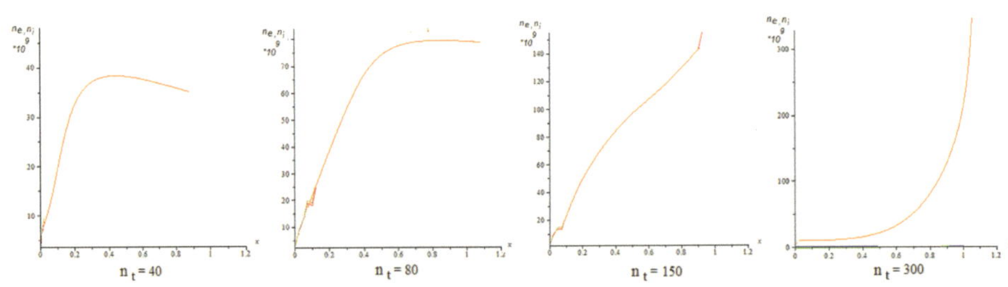

Figure 10.
Ion distribution in the gap on different time steps.

one-dimension on each special layer $z_i<z<z_i+1$ description for the electron ones. Thus, for ions and neutrals description we use Boltzmann kinetic equation:

$$\frac{\partial f_{i,n}}{\partial t} + \vec{v}_{i,n} \frac{\partial f_{i,n}}{\partial \vec{r}} + \frac{e}{M}\left(E + \frac{1}{c}[v \times B]\right)\frac{\partial f_i}{\partial v_i} = St\left\{ f_{i,n}\right\} \tag{18}$$

We solved this equation by splitting on the Vlasov equation for finding trajectories of ions and neutrals:

$$\frac{\partial f_{i,n}}{\partial t} + \vec{v}_{i,n} \frac{\partial f_{i,n}}{\partial \vec{r}} + \frac{e}{M}\left(E + \frac{1}{c}[v \times B]\right)\frac{\partial f_i}{\partial v_i} = 0 \tag{19}$$

and to correct the found trajectories considering the collision integral, in which we took into account the processes of ionization and elastic and inelastic collisions:

$$\frac{Df_{i,n}}{Dt} = St\left\{ f_{i,n}\right\} \tag{20}$$

The Vlasov equations were solved by the method of characteristics [20]:

$$\frac{d\vec{v}_k}{dt} = \frac{q_k}{M}\left(\vec{E} + \frac{1}{c}[v_k \times B]\right], \frac{d\vec{r}_k}{dt} = \vec{v}_k \tag{21}$$

To solve these equations the PIC method [21] with Boris scheme [22] was used to avoid singularities at the axis. For initial electric field distribution was taken electric field in the plasma absence: $E(r) = \frac{U_a}{rln(r_c/r_a)}$. The Monte-Carlo method was used for modeling of ionization in this field. The probability of a collision of a particle with energy ε_j during time Δt was found from expression [23]:

$$P_j = 1 - exp\left(-v_j \Delta t \sigma(\varepsilon_j) n_j\left(\vec{r}_j\right)\right) \tag{22}$$

here $\sigma(\varepsilon)$ – collision cross-section (elastic, ionization or excitation), n_j – density of similar particles at the point r_j. To determine the probability of collision a random number s is chosen from interval $[0, 1]$ with the help of a random number generator. If $s<P_j$, then assumed that collision has occurred. It is determined by the ratio of the cross-sections with the random number generator, which collision has occurred – elastic, excitation, or ionization. In dependence of this particle parameters change or new ion add in computational box. The evolution of all particles that are in the modeling region is traced at each time step. For this motion equations were solved, and new velocities and positions of the particles were found. Particles that move out the modeling box boundaries are excluded from consideration. After quite a long time particle density distribution was found. The ion charge density and current density are calculated from coordinates and velocities particles according to formulas:

$$\rho(r,t) = \frac{1}{V}\sum_j q_j R\left(\vec{r}, \vec{r}_j(t)\right), j(r,t) = \sum_j q_j v_j(t) R\left(\vec{r}, \vec{r}_j(t)\right) \tag{23}$$

where $R(r, r_j)$ – usual standard PIC – core, that characterizes particle size and shape and charge distribution in it. For a cylindrical coordinate system it has form [24]:

$$R\big((r_1,z_k),(r_j,z_j)\big) = \begin{cases} \dfrac{1}{V_i} * \dfrac{r_{i+1}^2 - r_j^2}{r_{i+1}^2 - r_i^2} * \dfrac{h_z - |z_k - z_j|}{h_z}, r_i < r_j < r_{i+1}, |z_k - z_j| < h_z \\ \dfrac{1}{V_i} * \dfrac{r_{i-1}^2 - r_j^2}{r_{i-1}^2 - r_i^2} * \dfrac{h_z - |z_k - z_j|}{h_z}, r_{i-1} < r_i < r_{i+1}, |z_k - z_j| < h_z \\ 0, i \end{cases} \tag{24}$$

here $V_i = 2\pi r_i h_r h_z$ – volume of the cell, h_r, h_z – steps in the spatial coordinates.

After that the Poisson equation was solved and new electric field distribution was found. Since electrons are magnetized, we consider their movement in radial plane only, thus can solve for electrons one-dimensional hydrodynamic equations on each layer at z separately. Solve it we find electron density, calculate electric field on each layer and correct particle trajectories. After that the procedure was repeated. Modeling time is large enough for establish of ion multiplication process. The formation of the sufficient number of ions is possible due to magnetic field presence, which isolates anode from the cathode. Ions practically do not feel the magnetic field action and move from anode to the axis, where create a space charge, first in the center of the system. Electrons move along the magnetic field strength

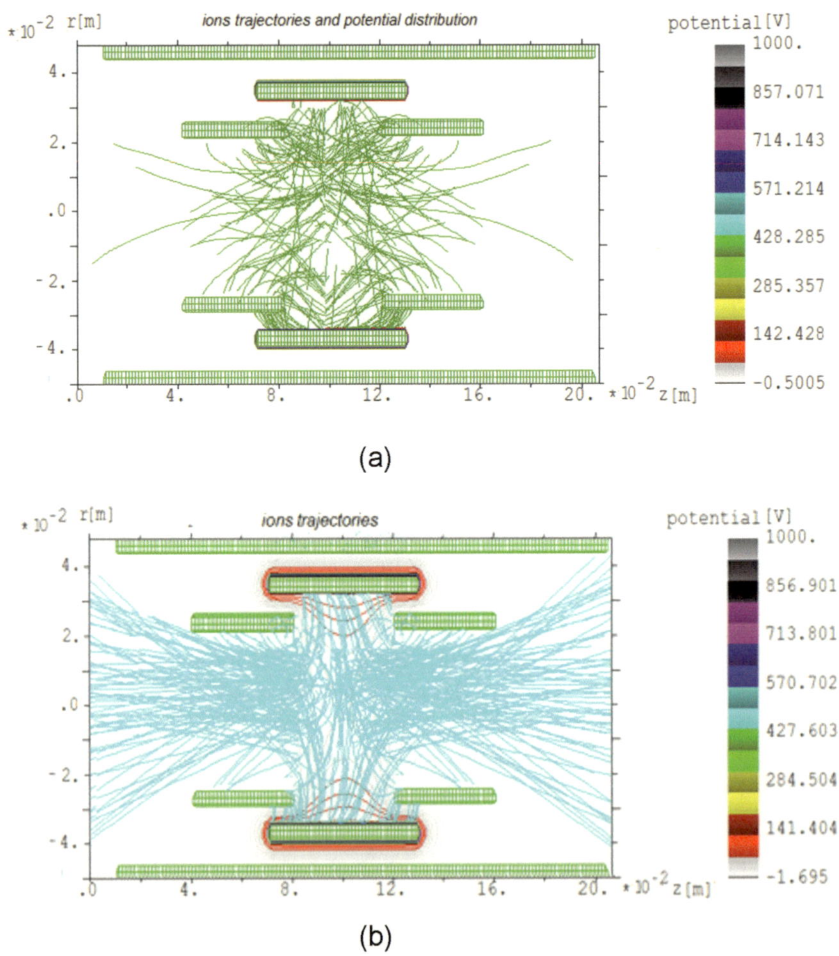

Figure 11.
Ion's trajectories (calculation for U_a = 1 kV, H = 0.03 T) for different time step a) N_t = 50, b) N_t = 200.

line, but due to collisions with neutrals, they start move across the magnetic field. An internal electric field is formed which slow down the ions and pushes out them from the volume along system axis. The **Figure 9b** shows results of modeling high-current mode (U_a = 1.2 kV, pressure 0.15 Pa and magnetic field at the axis is 0.03 T). In is shown how the ions number to axis increases when ionization process is steady-state. One can see that number of ions increase not only to axis but along axis from center to edge too. In **Figure 11** the calculated ion trajectories for different time steps are shown. One can see that the ions that appear due to ionization move to center of the system. Coinciding on the system axis, they accumulate and create a positive space charge, and then diverge along the axis in both directions under the action of created own electric field. The ion space charge distribution for these case is shown in **Figure 12**.

The electrons trajectories for this case are shown in **Figure 13a**. One can see that the electrons are magnetized, moving along magnetic strength lines, and their trajectories are almost parallel to the surface of the anode. The **Figure 14** shows the potential distribution for different time steps. One can see that at the beginning of ionization, the potential drop is not complete in the gap and even has a negative sign in the center of the system. With ions number increasing, they coincide in the

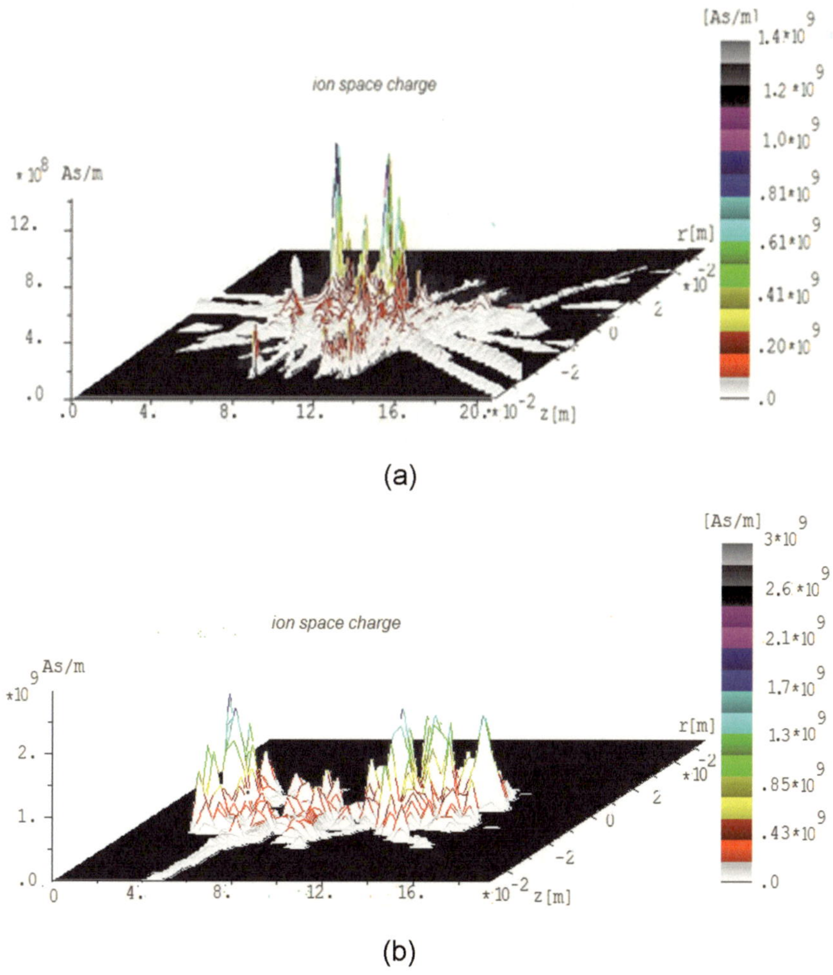

Figure 12.
Ion space charge for time step 70 (a) and 340 (b).

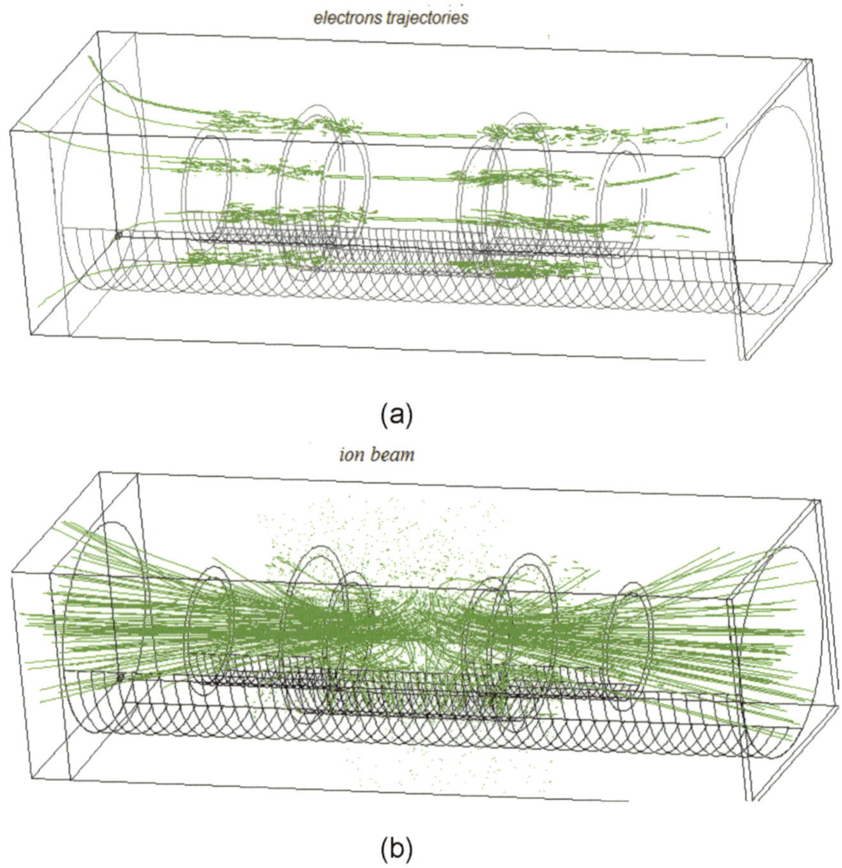

Figure 13.
Electron (a) and ions (b) trajectories in accelerator for H = 0.03 T.

center of the system and form a positive space charge cloud, the potential of which even exceeds the applied potential U_a. This creates an electric field under which ions begin to move along the axis of symmetry in both directions, from the center to the edges, taking out with them part of space charge, reducing it in the center and creating bulks of space charge at the ends of the system (see **Figure 12**).

If we look at the potential distribution along the z-axis, we see that the maximum potential with distance from the center first declines, but then gradually begins to increase (see **Figure 15**). One can see that at a distance of 0.16 m from system center maximum potential is 222 V($\sim$0.22 U_a), while at a distance 0.21 m it is already 396 V, which is equal to 0.4 U_a, and current density of ion beam reached 0.7 mA/cm^2. Thus, the formed ions, initially accumulating in the center of the system, under action of own created electric field can accelerated and create a powerful ion flux from both edges of the accelerator.

4. Conclusion

Our research of the accelerator with closed electron drift and open walls demonstrates its similarity to the accelerator with anode layer and metallic walls in the accelerator channel. The low-current operation mode with an anode layer confined by the interelectrode gap and the high-current operation mode with

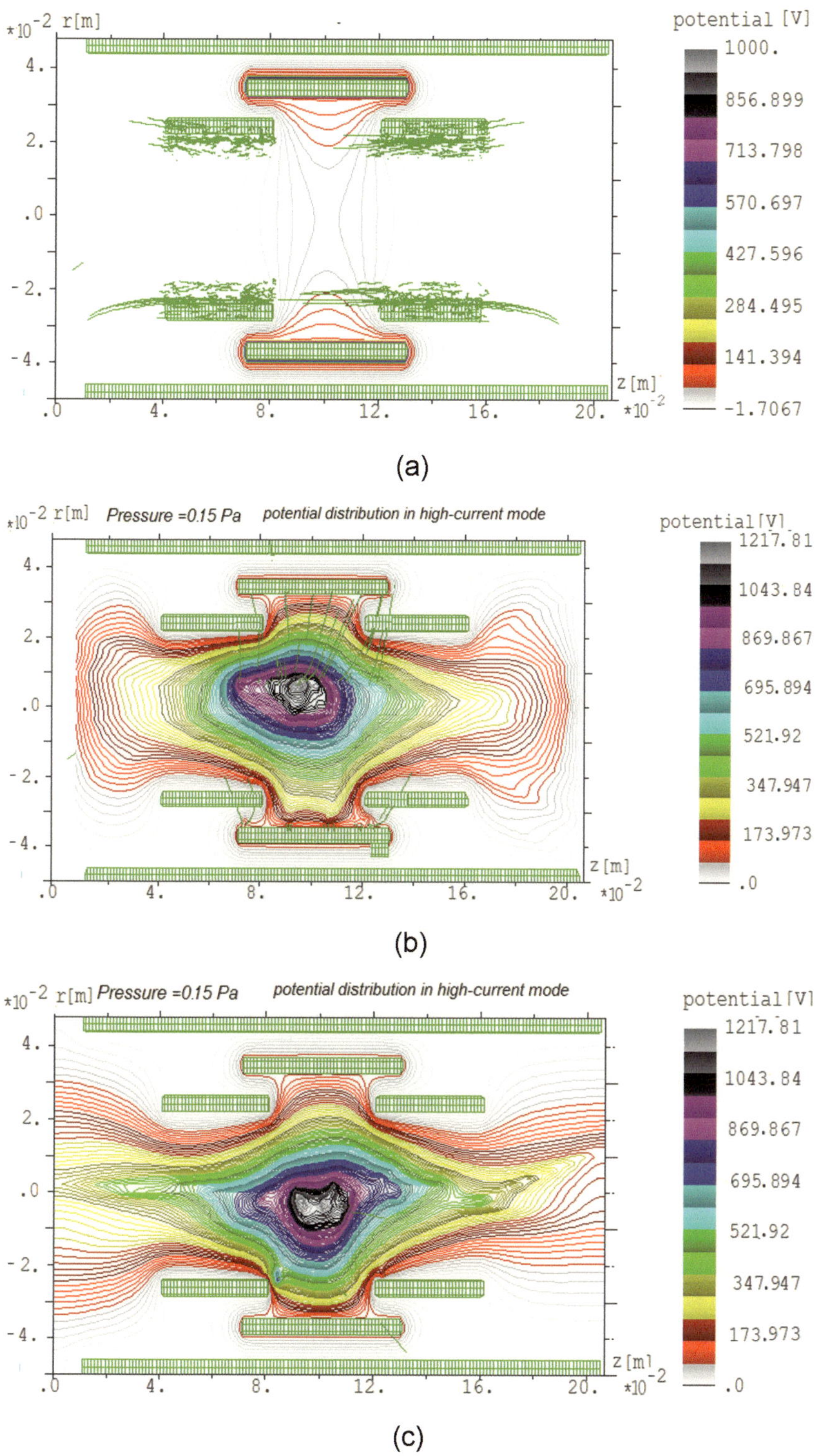

Figure 14.
Potential distribution for different time step: a) $N_t = 10$, b) $N_t = 100$, c) $N_t = 300$.

well-distinguishable plasma torches at system's ends are obtained. The jump-like transition between the modes is shown to occur under the influence of the anode potential and the working gas pressure. In the low-current mode, the discharge

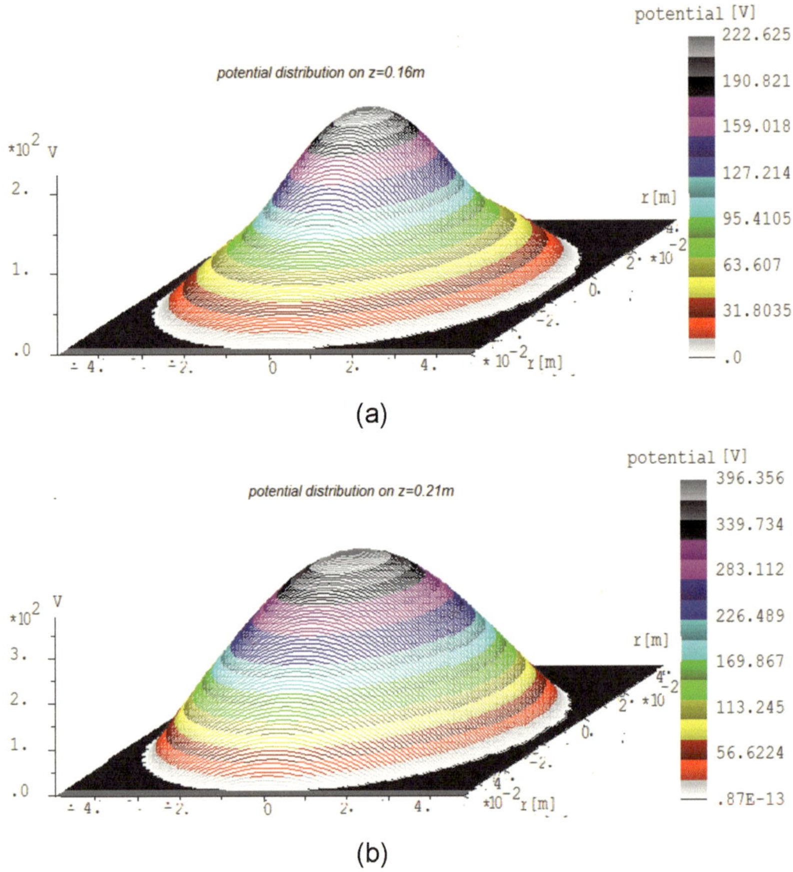

Figure 15.
Potential distribution on the system axis at different distances from the center of the system: a) z = 0.16 m., b) z = 0.22 m.

current depends much stronger on the voltage applied across the discharge gap than on the working gas pressure. A hybrid theoretical model was developed, and simulation results on its basis are obtained. In the framework of this model the possibility of creation a positive space charge at the system axes is shown. It is shown that the ions flow from the hump of electrical potential can lead to the creation of a powerful ion flow, which moves along the symmetry axis in both sides from the center.

Thus, it is shown the carried out theoretical and experimental explorations demonstrate attractive perspective for further development and elaboration new generation erosion-free Hall-type accelerators. This open up possibility to apply kind improved devices like the space electric propulsions and tools for High-Tech ion-plasma surface treatment of functional materials.

Acknowledgements

The work was performed within the framework of the targeted comprehensive program "Prospective research in the field of plasma physics, controlled fusion and plasma technologies" under the project PL-2021.

Author details

Iryna Litovko[1]*, Alexey Goncharov[2], Andrew Dobrovolskyi[2] and Iryna Naiko[2]

1 Institute for Nuclear Research NAS of Ukraine, Kiev, Ukraine

2 Institute of Physics NASU, Kiev, Ukraine

*Address all correspondence to: ilitovko@ukr.net

References

[1] Morozov A. and Lebedev S.; "Plasmaoptics" in Reviews of Plasma Physics, Edited by M. Leontovich (Consultants Bureau, New York, 1975).

[2] Grishin D., Leskov L., Kozlov N., Plasma Accelerators, Mashinostroenie (1983), Moscow-231 c.

[3] Kim V.P. Design features and peculiarities of working processes in modern stationary plasma engines of Morozov (in Russian). Journal of Technical Physics, 2015, v. 85, #3, pp. 45-59.

[4] Choueiri E. Y. A Critical History of Electric Propulsion: The First 50 Years (1906–1956). Journal of propulsion and power, Vol. 20, No. 2, March–April 2004

[5] Ding Y., Peng W., Sun H., Wei L., Zeng M., Wang F., and Yu D., "Performance characteristics of no-wall-losses hall thruster," Eur. Phys. J. Spec. Top. 226, 2945 (2017).

[6] Mazouffre, S., Vaudolon, J., Tsikata, S., er al. "Optimization of magnetic field topology and anode geometry for a Wallless hall Thruster," 51st AIAA/SAE/ASEE Joint Propulsion Conference, AIAA2015–4007, Cleveland, Ohio, 2015

[7] Goncharov A., Dobrovolsky A., Litovko I., Naiko I., Naiko L. Plasma accelerator with closed electron drift and open walls. Problems of atomic science and technology. – 2015. – №4 (98). – P. 26-31.

[8] Litovko I., Goncharov A., Dobrovolskiy A., Naiko L., Naiko I. Computer modelling new generation plasma optical devices (new results) Problems of atomic science and technology. – 2015. – №1(95). – P. 209-212.

[9] Goncharov A.A., Dobrovolskiy A.N., Litovko I.V., Naiko I.V., Naiko L.V. Modes of plasma-dynamical system with closed electron drift and open wall. 2017 IEEE International young scientists forum on applied physics and engineering (YSF-2017). Book of papers. – 2017. – P. 267-270. DOI: 10.1109/YSF.2017.8126633

[10] Goncharov A., Dobrovolsky A., Litovko I., Naiko L., Naiko I Novel modification of Hall-type ion source (study and the first results). Rev. Sci. Instrum. – 2016. – P. 87, 02A501.

[11] Morozov A. I. Intrduction in Plasmo Dynamics (CRC Press, 2012); (in Russian: – M.: Fismatlit, 2006.) – 576 p.

[12] Goncharov A.A., Dobrovol'skyi A. N., Evsyukov A.N., Protsenko I.M., Litovko I.V. New vision of the physics of gas magnetron-type discharges. Ukr. J. Phys, 2009, vol. 54, №1-2, pp. 63-67

[13] Physics and Applications of Plasma Accelerators, edited by A.I. Morozov. – Minsk: Nauka i technika, 1974.

[14] Goncharov A. A., Dobrovolskiy A. M., Dunets S. M., Litovko I. V., Gushenets V. I. et al. Electrostatic plasma lens for focusing negatively charged particle beams. Rev. Sci. Instrum. 83, 02B723 (2012); doi: 10.1063/1.3675387

[15] Goncharov, A. Dobrovolskiy, S. Dunets, A. Evsyukov, I. Litovko, V. Gushenets, E. Oks., Positive-Space-Charge Lens for Focusing and Manipulating High-Current Beams of Negatively Charged Particles. IEEE Trans. Plasma Sci., v. 39, № 6 (2011), pp. 1408-141113

[16] V. Gushenets, A. Goncharov, A. Dobrovolskiy, S. Dunets, I. Litovko, E. Oks., A. Bugaev, Electrostatic plasma lens focusing of an intense electron beam in an electron source with a vacuum arc plasma cathode. IEEE

Trans. Plasma Sci., v. 41, № 4, Part 3 (2013), pp. 2171-2174

[17] Goncharov, A. Dobrovolskiy, I. Litovko, V. Gushenets, E. Oks. Focusing intense electron beams using a positive space charge cloud plasma lens. Physica Scripta, v. 161, 014070 (2014).

[18] Litovko I.V., Dobrovolskiy A.N., Naiko L.V., Naiko I.V. a new type of plasma accelerator with closed electron drift. Ukr. J. Phys. 2018. Vol. 63, No. 2. Pp. 110–115.

[19] Litovko I.V., Goncharov A.A., Dobrovolskiy A.N., Naiko L.V., Naiko I. V., Modelling of new generation plasma optical devices//Nukleoika, 2016, vol. 61 (2), pp. 207-212.

[20] Pegoraro F. Theory and applications of the Vlasov equation. /F. Pegoraro, F. Califano, G. Manfredi, P. J. Morrison/ Eur. Phys. J. D. — 2015. — Vol. 69.— P. 68. — DOI:10.1140

[21] Potter D. Methods of Calculations in Physics (Mir, Moscow, 1975)

[22] Boris J.P., Lee R. Optimization of particle calculations in 2 and 3 dimensions, Commun. Math. Phys. 1969. № 12. p. 131

[23] V. Vahedi, M. Surendra, "A Monte Carlo collision model for the particle-in-cell method: Application for argon and oxigen discharges»//Computer Physics Communications 87 (1995), 179-198

[24] Birdsall C.K. Particle-in-Cell charged-particle simulation plus Monte Carlo collisions with neutral atoms, PIC-MCC/C.K. Birdsall//IEEE Transactions on Plasma Science. – 1991. – Vol. 19. – № 2. – P. 65-83.

Chapter 12

Gyrotron: The Most Suitable Millimeter-Wave Source for Heating of Plasma in Tokamak

Santanu Karmakar and Jagadish C. Mudiganti

Abstract

In this chapter, brief outline is presented about gyro-devices. Gyro-devices comprise of a family of microwave devices and gyrotron is one among those. Various gyro devices, namely, gyrotron, gyro-klystron and gyro traveling-wave tubes (gyro-TWT) are discussed. Gyrotron is the only microwave source which can generate megawatt range of power at millimeter-wave and sub-millimeter-wave frequency. Gyrotron is the most suitable millimeter wave source for the heating of plasma in the Tokamak for the controlled thermoneuclear fusion reactors. This device is used both for the electron cyclotron resonance heating (ECRH) as well as for the electron cyclotron current drive (ECCD). In this chapter, the basic theory of gyrotron operation are presented with the explanation of various sub-systems of gyrotron. The applications of gyrotrons are also discussed. Also, the present state-of-the-art worldwide scenario of gyrotrons suitable for plasma heating applications are presented in details.

Keywords: gyrotron, electron cyclotron resonance heating, fast wave device, gyrotron cavity

1. Introduction

The conventional microwave-tubes, such as traveling-wave tubes (TWSs) and klystrons follow the Pf^2 law. As per this law, the product of maximum power (P) and the square of frequency (f^2) is constant for a device. This limits the maximum power handling capability of a device at higher frequency, i.e., in the millimeter-wave and sub-millimeter-wave frequency regime. Hence, it was found to be highly difficult to develop a microwave source, capable of delivering megawatts of power at millimeter wave frequency. In the last few decades, a new class of microwave tube has emerged, called – Gyrotrons [1–9], which are based on cyclotron resonance maser (CRM) instability [10–12]. This class of device has the capability to produce very high power at millimeter-wave frequencies, much higher than other microwave devices.

The gyrotron is the most suitable source for the heating of plasma in Tokamak for controlled thermoneuclear fusion reactors. Gyrotron is being used as the heating source for electron cyclotron resonance heating (ECRH) as well as for the electron cyclotron current drive (ECCD).

IntechOpen

Gyro-devices comprise of a family of microwave devices and gyrotron is one among those. However, gyrotron being the most popular gyro-device, the entire gyro-device family is sometime referred as gyrotrons. Various other commercially available gyro-devices are: gyro-klystron, gyro-traveling wave tubes (gyro-TWT) and gyro-twistron (a combination of Gyro-TWT and Gyro-Klystron).

In a gyro-device, a hollow electron-beam is generated with the help of a special kind of electron gun, known as magnetron injection gun (MIG) operating in temperature-limited regime of thermionic emission. This hollow electron beam is made to gyrate at cyclotron frequency with the help of a strong axial magnetic field. Subsequently, this gyrating electron beam is passed through an interaction structure, where the electron-beam interacts with the electromagnetic-wave (EM-wave). In case of gyrotron, the interaction-structure is an open-ended cavity. In case of gyro-TWT, the interaction structure is waveguide with an input and output coupler. When the cyclotron frequency synchronizes with the frequency of the EM-wave (frequency of EM-wave supported by the cavity in case of gyrotron and frequency of the EM-wave fed at the input coupler in case of gyro-TWT) the beam-wave interaction takes place. The transverse kinetic energy of the electron-beam gets converted to electromagnetic energy. Hence, the EM-wave gets generated (in gyrotron) or amplified (in gyro-TWT).

Let us now briefly discuss the origin of gyrotrons. It has been well known since the mid-fifties that there appeared to be a limit to the upper frequency at which most vacuum microwave devices could be made to operate with sufficient power and efficiency, primarily due to the reduction of physical size of the components of the device with increase of frequency [13]. This problem can be explained as follows: As the frequency of operation of the device increases, the dimension of the waveguide or cavity or the loading elements inside the waveguide (such as helix in case of helix-TWT) become uncomfortably small, as their physical size is closely related to the operating wavelength of the device. Furthermore, since the depth of penetration of the field generated by the electromagnetic wave is proportional to the operating wavelength, the field penetration inside the loading-element reduces with the increase of operating frequency. Hence, in order to have a proper interaction between the electron-beam and electromagnetic wave, electron-beam needs to be placed closer to the structure carrying electromagnetic wave, if we wish to retain an acceptable efficiency of beam-wave interaction [1, 2]. All these awkward requirements clearly indicate an urgent need for a radical change of approach. In conventional vacuum electronic slow-wave devices (such as TWT), periodic loading elements are required for slowing down the phase velocity of the electromagnetic wave (slow wave interaction: $v_{ph} < c$) so that the phase velocity becomes synchronized with the velocity of the slow space-charge wave produced by a perturbed electron beam. In case of helix-TWT, helix acts as periodic loading element, alternatively known as slow-wave structure (SWS). Whereas in gyro-device, alternatively known as fast-wave devices, the periodicity in the propagating medium is removed and the periodicity is brought in-to the electron-beam. The interaction now takes place with an electromagnetic wave whose phase velocity is higher than the free-space velocity of light (fast wave interaction, $v_{ph} > c$) [1–3]. Instead of periodicity of the loading element, the periodicity of the electron-beam comes into play. This leads to a quasi-synchronism between the electromagnetic wave and the electron-beam.

Gyro-devices comes under the category of Bremsstrahlung radiation device [13]. Here, instead of periodic show-wave-structure, the electron beam is made periodic by generating a hollow electron-beam gyrating under the influence of a strong axial magnetic field. When this electron beam is perturbed, two cyclotron waves get generated, namely slow and fast cyclotron wave. When the velocity of the fast

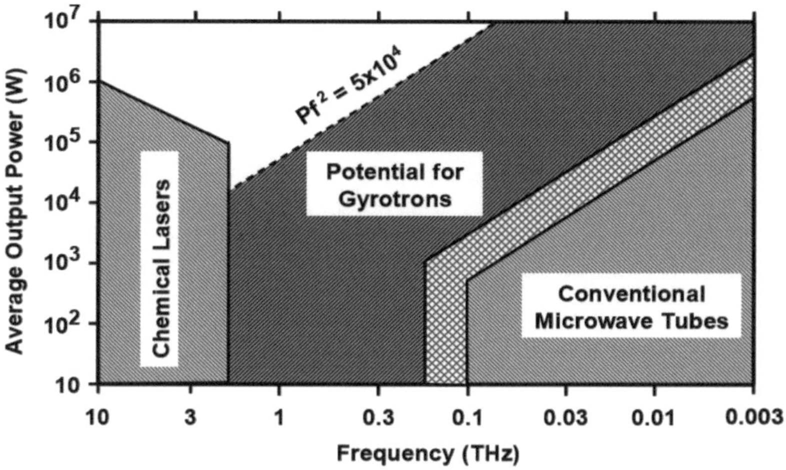

Figure 1.
Domain of microwave tubes/ laser devices.

cyclotron wave is synchronized with EM-wave, beam-wave interaction takes place. That's why these devices are known as fast-wave device.

A variety of interaction stricture geometries are proposed in the literature [9, 11–13] for gyro-devices. In case of gyrotron, the interaction structure is an open ended cavity. Well directed and concentrated efforts were made in the mid-seventies by Granastein and his team at the Naval Research Laboratory (NRL) [13], as well as Gapanov and his team at IAP, Russia [14] who, with some help from others, succeeded in mounting an extensive research effort in the whole area of Bremsstrahlung radiation device, which include free-electron lasers as well as gyrotrons. Since then, gyro-devices have developed very rapidly to offer prodigious amounts of power, and very high efficiency of the order of 50% or more. **Figure 1** shows capabilities of various vacuum electronic devices in terms of frequency and average power. It's evident from the figure, for frequencies above the Terahertz range, laser devices are most suitable source for generation of electromagnetic wave. Again, for frequencies below the millimeter-wave range, conventional microwave tubes are most suitable source for the generation of high power. Gyrotron fits in between these two frequency regimes. Gyrotrons are best suited when the operating wavelength is approximately 1 mm and output power requirement is between hundreds of kilowatts to few megawatts. That's why Gyrotron is found to be the most suitable source for the heating of plasma.

2. Basic types of gyro device

Ever since the advent of cyclotron-resonance maser (CRM) instability devices, a vast amount of research work has been carried out and a number of gyro-devices have been developed. Out of these, the most popular device is gyrotron (alternatively known as gyro-monotron). The other two commercially available gyro-devices widely used in radar applications are, gyro-TWT [3, 8], and gyro-klystron [3]. However, the entire class of gyro-devices are usually referred as gyrotron. There are few less popular gyro-devices, namely, Gyro backward-wave oscillator (gyro-BWO), gyro-twystron [3] (a combination of gyro-TWT and gyro-klystron), cyclotron autoresonance maser (CARM) and slow-wave cyclotron amplifier (SWCA) [13]. Technology for these devices are not as matured as for gyrotron, gyro-TWT and gyro-klystron. Some of the gyro-devices are oscillators and some are amplifiers. Same is brought out in the **Table 1.**

In **Table 1**, the most popular gyro-device names are written in red. As is evident from the **Table 1**, cyclotron autoresonance maaser (CARM) can be configured both as amplifier as well as oscillator.

In the following section, two most popular gyro-devices, namely, gyrotron, and gyro-TWT are discussed in brief with the schematic diagrams.

2.1 Gyrotron (gyro-monotron)

A schematic diagram of the gyrotron with axial output of cavity mode is shows in **Figure 2(a)**. The schematic view of high-power gyrotron with radial output of Gaussian beam is shown in **Figure 2(b)**. Here the beam-wave interaction take place in an open-ended cavity. The hollow electron-beam from the electron-gun (known as magnetron injection gun) is injected into a region with very strong axial magnetic field [3, 6–10]. Magnetic flux densities of the order of several Tesla are normally required and this usually necessitates the use of superconducting magnets [6].

The beam-wave interaction takes place in the interaction cavity region. In order to avoid the thermal issues, gyrotrons usually incorporate a highly overmoded cavity. The reported continuous wave (CW) and pulsed power capabilities of the gyrotron are three order of magnitude higher than the conventional microwave oscillators.

In case of axial output gyrotrons (**Figure 2(a)**), output millimeter-wave generated in the cavity propagates along the axis of the gyrotron and comes out of the gyrotron through an output-window. The spent-electron beam (the electron-beam after the beam-wave interaction) gets collected in the collector. In case of gyrotron with radial-output (**Figure 2(b)**), the cavity-resonator mode of EM-wave gets converted to Gaussian (TEM_{00}) mode with the help of a quasi-optical launcher (QOL) and 3 or 4 mirrors. The Gaussian beam comes out of the gyrotron radially (perpendicular to the axis of gyrotron) through the output-window and the spent electron-beam gets collected in the collector.

2.2 Gyro-traveling wave tube (gyro-TWT)

Gyro-TWT is a high power millimeter-wave amplifier [3, 8]. This is used in millimeter-wave radars. Gyro-TWT is also used for electron-cyclotron current drive (ECCD) for Tokamak. In this device, the interaction-cavity is replaced by a non-resonant structure (waveguide) to produce beam-wave interaction. This device has the potential of amplifying EM-powers of 2 order of magnitude higher than the

Oscillator	Amplifier
Gyrotron(Gyro-Monotron)	Gyro TWT
Gyro Backward Wave Oscillator	Gyro Klystron
Gyroton	Gyro Twistron
CARM	Cyclotron Autoresonance Maser (CARM)
	Slow Wave Cyclotron Amplifier (SWCA)
	Gyroton-TWT
	Magnicon

Table 1.
The complete family of gyro-device.

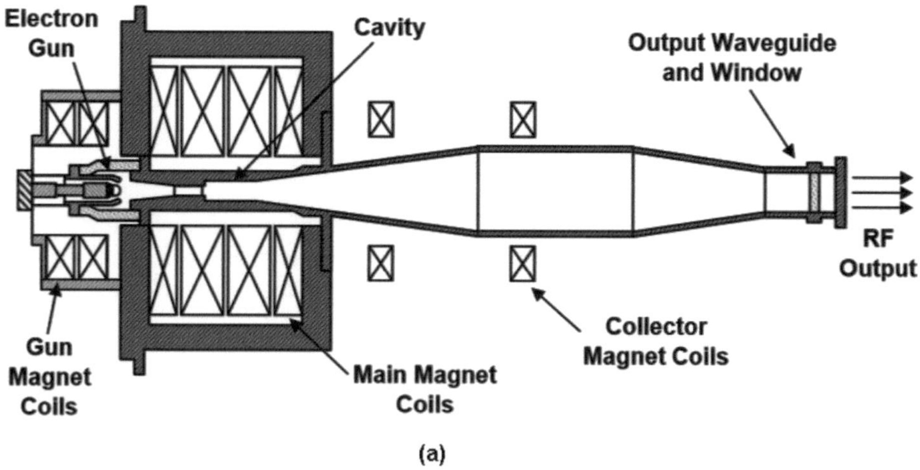

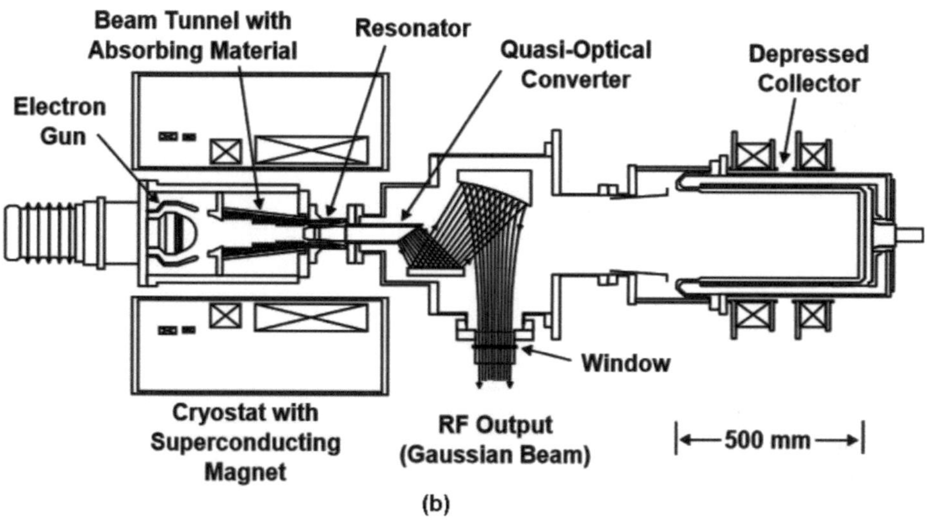

Figure 2.
(a): Gyrotron with axial output. (b): High power Gyrotron with radial output of Gaussian beam.

conventional TWT. Gyro-TWT provides a high spectral quality amplification over a narrow bandwidth. The device interaction essentially involves a narrow band resonance between the electron-beam and the electromagnetic-wave near the waveguide cut-off due to the dispersive nature of the waveguide interaction structure. However, wideband coalescence is possible by proper dispersion shaping of the waveguide. Axial phase synchronism is required between the traveling wave and the gyrating electron. Techniques are being used to increase the band-width by

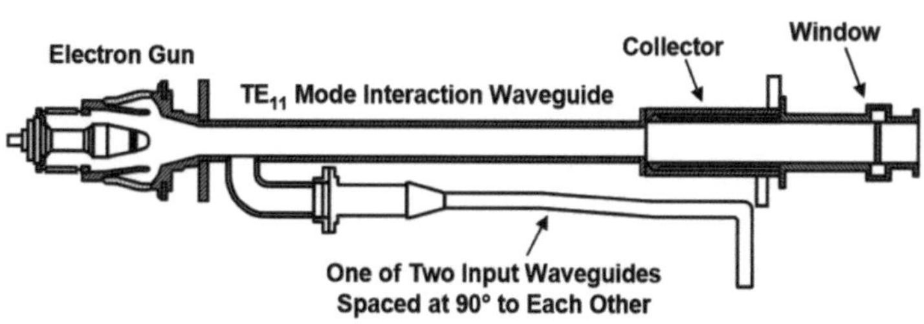

Figure 3.
Gyro TWT.

tapering the magnetic field or by periodically loading the waveguide structure. The cross sectional view of gyro-TWT is shown in the **Figure 3**.

3. Basic principle of gyrotron

In a gyrotron, the electron beam, which is normally in the shape of a thin hollow cylinder, is injected into a region with strong axial magnetic field and passed through a cylindrical cavity or waveguide region containing an electromagnetic wave with an azimutal component of electric field [1–3]. The rotational velocities of the electrons are normally 1.2 to 2 times the axial velocity. So, majority of the electron energy is rotational.

Because the magnetic field is very large, the orbit diameter for the electrons is very small. As a result, the thickness of the hollow electron beam is several times the diameter of the electron orbit as shown in **Figure 4**, and in effect, the hollow electron-beam contains a large number of small beams, referred as beamlets [2, 6]. **Figure 4** shows the thickness of the hollow electron-beam as twice the diameter of beamlet.

The basic operating mechanism of gyrotron can be explained by considering the interaction of a single beamlet of electrons with the electric field. In **Figure 5** it is assumed that electrons in a single beamlet are initially uniformly distributed along a single helical path prior to interaction with the RF electric field. The electrons are assumed to rotate in the counter clockwise direction as they move through the RF field. The rotational frequency of electrons is the cyclotron frequency, which is given by

$$\omega_c = \frac{e}{m} B_0 \tag{1}$$

Where, B_0 is the externally applied axial magnetic field, e is the electron-charge, and m is the mass of the electron. In a gyrotron, the velocity of electrons is a significant fraction of the free-space velocity of light, hence, the relativistic effect comes into play. So, the mass of the electrons is significantly greater than their rest-mass, i.e.,

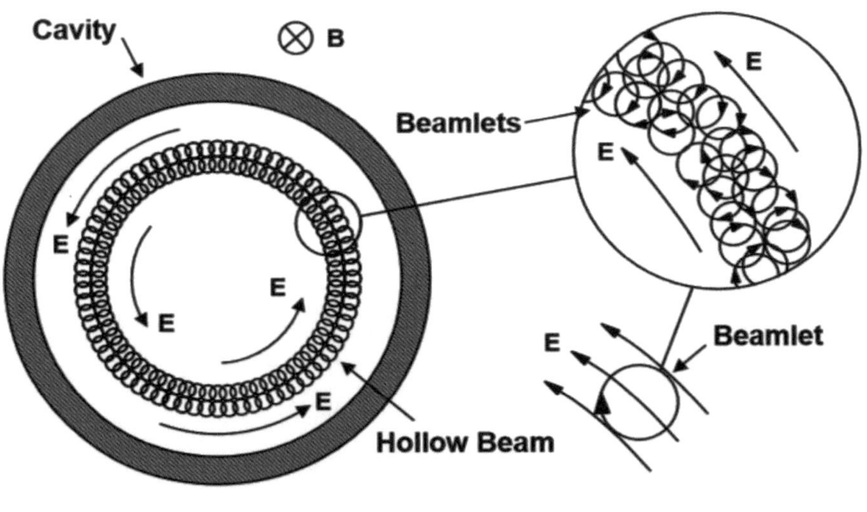

Figure 4.
Gyrotron cross section showing electron trajectories.

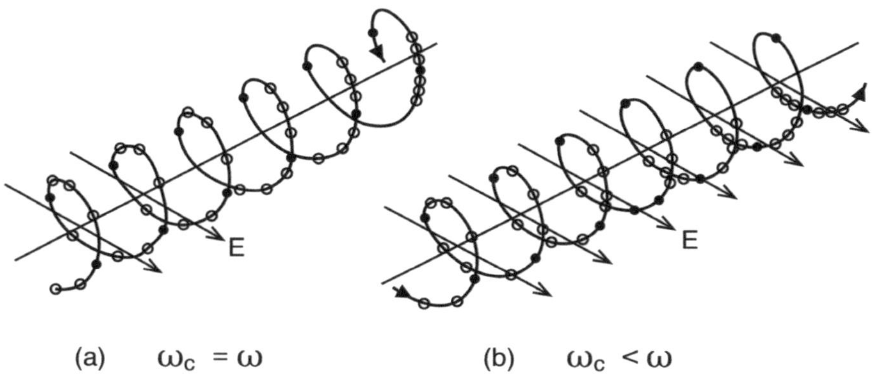

Figure 5.
Bunching of electrons in a Gyrotron.

$$m = \gamma m_0 \tag{2}$$

Where m_0 is the rest-mass of the electron and γ is the relativistic mass factor, which is given by

$$\gamma = \frac{1}{\sqrt{1 - (v/c)^2}} \tag{3}$$

Where, v is the velocity of electron and c is the free space velocity of light. The value of γ increases when an electron is placed in an accelerating field.

The radius of the gyrating orbit, alternatively known as Larmor radius (r_L), may be obtained by the following equation

$$\omega_c r_L = v_t \tag{4}$$

Where, v_t is the transverse velocity of the electron-beam.

Now, referring again to **Figure 5(a)**, when the electric field is such that it tends to accelerate electrons (top of the orbits), the electron mass is increased and so the cyclotron frequency (ω_c) decreases for these electrons. Similarly, when the electric field tends to decelerate electrons (bottom of the orbits), the electron mass is decreased, and so ω_c is increased. Since the rate of rotation is decreased for some electrons and is increased for others, orbital bunching occurs if the electrons are permitted to drift, as indicated at the right-hand side of **Figure 5(a)**.

If the cyclotron frequency (ω_c) is somewhat lower than the frequency of the electromagnetic wave (ω), then the position of the bunches along the helical orbit is delayed with respect to the phase of the applied field as indicated in **Figure 5(b)**. Hence, the bunched electrons face a decelerating field and give-up their kinetic energy to the field. As the electron bunches rotate in near synchronism with the alternating RF-field, they continue to give-up energy on each half-cycle of rotation.

The interaction that has just been described for the electrons in a single beamlet in a gyrotron also takes place in the other beamlets. Thus, the electron distribution becomes as indicated in **Figure 6**. As the direction of the electric field alternates, the direction of motion of electron also alternates, and so the electrons throughout the transit period of the electron-beam give-up energy on each half cycle of operation. This is how the beam-wave interaction happens.

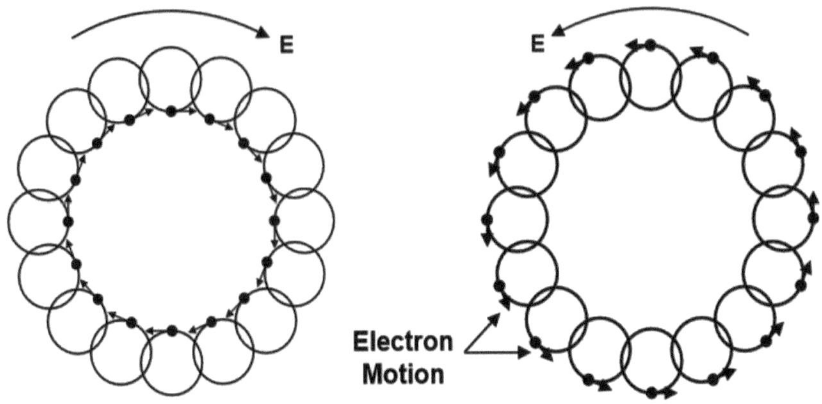

Figure 6.
Electron motion in relation to direction of electric field in a gyrotron.

3.1 Operation at harmonics of cyclotron frequency

With the proper shape of the RF-field, it is possible to excite harmonic mode of interactions with the electrons [1, 2, 8, 13–15]. As shown in **Figure 7**, the EM-wave field oscillates at a frequency twice the cyclotron frequency [2], i.e., $\omega = 2\omega_c$. The direction of the field reverses in the center of the electron orbit. Thus, an electron that is initially decelerated by the field is moving transverse to the field when the field reverses, and so does not have its orbital energy changed. By the time the field reverses again, the electron has moved 90^0 around its orbit and is again in a decelerating field. Thus, during each full orbital motion of electron, the RF-field goes through two complete cycles. Hence, for harmonic mode of operation, for a given operating frequency, the cyclotron frequency is half the value used in fundamental mode of operation. As a result, the magnetic field is reduced by a factor of two. Operation at frequencies higher than the second harmonic are also being examined [15], but the intensity of the interaction is reduced, making the efficiency of gyrotron low.

For harmonic mode operation, the frequency of operation of the gyrotron is approximately given by

$$\omega \cong s\left(\frac{e}{\gamma m_0}\right)B_0 \tag{5}$$

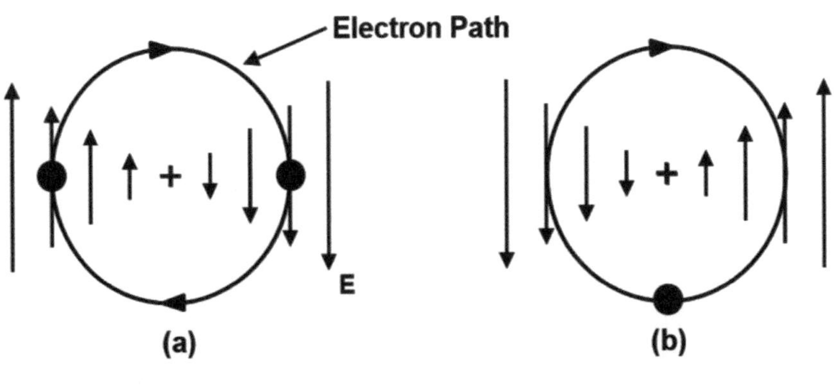

Figure 7.
Harmonic interaction of an electron and a field varying at twice the cyclotron frequency. (a) At an arbitrary time T, (b) At half RF cycle after T.

where, s is an integer, representing the harmonic number. Value of s equals to 2 corresponds to second harmonic operation. It signifies that the electromagnetic-wave frequency of the gyrotron is chosen to be twice the cyclotron frequency. Harmonic operation reduces the magnetic field requirements by factor of s (B_0/s). For gyrotrons operating at W-band or above, magnetic field requirement is very high (beyond 3 Tesla), it's not possible to obtain such magnetic field from a normal solenoid magnet. This necessitates the use of superconducting-magnets. Harmonic operation is a suitable choice for such class of gyrotrons, as second harmonic operation reduces the magnetic field requirement to half. Hence, such magnets can be built with non-superconducting solenoid coils. However, the harmonic operation reduces the efficiency of the gyrotron.

4. Main subsystems of gyrotron

In a gyrotron, a hollow electron beam gyrating at cyclotron frequency under the influence of a strong axial magnetic field interacts with the transverse electric field excited inside the cavity. If the cyclotron frequency is synchronized with the frequency of millimeter wave supported by the cavity (cut-off frequency of the cavity) for the selected higher order mode, millimeter wave gets generated. This phenomenon is known as cyclotron resonance maser (CRM) interaction. The hollow gyrating electron-beam is generated with the help of magnetron injection gun (MIG). The gyrating electron-beam is passed through a beam-tunnel and fed into an open ended interaction-cavity. The millimeter wave generated in the cavity region diffracts out with the help of a non-linear taper (NLT). The waveguide mode of electromagnetic-wave is covered to Gaussian mode with the help of a quasi-optical launcher (QOL) and mirror units. The millimeter wave is taken out of the gyrotron with the help of a high power millimeter-wave window. The spent electron-beam is collected in a collector. The required axial magnetic field throughout the gyrotron, starting from the MIG to collector is provided by a magnetic-system consisting of a main superconducting-magnet along with a number of non-superconducting solenoid magnets. Out of all these subsystems, MIG and interaction-cavity are the most important subsystems of gyrotron. The following section describes some of these main subsystems of gyrotron.

4.1 Magnetron injection gun (MIG)

Most high power gyrotrons use magnetron injection guns (MIGs), which produce annular electron-beams in which electrons gyrates in cyclotron frequency. The gyrating frequency is so chosen that the beam-wave interaction at desired mode can take place. For good interaction-efficiency, the transverse velocity component of electron should be as large as possible. A spread in transverse velocity results in a spread in axial velocity, and eventually reduces the efficiency of the gyrotron. Hence, the electron velocity spread should be kept as small as possible [8]. The cut-section view of a typical MIG with anode is shown in **Figure 8** indicating various parts of MIG.

The electrons are emitted from a annular cathode operating in temperature limiting regime of thermionic emission [2, 3]. MM-type dispenser cathode is used as emitter. The electron motion is taking place in crossed electric and magnetic fields so that the electrons follow helical trajectories around the magnetic flux lines with the electrons gyrating in cyclotron frequency. The accelerating potential of 20–70 kV is applied between the cathode and the anode. The MIG can have a diode or a triode configuration. In the triode configuration, there are two anodes, namely

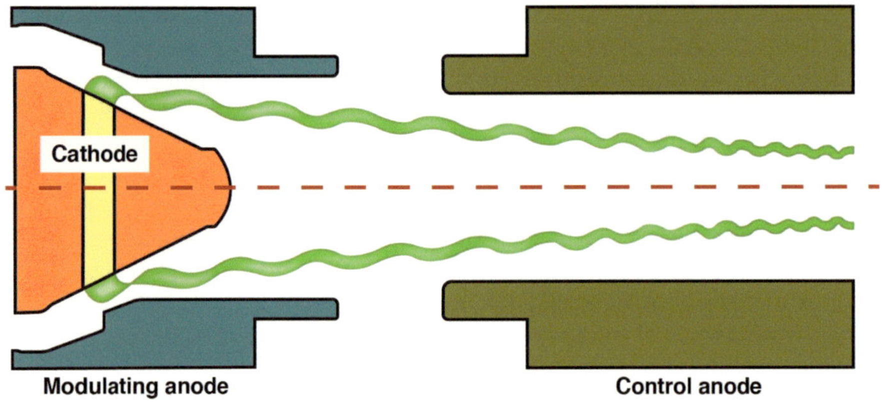

Figure 8.
Cut-section view of a typical MIG.

modulating anode and accelerating anode. In triode configuration, second anode provides the main accelerating potential. Whereas, the first anode (which is closer to cathode) is used to fine-tune the velocity pitch-factor of the beam (ratio of transverse to axial beam-velocity) as well as for pulsing the beam (i.e., for switching the beam ON and OFF). Gyro-TWT's usually incorporate triode MIG. In diode configuration, there is only one anode. Diode MIG needs much simpler power-supply for providing the necessary voltages. However, on the flipside, gyrotrons with diode MIG have lesser control over the beam.

4.2 Interaction cavity and nonlinear taper

The gyrating electron beam enters the interaction cavity, where the beam-wave interaction takes place [3, 6, 8–13]. This is an open ended overmoded cavity operating near cut-off [3, 6]. The interaction cavity generally consist of 3 sections, namely downtaper-section, straight section and uptaper-section. The shape of the cavity is dependent on the mode of the electromagnetic field with which the beam is intended to interact and also the harmonic number of interaction. The required diffractive quality factor of the cavity is achieved by proper fine tuning of the cavity shape. Schematic drawings of a typical Gyrotron cavity is presented in **Figure 9**.

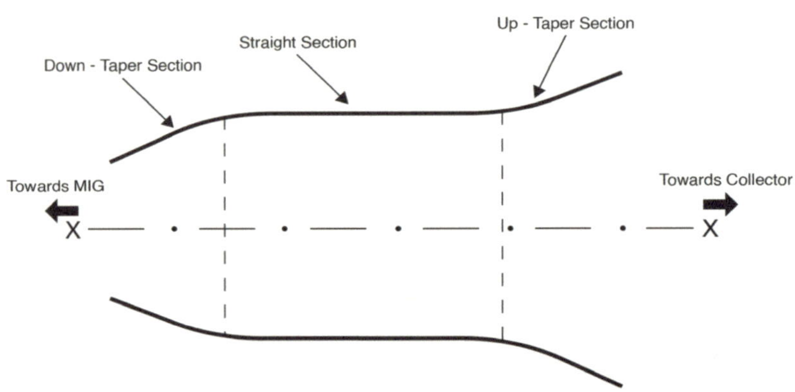

Figure 9.
A typical Gyrotron cavity.

The down-tapering is offered to the input-end of the cavity. This prevents the millimeter wave from back-traveling towards the MIG. The up-tapering is offered in the output-end of the cavity. The up-tapering helps the millimeter-wave to diffract out of the cavity. In case of coaxial gyrotrons, a coaxial insert is placed at the center of the cavity. The main beam-wave interaction takes place at the straight section of the cavity.

The millimeter-wave signal generated in the cavity needs to diffract out of the cavity. The same is achieved by the non-linear taper (NLT). This NLT is basically a tapered waveguide section with a specific tapering profile. A raised-cosine profile is incorporated in the NLT region to avoid reflection of electromagnetic-wave. This section acts as an interface between the interaction cavity and the QOL [6, 16–20]. Generally interaction cavity operates at a mode much higher than the dominant mode of the cavity. This enable the use of much higher cavity dimension and volume and this in-turn eliminates the bearing on the maximum power handling capacity at higher frequencies of millimeter-wave and sub-millimeter-wave regime. Broadly, the cavity operating modes are divided into three categories, namely, $TE_{0\ n}$ mode, $TE_{m\ n\ (m > n)}$ mode and $TE_{1\ n}$ mode. $TE_{m\ n}$ mode, when m> > n, is called the whispering gallery mode. This mode is most widely used in gyrotrons for plasma heating applications. The relative merits and demerits of these modes are presented in the **Table 2**.

4.3 Beam-tunnel

The radius of the hollow electron beam generated by MIG is generally much larger than the required hollow beam radius at the cavity region. The purpose of the beam-tunnel is to gradually bring down the beam radius to the value needed in the cavity region. Beam-tunnel is basically a cylindrical waveguide structure placed between the anode and interaction-cavity. The inner radius of the beam-tunnel at the anode end is matched to the anode inner radius and at the cavity end is matched to the input inner radius of the cavity. In order to ensure that the beam-tunnel does not take part in interaction, lossy dielectric material is placed inside the beam-tunnel. One of the popular configuration of beam-tunnel is a stack of alternate metal (OFHC copper) and lossy ceramic (AlN, SiC) rings stacked inside the cylindrical waveguide of beam-tunnel (**Figure 10**). The axial length of the beam-tunnel is so chosen that that the electron beam undergoes an adiabatic compression as it propagates from the MIG to the cavity, i.e., the beam trajectories follows the magnetic flux lines. This configuration ensures maximum beam laminarity and minimum beam-turbulence. Cavity.

4.4 Quasi-optical launcher

The purpose of the quasi-optical launcher (QOL) is to convert the cavity mode of EM-wave into a Gaussian (TEM_{00}) mode. This is accomplished with the help of a helically-cut waveguide section (QOL) followed by 3 or 4 toroidal mirrors system. QOL consist of a mildly tapered waveguide structure with helically cut end (known as Vlasov launcher) with dimple patterned inner surface (Denisov type surface deformation). Millimeter wave is launched from the QOL to the mirror system [16]. After passing through the mirror system, a Gaussian beam (with more than 98% Gaussian mode purity) is emerged. A typical QOL and 3 mirrors for converting $_{\text{cavity}}$ mode to Gaussian (TEM_{00}) mode is shown in **Figure 11**. The Gyrotron with Gaussian output is most suited for plasma heating applications. Because, the Gaussian

TE_{0n} Mode

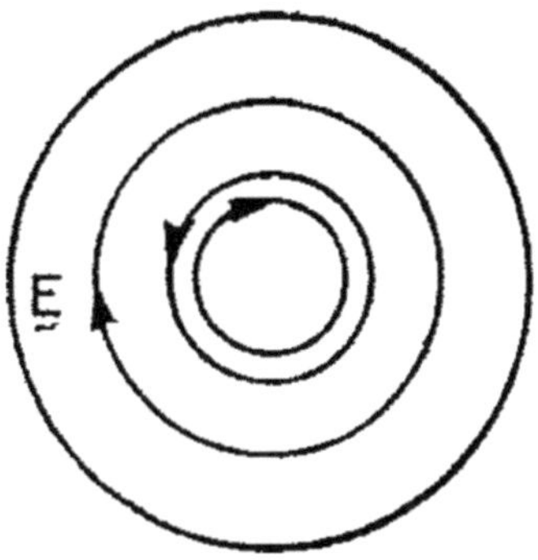

Advantage

- Smallest Ohmic Loss
- Easy to excite

Disadvantage

- Suffers from mode competition with TE_{2n} mode
- Difficult to convert to Gaussian (TEM_{00}) mode

Whispering Gallery Mode ($TE_{m\,n}$) m > > n

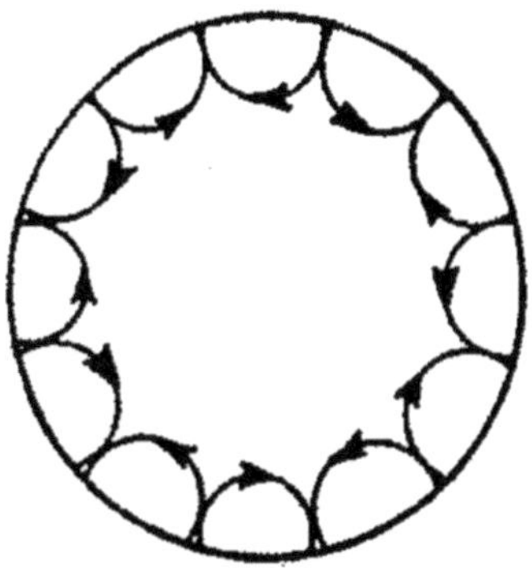

Advantage

- Beam need to be placed near the cavity wall. Reduces space charge depression
- Suitable for high harmonic operation (reduced magnetic field requirement)
- Large Cavity dimension leads to high power handling capability

Disadvantage

- Prone to higher beam interception on the cavity wall

$TE_{1\,n}$ mode

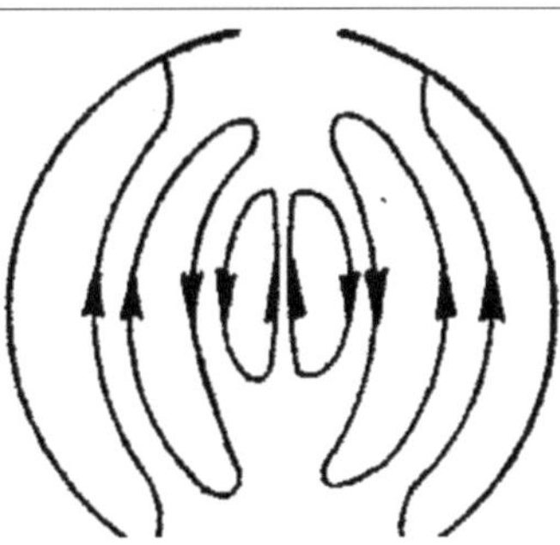

Advantage

- Mode competition can be avoided by splitting the cavity wall
- Beam need to be placed near the center of the cavity. Hence less beam interception on the cavity wall

Disadvantage

- Higher space-charge depression inside the cavity

Table 2.
Gyrotron cavity modes.

millimeter-wave beam can be transmitted through a waveguide over a very long distance with very little attenuation. Hence, the gyrotron can be placed away from the plasma vessel. Sometimes, the Gaussian beam is further converted to HE_{11} mode with the help of a matching optic unit (MOU) placed external to gyrotron and then transmitted to the plasma vessel. This arrangement further reduces the attenuation of the beam.

4.5 High power output window

The millimeter-wave signal is finally taken out of the gyrotron with the help of the high-power output window. This window consists of a ceramic disc which isolates the ultra-high vacuum environment inside the gyrotron enclosure from the outside

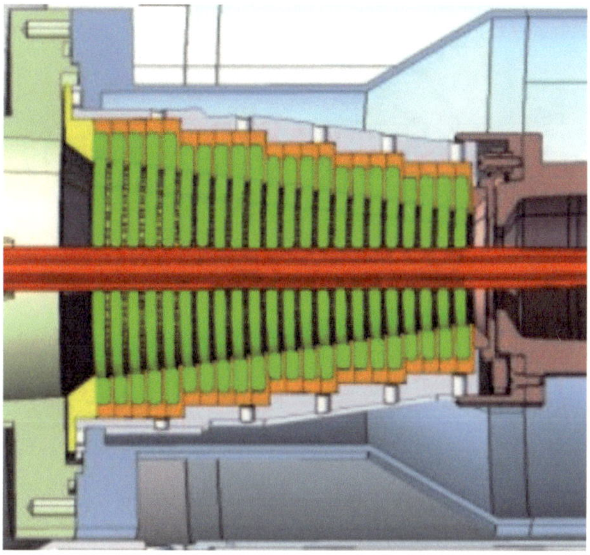

Figure 10.
Cut section view of a beam tunnel.

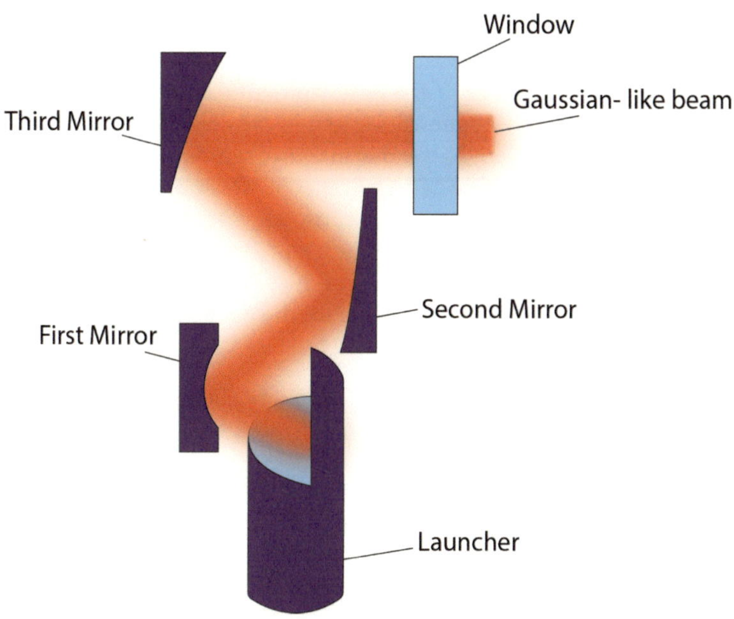

Figure 11.
A typical QOL for converting $TE_{6,2}$ cavity mode to Gaussian mode.

atmosphere. The ceramic disc material and thickness is so chosen that it appears almost transparent to the electromagnetic-wave and the millimeter-wave comes out of the gyrotron through the window with minimum attenuation. Also, a material, which is a good thermal conductor but bad electrical conductor, is chosen for window. Usually single disc window is used. However, for the purpose of VSWR matching, sometime double disc window may be incorporated. For short-pulse operation, sapphire, beryllium-oxide (BeO) or boron-nitride may be used as window ceramic materials. For long pulse high-power operation, chemical vapor deposition (CVD)

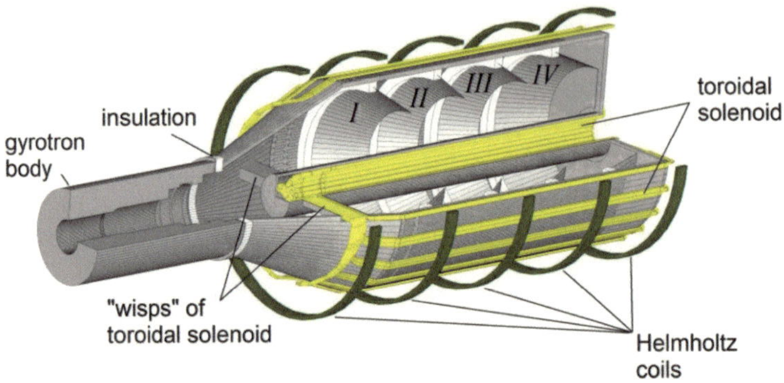

Figure 12.
Cut-section view of a typical collector.

diamond is generally used as the window material due to its very high thermal conductivity.

4.6 Collector

After the electron-beam comes out of the interaction-cavity, the spent electron-beam gets collected in the collector. The kinetic energy of the spent-electron beam (the electron-beam which has already undergone beam-wave interaction) gets dissipated in the collector. Hence, if the electronic-efficiency of a gyrotron is 40%, 60% of the electron beam power gets dissipated in the collector. Since the dissipated power in the collector is very high, the thermal management of collector is a very critical issue. The cut-section view of a typical collector is presented in the **Figure 12**. In some of the high power gyrotrons, a low frequency magnetic sweeping coil is used for sweeping the electron-beam along the length of collector to avoid creation of hot-spots. For the enhancement of overall efficiency of gyrotron, multi-stage depressed collector (MDC) is used, where the collector is kept at a negative potential with respect to cavity.

4.7 Magnetic system

The purpose of the magnetic system is to generate required axial magnetic field profile needed for the cavity as well as the MIG and collector. For lower frequency operation, non-superconducting air-cooled solenoids are preferred. However, for higher frequency operation (i.e., for gyrotrons operating at W-band or beyond), superconducting magnets are being used. The state-of-the-art magnets employ cryogen free superconducting magnet technology. Which eliminates the need of re-filling of liquid helium. Some researchers have reported gyrotrons developed with Samarium-Cobalt (Sm_2Co_{17}) permanent magnets and special type of room temperature solenoid made out of copper foil. Such gyrotrons usually operate at higher harmonic mode of interaction.

5. Applications of gyrotrons

Gyrotrons have wide range of applications. These applications coves the domain of scientific research, industrial heating, homeland security and defense. Same is shown in tree diagram (**Figure 13**).

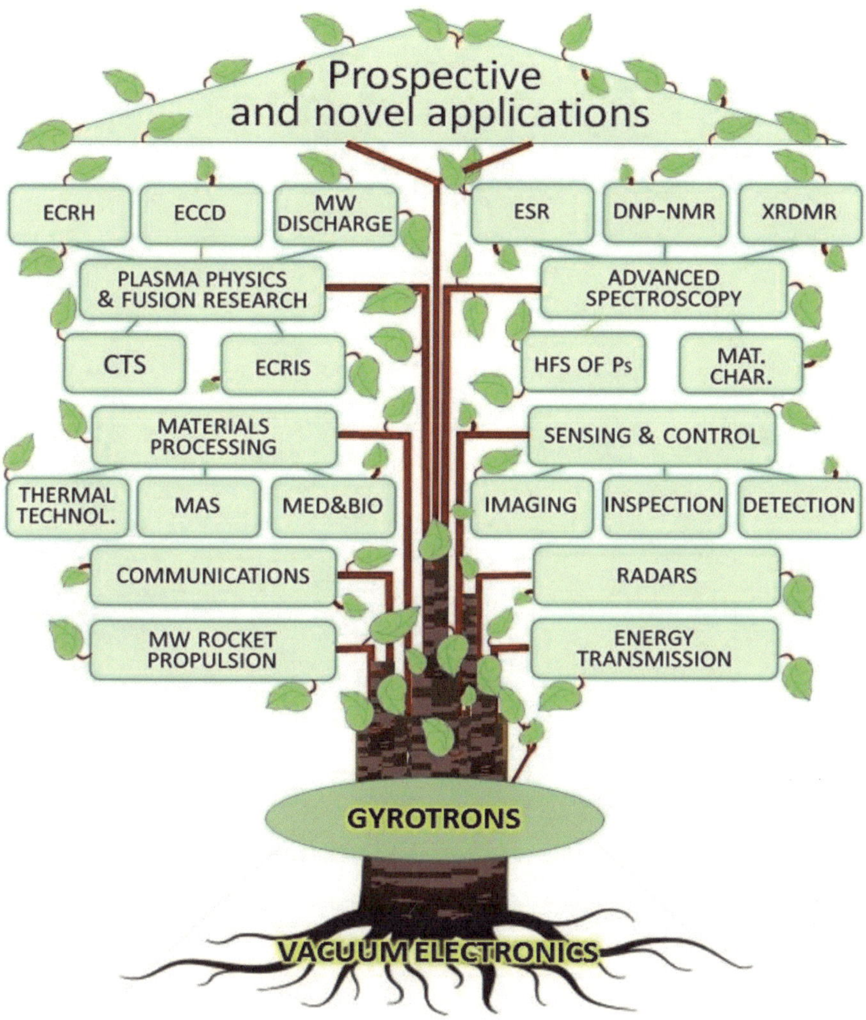

Figure 13.
Applications of gyrotron.

5.1 Major application

Majority of gyrotrons developed worldwide are being used for the electron cyclotron resonance heating (ECRH) of plasma in the controlled thermoneuclear fusion reactor [7]. The plasma is kept confined in the plasma vessel with the help of very high value of superconducting magnetic field (magnetic confinement). The magnetically confined plasma is then exposed to very high power millimeter wave beam generated with the help of a gyrotron. This elevates the temperature of the plasma to 100000^0 C. At this temperature, fusion reaction takes place. Millimeter-wave beam generated by a gyrotron is also used for the electron cyclotron current drive (ECCD), electron cyclotron resonance ion source (ECRIS) and also for the diagnostics cooling tower system (CTS). For the ITER (international thermoneuclear experimental reactor) project, it's proposed to use 20 numbers of 170 GHz long-pulse gyrotrons to generate combined heating power of 24 MW. For this purpose, till now, the highest order mode number attempted is $TE_{34,19}$ for the generation of 2 MW of continuous

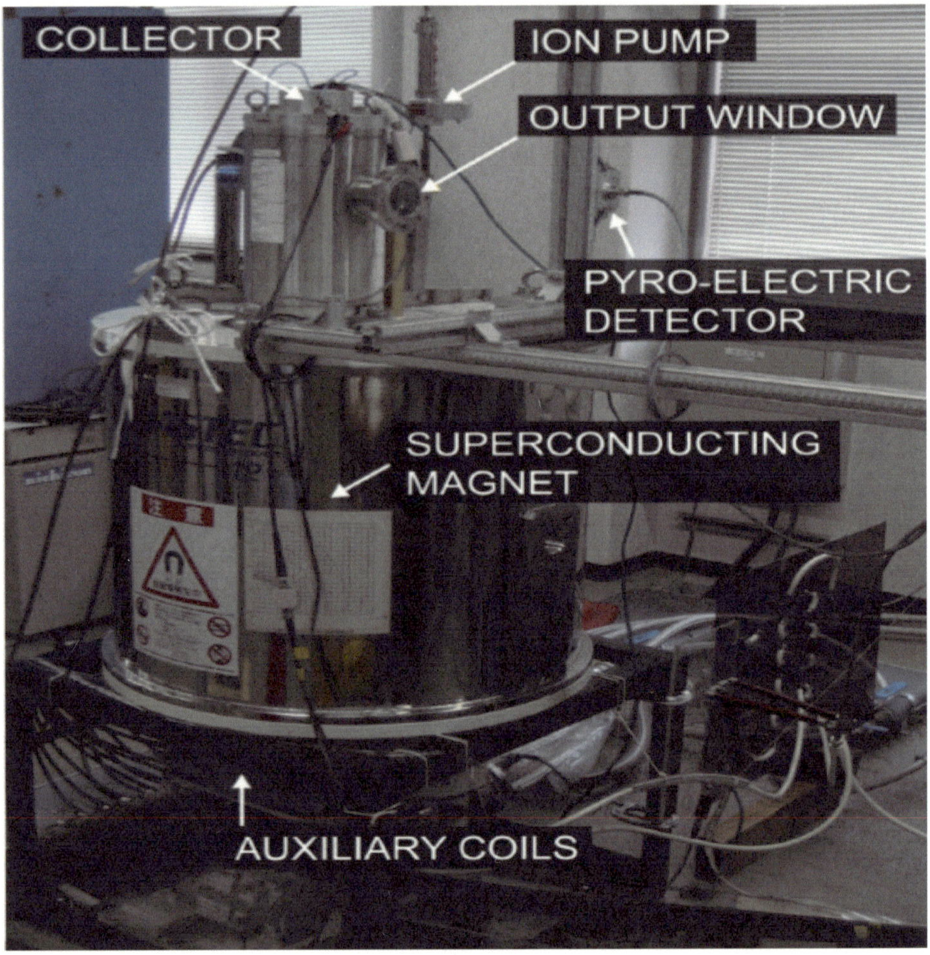

Figure 14.
2 MW Gyrotron for ECRH application.

Figure 15.
140 GHz Gyrotron based ECRH system (Stellarator W7-AS).

power at 170 GHz. The photograph of a 2 MW gyrotron for ECRH application is shown in **Figure 14**. The photograph of a ECRH System (Stellarator W7-AS) with 140 GHz gyrotron is shown in **Figure 15**.

5.2 Other applications of gyrotron

Other important scientific research application of gyrotron is in the area of spectroscopy. This includes, electron spin resonance (ESR) spectroscopy, dynamic nuclear polarization – nuclear magnetic resonance (DNP-NMR) Spectroscopy, X-ray diffraction magnetic resonance (XRDMR) spectroscopy etc. Also, W- band frequency of 95 GHz being an atmospheric window, gyrotrons operating at W-band are having special significance in connection with defense and homeland-security [17, 18]. The active denial systems (ADS) for controlling low-intensity conflicts, uses 95 GHz gyrotron. Also a number of millimeter-wave radar systems, such as space surveillance radar, space derby radar, imaging radar and weather radar uses W-band gyrotron as well as gyro-TWT/ gyro-klystrons.

6. Worldwide scenario of gyrotron for plasma heating

Worldwide, a number of research institutions, academic institutions and industries are working in the field of gyrotron, with frequency varying from lower end of microwave range (8 GHz) to 1 THz. The output power of these gyrotrons also ranges from 100 s of kW to few MW. Pulse duration also varies from few milliseconds to full continuous wave (CW) operation. The efficiency of gyrotron varies from 10–70%. The worldwide scenario of Gyrotron for plasma heating purpose are presented in the **Tables 3** and **4** [7, 11–13, 19–21].

Institute	Frequency (GHz)	Mode		Power (MW)	Efficiency (%)	Pulse length (Sec)
		Cavity	Output			
CPI, USA	28,35	TE_{02}	TE_{02}	0.2	37	CW
CPI, USA	53.2,56,60,70	$TE_{01/02}$	TE_{02}	0.23	37	CW
CPI, USA	70.15	$TE_{10,3}$	TEM_{00}	0.6	47 (SDC)	2.25
CPI, USA	84	$TE_{15,4}$	TEM_{00}	0.56	44 (SDC)	2.0
CPI, USA	94.9	$TE_{6,2}$	TEM_{00}	0.12	50 (SDC)	CW
Gycom, Russia	68 (70)	$TE_{9,3}$	TEM_{00}	0.5 (0.68)	50 (48) (SDC)	1.0 (3.0)
Gycom, Russia	75	$TE_{11,5}$	TEM_{00}	0.8	70 (SDC)	0.1
Gycom, Russia	82.7	$TE_{10,4}$	TEM_{00}	0.65	53 (SDC)	0.3
Gycom, Russia	82.7	$TE_{10,4}$	TEM_{00}	0.9	32	0.3
Gycom, Russia	82.7	$TE_{10,4}$	TEM_{00}	0.2	52 (SDC)	CW
Gycom, Russia	84	$TE_{12,5}$	TEM_{00}	0.88	54 (SDC)	3.0
Gycom, Russia	84	$TE_{12,5}$	TEM_{00}	0.5 (0.2)	50 (SDC)	10 (CW)
Hughes	60	TE_{02}	TE_{02}	0.2	35	0.1
IECAS, China	24.1	TE_{01}	TE_{01}	0.15	24	0.02
IECAS, China	34.3(2Ωc)	$TE_{02/03}$	TE_{03}	0.2	30	0.02
Mitshubishi, Japan	88	$TE_{8,2}$	TEM_{00}	0.35	29	0.1
NEC, Japan	35	TE_{01}	TE_{01}	0.1	30	0.001
NRL, USA	35	TE_{01}	TE_{01}	0.15	31	0.02
Philips, Germany	70	TE_{02}	TE_{02}	0.14	30	CW

Institute	Frequency (GHz)	Mode		Power (MW)	Efficiency (%)	Pulse length (Sec)
		Cavity	Output			
Toshiba, Japan	77	$TE_{18,6}$	TEM_{00}	1.2	38 (SDC)	10.0
Toshiba, Japan	77	$TE_{18,6}$	TEM_{00}	0.3	36 (SDC)	900
UESTC, China	70(2Ωc)	$TE_{02/03}$	TE_{03}	0.1	20	0.0001
UESTC, China	94(2Ωc)	$TE_{02/03}$	TE_{03}	0.12	20.5	0.0001
	94	$TE_{61/62}$	$TE_{61/62}$	0.09	43	CW

SDC: Single-Stage Depressed Collector; CW: Continuous Wave Operation.

Table 3.
Gyrotrons for electron cyclotron resonance heating, 28–95 GHz.

Institute	Frequency (GHz)	Mode		Power (MW)	Efficiency (%)	Pulse length (Sec)
		Cavity	Output			
CPI, USA	140	$TE_{02/03}$	TE_{03}	0.1	27	CW
CPI, USA	140	$TE_{15,2}$	$TE_{15,2}$	0.32	31	3.6
CPI, USA	140.2	$TE_{28,7}$	TEM_{00}	0.9	33 (SDC)	1800
KIT, Germany	140.8	TE_{03}	TE_{03}	0.12	26	0.4
KIT, Germany	162.3	$TE_{25.7}$	TEM_{00}	1.48	35	0.007
KIT, Germany	139.8	$TE_{28,8}$	TEM_{00}	1.0	50 (SDC)	12
KIT, Germany	139.8	$TE_{28,8}$	TEM_{00}	0.92	44 (SDC)	1800
Gycom, Russia	140	$TE_{22,6}$	TEM_{00}	0.96	36	1.2
Gycom, Russia	140	$TE_{22,6}$	TEM_{00}	0.54	36	3.0
Gycom, Russia	140	$TE_{22,6}$	TEM_{00}	0.1	35	80
Gycom, Russia	170	$TE_{25,1}0$	TEM_{00}	1.0	53 (SDC)	570
Gycom, Russia	170	$TE_{25,1}0$	TEM_{00}	0.8	55 (SDC)	1000
Gycom, Russia	140	$TE_{22,6}$	TEM_{00}	0.8	32	0.8
Gycom, Russia	140	$TE_{22,6}$	TEM_{00}	0.88 50	50.5(SDC)	1.0
Toshiba, Japan	170	$TE_{31,8}$	TEM_{00}	1.3	32	0.003
Toshiba, Japan	170	$TE_{31,12}$	TEM_{00}	1.56	27	0.1
Toshiba, Japan	168	$TE_{31,8}$	TEM_{00}	0.52	19	1.0
Toshiba, Japan	168	$TE_{31,8}$	TEM_{00}	0.52	30 (SDC)	1.0

Table 4.
Gyrotrons for electron cyclotron resonance heating, above 140 GHz.

Acknowledgements

The authors are thankful to Dr. M.V. Kartikeyan, professor, Indian Institute of Technology (IIT)-Tirupati, India, for his valuable suggestions and inputs. The authors are also thankful to Sri G. Viswam, Center Head, Microwave Tube Research & Development Center (MTRDC), Bangalore, India, for his necessary help and support. Also, the authors are thankful to Shroyonaa Karmakar, Jain University School of Design, for the preparation of the diagrams.

Author details

Santanu Karmakar[1,2]* and Jagadish C. Mudiganti[3]

1 VIT University, Vellore, India

2 MTRDC, Bangalore, India

3 VIT AP University, Amaravati, India

*Address all correspondence to: santanukarmakar@yahoo.com

References

[1] Basu B.N., "Electromagnetic Theory and application in Beam-Wave Electronics", World Scientific, Singapore, 1996

[2] Gilmour, A.S. Jr. "Microwave Tubes", Artech House, USA, 1986

[3] Gilmour, A.S. Jr. "Klystrons, Traveling Wave Tubes, Magnetrons, Crossed-Field Amplifiers, and Gyrotrons", Artech House, USA, 2011

[4] Rowe, J.E., "Non-linear Electron wave interaction phenomenon", Academic Press, New York, 1965.

[5] D. E. Baldwin, "Advances in Plasma Physics", vol. 3, John Wiley & Sons, New York, 1969.

[6] M. V. Kartikeyan, E. Boris, M.K.A. Thumm, "Gyrotrons: High Power Microwave and Millimeter Wave Technology", Springer-Verlag, Berlin, Heidelberg, 2004

[7] M K.A. Thumm, State-of-the-art of high power gyro- devices and free electron masers, Update 2016, KIT, Karlsruhe, Germany, 2017.

[8] G. S. Nusinovich, Introduction to the Physics of Gyrotron, Maryland: JHU, USA, 2004.

[9] C. J. Edgecombe, Gyrotron Oscillators: Their principles and practice, Taylor & Francis Ltd., London, 1993.

[10] Shulim E. Tsimring, "Electron Beams and Microwave Vacuum Electronics", Wiley-Interscience, 2007

[11] Barker, R.J. and E. Schamiloglu, "High Power Microwave Sources and Technologies", IEEE Press, New York, 2001

[12] Barker R.J., J.H. Booske, N.C. Luhmann and G.S. Nusinovich, "Modern Microwave and Millimeter Wave Power Electronics", IEEE Press, NJ, 2005

[13] Victor L. Granstein and Igor Alexef, "High power microwave sources", Artech House, London

[14] Flyagin, V. A., Gaponov, A.V., Petelin, M.I. and Yulpatov, V.K., The Gyrotron, IEEE. Trans. Microwave Theory and Tech., vol. 25, pp 514 – 512, 1977

[15] Idehara T., I. Ogawa, S. Mitsudo, "A high harmonic Gyrotron with an axis-encircling electron beam and a permanent magnet", IEEE Trans. Plasma Sci., vol. 32, pp 903, 2004

[16] Shahana K, Vishal Kesari, Shantanu Karmakar, R. Seshadri, "Simulation of TE6,2 to Gaussian Mode Converter for a 95 GHz Gyrotron", IEEE Trans. Plasma Science, vol. 46, Issue 1, pp 84-89, 2018.

[17] Alastair D. McAulay, Military Laser Technology for Defense, John Wiley & Sons, 2011.

[18] Tolkachev A.A., B.A. Levitan, G.K. Solovyev, "A megawatt power millimeter wave phased-array radar", IEEE Trans. Aerosp. Electron. Syst., vol. 25, pp25, 2000

[19] Glyavin M., A.L. Goldenberg, A.N. Kuftin, "Experimental studies of Gyrotron electron beam systems", IEEE Trans. Plasma Sci., vol 27, pp 474, 1999

[20] C. D. Marchewka, E. M. Choi, M. A. Shapiro, J. R. Sirigiri, and R. J. Temkin, "Observation of low-frequency paracitic oscillations in 1.5 MW, 110 GHz gyrotron," Proc. IEEE Int. Vacuum Electronics Conf., p. 535, May 2006.

[21] S. Sabchevski and I. Zhelyazkov, "Modeling and simulation of new generation powerful gyrotrons for the fusion research," J. Phys. Conf. Series, vol. 63, pp. 012003-1-11, Jan. 2007.